Science and Engineering Mathematics with the HP 49 G

VOLUME I
Introduction, Pre-calculus Mathematics, Graphics, and Linear Algebra

Copyright © 2000 Gilberto E. Urroz, Ph.D., P.E.
All Rights Reserved.

ISBN 1-58898-043-X

Science and Engineering Mathematics with the HP 49 G

VOLUME I
Introduction, Pre-calculus Mathematics, Graphics, and Linear Algebra

Gilberto E. Urroz, Ph.D., P.E.
greatunpublished.com
No. 43

Science and Engineering Mathematics with the HP 49 G

VOLUME I
Introduction, Pre-calculus Mathematics, Graphics, and Linear Algebra

To my math High-School teachers at *Colegio Calasanz*, Managua, Nicaragua (1971-1974):

- Mr. Lucinio Sanmillán (mathematics),
- Mr. Jorge Krüger (mathematics),
- Mr. Humberto Bermúdez (science, physics, and chemistry) and
- Mr. Alejandro Robles (mathematics and physics),

for they awakened in me the love for mathematics and the sciences.

To my math, physics, and engineering mechanics college professors at *Universidad Nacional Autónoma de Nicaragua – Recinto Universitario "Rubén Darío,"* Managua, Nicaragua (1975-1980):

- Mr. Antonio Rodríguez (mathematics),
- Mr. Rafael Sánchez Richardson (mathematics),
- Mr. Douglas López (mathematics),
- Mr. Aníbal Fonseca (physics),
- Mr. Max García (statistics),
- Dr. Eduardo Gómez (physics),
- Mr. Douglas Quinn (physics),
- Dr. Róger Argeñal (physics), and
- Dr. Moisés Hassan (continuum mechanics),

for they taught me the right way to apply mathematics to the hard sciences.

Table of Contents

Preface 1

Preface to Volume I 5

1 Introduction 7

Notation used in this book 7
The HP 49 G operating system or ROM 7
Keeping up to date with the HP 49 G 8
Programs for the HP 49 G 9
Some HP 49 G programming concepts 9
    Programming Languages 9
    Libraries 10
Calculator operating modes 10
    Changing the calculator mode 10
    Comparing algebraic mode with RPN mode 10
Flags 13
    Example of flag setting: general solutions vs. principal value 13
    Other flags of interest 14
Objects and their types in the HP 49 G calculator 15
Organizing data in your calculator 16

2 The HP 49 G keyboard 18

Primary function of each key *18*
Alphabetic characters *18*
Left-shift and right-shift functions *18*
Secondary functions of the soft menu (F1-F6) keys *19*
The arrow keys... *19*
Utility keys *19*
Mathematical operations keys *24*
The ALPHA key *26*
Numeric keypad.. *26*
Arithmetic operation keys *28*

3 Basic calculator operation 30

Undo, Arg, and Cmd *30*
Deleting variables *30*
Transferring data between two HP 49 G calculators *31*
Transferring data from the HP48 to the HP 49 .... *31*
Transferring data from the HP49 to the HP 48 .... *31*
How to type Greek letters and other characters *32*
    Keyboard shortcuts for special characters 32
Changing the display format *33*
    Exercises using different number formats 33
Entering numbers as powers of ten *35*
Changing the angle mode and coordinate system *35*
    Important relationships between angle units 36
    Quick conversions from degrees to radians and vice versa 37
    Important relationships between coordinate systems 37
    Additional examples on angle measure and coordinate system conversions 37
HP 49 G standard mathematical constants *38*
Physical constants available in the HP49 G calculator *39*
Utility menus adapted from the HP 48 G *39*
Accessing the EQ LIB menus *40*
    Accessing the constants library 40
    The function CONST 41
    Accessing the constant library through the command catalog 41
Using the command catalog *41*
Working with units in the HP 49 G calculator *42*
    Universal gas law 43

4 Calculations with real numbers........ 45

What is the CAS in the HP 49 G calculator? 45
   CAS options 45
   Checking current CAS settings 46
   Checking the calculator mode 46

Real number calculations 46
   Modifying the display to obtain a textbook-like appearance 48
   Changing from complex to real mode 48
   Some functions accessed through the MTH menu 49
      Examples of functions from the REAL menu 50

Combinatorics, random numbers, and probability functions 52
   Factorials, permutations, and combinations 52
   The Gamma function 53
   Generating random numbers 53
   Examples of probability calculations for continuous random variables 54
      Normal distribution pdf 54
      Normal distribution cdf 55
      The Student-t distribution 56
      The Chi-squared (χ^2) distribution 57
      The F distribution 57

FFT menu: Fast Fourier Transform - brief introduction 58

5. Calculations with complex numbers .. 59

Introduction to complex numbers 59
The CMPLX menus - basic complex number applications 60
    CMPLX menu through MTH 60
    CMPLX menu defined in the keyboard 60
Polar representation of a complex number 61
Examples of basic complex number operations 61
    Polar representation 62
Complex number calculations 62
    Examples of operations with complex numbers 63
Functions of a complex variable 64
    Example: Expanding ln(z) using the HP 49 G calculator 65
    Example: Expanding z^2 using the HP 49 G calculator 65
Functions of complex numbers in the HP 49 G calculator 66

6 Lists, functions, and programs 67

Lists 67
   Entering a list of objects 67
   Operations with lists of numbers 67
   Composing and decomposing lists 69

Functions 69
   Defining a function of one variable 69
   Evaluating the function 70
   When list arguments fail 70

An example of programming 71
   Global and local variables 72
   Global Variable Scope 73

Functions defined by more than one expression 74
   The IFTE function 74
   Combined IFTE functions 75

Generating a table of values for a function 75
   The EQ variable 76
   The TPAR variable 76

Functions of more than one variable 77

Applications of function definitions - probability distributions 78
   Discrete probability distributions 78
   Defining summations in the HP 49 G 79
      Binomial probability distribution function 79
      Binomial cumulative distribution function 80
      Other discrete probability distributions and distribution functions 80
   Continuous probability functions 81
      The gamma distribution 81
      The exponential distribution 82
      The beta distribution 82
      The Weibull distribution 82
      Defining integrals in the HP 49 G - brief introduction 82

Manipulating elements of lists 85
   Determining list size 85
   Extracting and inserting elements in a list 85
   Element position in the list 86
   HEAD and TAIL functions 86

Applications of list operations 86
   Mean, variance, and standard deviation of a sample 86
   Calculating statistics from grouped data 87
   Mean and variance of a discrete probability distribution 88

vii

© 2000 – Gilberto E. Urroz
All rights reserved

Programs for generating lists of numbers *89*
Applications of list-generating programs *90*
    Generating tables of mass and cumulative distribution functions for discrete random
    variables 90
        Binomial distribution 91
        Poisson distribution 91
        Geometric distribution 92
Functions and operations applied to lists with DOLIST, DOSUBS, STREAM, and SEQ *92*
    DOLIST 93
    Defining a function instead of using DOLISTS 93
    DOSUBS 94
    Moving averages 94
    STREAM 94
    SEQ 95

7 HP 49 G programming basics 96

Examples of sequential programming *96*
    Programs generated by defining a function. 96
    Programs that mimic a sequence of calculator operations....................................... 98

Interactive input in programs *100*
    Prompt with an input string 102
    A function with an input string 103
    Debugging the program 103
    Fixing the program 104

Input string for programs requiring two or three input values *105*
    Input string program for two input values 106
    Input string program for three input values 108

Identifying output in programs *109*
    Tagging a numerical result 109
    Decomposing a tagged numerical result into a number and a tag 109
    "De-tagging" a tagged quantity.. 110
    Using a message box for output 112
    Including input and output in a message box - string concatenation 113

Incorporating units within a program *114*

Relational and logical operators *117*
    Relational operators 117
    Logical operators 118

Program branching *119*
    What IF...? 119
        IF...THEN...END 119
        IF...THEN...ELSE...END 121
        Nested IF...THEN...ELSE...END constructs 122
    Just in CASE 124

Program loops *125*
    START 125
        START...NEXT 126
        START...STEP 128
    FOR 129
        FOR...NEXT 129
        FOR...STEP 130
    DO 130
    WHILE 131

Procedural programming and object-oriented programming *132*

Concluding remarks on programming *133*

8 Algebra and related subjects.......... 134

The CAS and RPN mode 134
Some transcendental functions 134
   Trigonometric functions 134
   Trigonometric identities 135
   What is a logarithm? 136
   Logarithms of base 10 137
   Natural logarithms and the exponential function 137
   Properties of the exponential function 138
   Properties of logarithms 138
   Converting logarithms of different bases 138
   Hyperbolic functions 139
   Euler formula and complex arguments for logarithmic, exponential, and hyperbolic functions 140

Operations with algebraic objects 142
   Entering algebraic objects 142
   Basic operations 142
   Expanding expressions 143
   Factoring expressions 144
   Complex mode factorization 145
   Substituting expressions in algebraic objects 145
   Substitution by using HP variables 146

Expansion and factorization using transcendental functions 147
   Expansion and factorization in terms of exponential and logarithmic functions 147
   Expansion and factorization in terms of trigonometric functions 149

Fraction Manipulation 152
   SIMP2 152
   PROPFAC 153
   PARTFRAC 153
   FCOEF 153
   FROOTS 154

Modular arithmetic 154
   Operations in modular arithmetic 154
   Formal definition of a finite arithmetic ring 155
   Finite arithmetic rings in the HP 49 G 155
   Modular arithmetic in the HP 49 G 156
      Setting the modulus (or MODULO) 156
   Modular arithmetic operations with numbers 156
      The modular inverse of a number 158
      The MOD function 158
      Other modular arithmetic functions 158

Polynomials *159*
   Modular arithmetic with polynomials 159
   ABCUV 159
   CHINREM 160
   DIV2 160
   EGCD 160
   FACTOR 161
   FCOEF 161
   FROOTS 161
   GCD 161
   HERMITE 161
   HORNER 162
   The variable VX, or "Why do you use only X in your examples?" 163
   LAGRANGE 163
   Entering matrices directly in the stack 164
   LCM 164
   LEGENDRE 164
      Checking the solution to Legendre's equation 165
   PARTFRAC 165
   PCOEF 165
   PROOT 166
   Direct access to polynomials numerical solution using NUM.SLV 166
   PTAYL 166
   QUOTIENT and REMAINDER 167
   EPSX0 and the CAS variable EPS 167
   PEVAL 167
   TCHEBYCHEFF 167

9 Vectors 169

Entering vectors *169*
   Typing a vector directly into the stack 169
   Using the matrix writer (MTRW) 169
   Vectors vs. matrices 170
   Moving to the right vs. moving down in the matrix writer 170
   Building a vector with →ARRY 172

Identifying, extracting, and inserting elements of a vector *172*
   Arrays, vectors, and matrices 174

Simple operations with vectors *174*

The VECTR menu *174*

Vectors as physical quantities *176*
   Operations with vectors 176
   Vectors in Cartesian coordinates 179
   Calculating 2×2 and 3×3 determinants 180
   Cross product as a determinant 182

Operations with 2- and 3-dimensional vectors in the HP 49 G calculator *182*
   Example 1 - Position vector 182
   Example 2 - Center of mass of a system of discrete particles 183
   Example 3 - Resultant of forces 184
      Writing the vector that joins two points in space 186
      Relative position vector 186
   Example 4 – Equation of a plane in space 187
   Example 5 – Moment of a force 188
   Example 6 – Cartesian and polar representations of vectors in the x-y plane 189
   Example 7 - Planar motion of a rigid body 191
      Angular velocity and acceleration... 192
      Relative position vector 192
      Velocity 192
      Solution of equations – one at a time 193
         Acceleration 194
      Solution of a system of linear equations using matrices 195

Row vectors, column vectors, and lists *196*
   Transforming a row vector into a column vector 196
   Transforming a column vector into a row vector 197
   Transforming a list into a vector 197
   Transforming a vector (or matrix) into a list.. 198

10 Matrices and linear algebra......199

Definitions 200
   Matrices as tensors and the Kronecker's delta function 201
Matrix operations 201
   Einstein's summation convention for tensor algebra 202
Entering matrices in the HP 49 G stack 203
   Using the matrix editor 203
   Typing in the matrix directly into the stack 204
Examples of matrix operations 204
   Multiplication by a scalar 207
   Matrix multiplication 207
   Inverse matrices 208
   Verifying properties of inverse matrices 209
Creating matrices using calculator functions 209
   Functions GET and PUT 209
   Functions GETI and PUTI 210
   The function SIZE 211
   The function TRN 211
   The function CON 212
   The function IDN 212
   The function RDM 212
   The function RANM 213
   The function SUB 213
   The function REPL 214

   The function →DIAG 214

   The function DIAG→ 214
   The function VANDERMONDE 215
   The function HILBERT 216
Using the matrix CREATE menu 216
A program to build a matrix out of a number of lists 217
   Lists represent columns of the matrix 217
   Lists represent rows of the matrix 218
Manipulating matrices by columns 218

   The function →COL 219

   The function COL→ 219
   The function COL+ 219
   The function COL- 220
   The function CSWP 220

Manipulating matrices by rows 220
    The function →ROW 221
    The function ROW→ 221
    The function ROW+ 221
    The function ROW- 221
    The function RSWP 222
    The function RCI 222
    The function RCIJ 222

Symmetric and anti-symmetric matrices 223

Matrices and solution of linear equation systems 224
    Solution to a system of linear equations using the numerical solver (NUM.SLV) 224
    Example 1 - A system with more unknowns than equations 224
    Example 2 - A system with more equations than unknowns 226
    Example 3 - A system with equal number of equations and unknowns 227
    Direct solution of a linear system in the stack 228
    Solution using the inverse matrix 228

Characterizing a matrix 228
    The matrix NORM menu 228
    The function ABS 229
    Matrix decomposition 229
    Singular value decomposition and rank 230
    The function SNRM 230
    Row norm and column norm of a matrix 230
    The functions RNRM and CNRM 230
    Eigenvalues and eigenvectors of a matrix 231
    The function SRAD 231
    Determinants, singular matrices, and conditions numbers 231
    The function COND 231
    The rank of a matrix 232
    The function RANK 232
    The determinant of a matrix 233
    The function DET 233
    Properties of determinants 233
    Cramer's rule for solving systems of linear equations 235
    The function TRACE 236
    The function TRAN 237

Additional matrix operations 237
    The matrix OPER menu 237
    The function AXL 237
    The function AXM 237
    The function HADAMARD 237
    The function LSQ 238
    The function MAD 239
    The function RSD 239
    The function LCXM 240

Gaussian and Gauss-Jordan elimination 241
   Gaussian elimination using a system of equations 241
   Gaussian elimination using matrices 242
   Gauss-Jordan elimination using matrices 243
   Pivoting 244
   Example of Gauss-Jordan elimination with full pivoting 244

Step-by-step calculator procedure for solving linear systems 248
   Solving multiple set of equations with the same coefficient matrix 249
   Calculating the inverse matrix step-by-step 250
   Inverse matrices and determinants 250

Solution to linear systems using calculator functions 251
   The function LINSOLVE 251
   Functions REF, rref, RREF 251

Eigenvalues and eigenvectors 253
   The function PCAR 253
   The function EGVL 253
   The function EGV 254
   The function JORDAN 254
   The function MAD 255
   Matrix factorization 255
   The function LU 256
   Orthogonal matrices and singular value decomposition (SVD) 256
   The function SVD 256
   The function SVL 257
   The function SCHUR 257
   The function LQ 257
   The function QR 257

Matrix Quadratic Forms 258
   The QUADF menu 258
   The function AXQ 258
   The function QXA 258
   Diagonal representation of a quadratic form 259
   The function SYLVESTER 259
   The function GAUSS 259

Matrix applications 260
   Electric circuits 260
   Structural mechanics 261
   Dimensionless numbers in fluid mechanics 263
   Stress at a point in a solid in equilibrium 266
   Principal stresses at a point 270
   Multiple linear fitting 271
   Polynomial fitting 273
   Selecting the best fitting 277
   Keystroke combinations for writing the programs POLY and POLYR 279

11 Graphics and character strings ... 281

Graphs options in the HP 49 G ... 281

Plotting an expression of the form y = f(x) ... 282
    Some useful PLOT operations for FUNCTION plots ... 284

Saving a graph for future use ... 286

Graphics of transcendental functions ... 286
    Graph of ln(X) ... 286
    Graph of the exponential function ... 288
    The PPAR variable ... 288
    Inverse functions and their graphs ... 289

Summary of FUNCTION plot operation ... 290

Plots of trigonometric and hyperbolic functions and their inverses ... 293

Plots in polar coordinates ... 294

Plotting conic curves ... 295

Parametric plots ... 296
    Creating a table of results ... 299

Plotting the solution to simple differential equations ... 299

Truth plots ... 301

Plotting histograms, bar plots, and scatter plots ... 302
    Bar plots ... 302
    Scatter plots ... 303

Slope fields ... 305

Fast 3D plots ... 306

Wireframe plots ... 307

Ps-Contour plots ... 309

Y-Slice plots ... 310

Gridmap plots ... 311

Pr-Surface plots ... 312
    The VPAR variable ... 313

Programming with graphics ... 313
    Example: Residual plots for polynomial fitting ... 313

Plotting commands for programming 315
    User-defined key for the PLOT menu 315
    The PLOT menu 316
    The function LABEL (10) 316
    The function AUTO (11) 316
    The function INFO (12) 317
    The variable EQ (3) 317
    The function ERASE (4) 317
    The function DRAX (5) 317
    The function DRAW (6) 317
    The PTYPE menu under PLOT (1) 318
    The PPAR menu (2) 318
    INFO (n) and PPAR (m) 318
    INDEP (a) 319
    DEPND (b) 319
    XRNG (c) and YRNG (d) 319
    RES (e) 320
    CENTR (g) 320
    SCALE (h) 320
    SCALEW (i) 320
    SCALEH (j) 320
    ATICK (l) 320
    AXES (k) 321
    RESET (f) 321
    The 3D menu within PLOT (7) 321
    The PTYPE menu within 3D (IV) 322
    The VPAR menu within 3D (V) 322
    INFO (S) and VPAR (W) 322
    XVOL (N), YVOL (O), and ZVOL (P) 323
    XXRNG (Q) and YYRNG (R) 323
    EYEPT (T) 323
    NUMX(U) and NUMY (V) 324
    VPAR (W) 324
    RESET (X) 324
    The STAT menu within PLOT 324
    The PTYPE menu within STAT (I) 325
    The DATA menu within STAT (II) 325
    The ΣPAR menu within STAT (III) 325
    INFO (M) and ΣPAR (K) 325
    XCOL (H) 326
    YCOL (I) 326
    MODL (J) 326
    ΣPAR (K) 326
    RESET (L) 326
    The FLAG menu within PLOT 326

So, how are plots generated in a program? 327
    Two-dimensional graphics 327
    Three-dimensional graphics 327
    The variable EQ 328
    Examples of interactive plots using the PLOT menu 328
    Examples of program-generated plots 329

Interactive drawing with the HP 49 G calculator *331*
    DOT+ and DOT- 332
    MARK 332
    LINE 332
    TLINE 332
    BOX 332
    CIRCL 333
    LABEL 333
    DEL 333
    ERASE 333
    MENU 333
    SUB 333
    REPL 334
    PICT→ 334
    X,Y→ 334

Drawing commands for use in programming *334*
    PICT 334
    PDIM 335
    PICT and the graphics screen 335
    LINE 335
    TLINE 336
    BOX 336
    ARC 336
    PIX?, PIXON, and PIXOFF 336
    PVIEW 336
    PX→C 336
    C→PX 336

Programming examples using drawing functions *337*
    Example 1 - A program that uses drawing commands 337
    Example 2 - A program to plot a natural river cross-section 337
    Example 3 - A program to visualize a polynomial fitting 341

Binary, octal, and hexadecimal number systems - a brief introduction *342*
    The BASE menu 343
    Writing non-decimal numbers in the calculator 343
    Conversion between number systems 343
    The LOGIC menu 344
    The BIT and BYTE menus 344
    Hexadecimal numbers for pixel references 344
    Pixel coordinates 345

Zooming in and out in the graphics display *345*
    ZFACT, ZIN, ZOUT, and ZLAST 345
    BOXZ 346
    ZDFLT, ZAUTO 346
    HZIN, HZOUT, VZIN and VZOUT 346
    CNTR 346
    ZDECI 346
    ZINTG 346
    ZSQR 347
    ZTRIG 347

Special function for gas dynamics: ZFACTOR(Tr, Pr) *347*

Animating graphics *347*
   Animating a collection of graphics 348
   More information on the ANIMATE function 349

Grabbing GROBs *350*
   The GROB menu 351
   →GROB 351
   BLANK 351
   GOR 351
   GXOR 351
   →LCD 352
   LCD→ 352
   SIZE 352
   An example of a program using GROB 352

Strings attached *353*
   The CHARS menu 353
   The functions NUM and CHAR 354
   The characters list 354
   Parting thoughts about character strings 355

A program with plotting and drawing functions - Generating Mohr's circle for two-dimensional stress *355*
   Background 355
   Modular programming 356
   Running the program 361
   A program to calculate principal stresses 362
   Ordering the variables in the sub-directory 363
   A second example 363

12 Solution to equations 365

Symbolic solution of algebraic equations 365
   The functions ISOL and SOLVE 365
   The S.SLV menu 366
   The function SOLVEVX 366
   The function ZEROS 366

Numerical solver menu 367

Polynomial equations 367
   Finding the solutions to a polynomial equation 368
   Generating polynomial coefficients given the polynomial's roots 368
   Generating an algebraic expression for the polynomial 369
   Evaluating a polynomial (or any) expression 369
   Replacing the variable in the polynomial's algebraic expression 370

Financial calculations 370
   Definitions 370
   Payment at beginning or end of a period... 370
   Examples 371

Solving equations with one unknown through NUM.SLV 373
   How does the numerical solver for single-unknown equations work?....... 373
      Example 1 – Hooke's law for stress and strain. 374
      Example2 – Specific energy in an open channel flow........................ 375

Special function for pipe flow: DARCY(ε/D, Re) 377
   The function FANNING(ε/D, Re) 377
   Example 3 – Flow in a pipe 377

Functions UTPN, UTPT, UTPC, and UTPF 379
   Example 3 – Upper tail probabilities for Normal, Student-t, χ^2, and F distributions 380
   Example 4 – Universal gravitation 382
   Example 5 – Energy losses in three pipelines in series........................... 383
   Changing the order of the variables 384

Graphical solution of single-unknown equations 385
   Example 1 – Solution of a cubic equation.. 385
   Example 2– Clausius-Clapeyron equation 386

Solving multiple equations 387
   Linear equation systems 387
   Rational equation systems 387
      Example 1 – Projectile motion.. 387
      Example 2 – Stresses in a thick wall cylinder 388

Using the Multiple Equation Solver (MES) 389

© 2000 – Gilberto E. Urroz
All rights reserved

Solution of triangles using the MES *389*
   Trigonometric functions in a right triangle 390
   Solution of triangles 391
   Calculating the triangle's area with Heron's formula 392
   Triangle solution using the HP 49 G's Multiple Equation Solver 392
      Creating a working directory 393
      Entering the list of equations 393
      Entering a window title 393
      Creating a list of variables 394
      Preparing to run the MES 394
      Running the MES interactively 395
      Organizing the variables in the sub directory 396
   Programming the MES triangle solution using User RPL 397
   Running the program - solution examples 397
   Adding an INFO button to your directory 399

Velocity and acceleration in polar coordinates - solution using the calculator's MES *399*

Using the SOLVESYS library for simultaneous equations *402*
   Example 1 - two conic equations 402
   Example 2 - Manning's equation for circular cross-section 403

Graphical solutions of two simultaneous equations *405*

REFERENCES - Vol. I only 406

REFERENCES - Vol. II only 406

REFERENCES - For both Vols. I and II 406

INDEX... 407

Preface

This book covers a variety of Analysis-based mathematics utilizing the amazing algebraic, numerical, and graphical capabilities of the HP 49 G calculator. The book emphasizes the practical applications of mathematics to engineering and the physical sciences. Each chapter includes a review of the mathematical concepts used, a description of the appropriate calculator features, and examples showing their application.

This book is the result of many years of experience in teaching courses on:

- Engineering Mechanics II (Dynamics),
- Uncertainty in Engineering Analysis (i.e., probability and statistics applied to engineering),
- Hydraulics,
- Fluid Mechanics, and
- Numerical Methods in Engineering

in which the use of the programmable calculators HP 48 G/G+/GX and HP 49 G has been emphasized. Many of the examples presented throughout the book had been published previously as class handouts and class notes. In the preparation of my courses, and even in some research activities, I have developed many User RPL programs for the HP 48 G/G+/GX and the HP 49 G, a good number of which are also included in this book.

The book, in its present form, developed from a set of class notes I prepared in the Spring Semester of the year 2000 for a colleague's Engineering Freshmen Seminar class. The HP 48 series calculator has been required from our engineering students since the HP 48 SX was introduced back in the early 1990's. This year, the HP 49 G made its debut among the freshmen class, therefore, there was a need to produce some training material using this calculator. The task fell on me, given my extensive experience with the HP 48 G series calculator, and my familiarity with the HP 49 G calculator since it first came out in August of 1999. The 23-page-long handout I produced for my colleague's class, together with the class notes I have prepared for the HP 48 G series calculator through the last six years, plus long hours of typing away in my computer, have developed into the book you now have in your hands.

The reader should think of this book as a mathematics handbook that emphasizes the extraordinary capabilities of the HP 49 G calculator in demonstrating different mathematical concepts. While I have made the effort of introducing those concepts before using them in each chapter, detailed explanation of mathematical concepts and proofs of theorems used in this book is to be found in more traditional mathematical textbooks. A list of references is provided in the book for that purpose.

Although the calculator includes features useful in number theory and in operations with number bases other than decimal, the book does not expand on these subjects beyond some basic description of the appropriate functions. The reason for this omission is the lack of experience of the author in those subjects. Please keep in mind that the author's training is in Civil and Environmental Engineering, where the emphasis is in Analysis-based mathematics.

Get yourself a notebook: I recommend that you go through the book armed with your calculator and a notebook. You want to have a notebook handy because sometimes the calculator display is not large enough to hold all the information you want to see when solving a given problem. Also, you may want to keep your own notes on particular types of operations with the calculator that are of interest to you.

A note about RPN: While the calculator uses the algebraic mode by default, I make it clear from the start that the book emphasizes the Reverse Polish Notation (RPN) mode. The emphasis on RPN mode is not only because it is the mode that most HP 48 G series users are familiar with, but also because it is more efficient than the algebraic mode in using the calculator display. I should also point out that the HP 49 G converts function calls and programs into RPN mode when performing operations. Therefore, it is useful that the user learn the RPN mode to better understand the workings of the calculator, and to be able to communicate with the wide community of HP 48/49 calculator users around the world.

Preferred calculator settings: When you take the calculator out of the box, or when it recovers after a system crash, the original calculator settings are such that the calculator's `Operating Mode` is set to `Algebraic`, the `beep` option is selected, the calculator's display is set to `Textbook` mode, and system flag 117 is cleared (i.e., CHOOSE boxes, rather than Soft MENU keys are selected), among other default settings. For the applications in this book I prefer that you change your settings as follows:

⬇ Press [MODE][+/-] to change to RPN mode. Change other settings so that the CALCULATOR MODES screen should looks like this:

```
▓▓▓▓▓▓▓ CALCULATOR MODES ▓▓▓▓▓▓▓
Operating Mode..RPN
Number Format....Std          _FM,
Angle Measure....Radians
Coord System......Rectangular
 _Beep   _Key Click  ✓Last Stack

Choose calculator operating mode
FLAGS CHOOS  CAS  DISP CANCL  OK
```

⬇ Press [CAS] (i.e., the F3 key). [CAS stands for *Computer Algebraic System*, a generic name given to programs that lets you produce algebraic and calculus operations in a computer or, in this case, a calculator]. Change settings, if needed, so that the CAS MODES screen looks like this:

```
▓▓▓▓▓▓▓▓▓ CAS MODES ▓▓▓▓▓▓▓▓▓
Indep var:'X'
Modulo:   3
 _Numeric  _Approx    _Complex
 _Verbose  _Step/Step _Incr Pow
 ✓Rigorous✓Simp Non-Rational
Enter modulo value
 EDIT                  CANCL  OK
```

⬇ Press [OK] to return to the CALCULATOR MODES screen. Within that screen press [FLAGS] (i.e., the F1 key). Next, press [▲] to access the last flags in the list. Press [▲][▲] to highlight system flag 117, then press [✓CHK] (i.e., the F3 key) to change the setting to Soft MENU. The SYSTEM FLAGS screen should look like this:

```
░░░░░░░░░ SYSTEM FLAGS ░░░░░░░░░
   111 Simp non rat.              ↑
   113 Linear simp on
   114 Disp 1+x → x+1
   116 Prefer cos()
 ✓ 117 Soft MENU
   119 Rigorous on
   120 Silent mode off
 ░░░░░░░ ✓CHK ░░░░░░  CANCL  OK
```

⬇ Press [OK][OK] to return to normal (that is, RPN mode) calculator display.

A note on CAS modes: One of the greatest features of the HP 49 G calculator is its CAS (Computer Algebraic System). The CAS is used in almost every operation in the calculator that involves algebraic or calculus manipulations. The CAS prefers that you use the Exact mode for most operations in order to provide the most accurate result. You will know that the Exact mode is selected if you see that the Approx mode is *not* selected in the CAS MODES screen (see above). Make sure that your calculator is set to Exact mode all the time. While in the stack, in RPN mode, this can be quickly accomplished by clearing system flag 105:

[1][0][5][+/-][ALPHA][ALPHA][C][F][ENTER].

Many errors produced when operating the calculator can be traced to not having it set to Exact mode. On the other hand, whenever the calculator, set to Exact mode, tries to evaluate expressions involving floating point numbers (i.e., numbers with decimals), it will request that the CAS mode be changed to Approx. Accept the request for changing the CAS mode, but, when done, make sure that you return the calculator to Exact mode.

Many other operations will request you to change the mode to Complex. Accept the changes when requested to obtain complex results. Within the stack, in RPN mode, if you want to return to Real (i.e., not Complex) mode, clear system flag 103:

[1][0][3][+/-][ALPHA][ALPHA][C][F][ENTER].

In normal calculator display you can check what the current CAS settings, and other calculator settings, are by checking the characters in the upper left corner of the display. The settings I prefer, unless otherwise noticed, will look like this:

```
RAD XYZ HEX R= 'X'
{HOME}
5:
4:
3:
2:
1:
EDIT VIEW STACK RCL PURGE CLEAR
```

The items in the upper left corner are interpreted as follows:

- RAD stands for radians for angular measure
- XYZ stands for rectangular (i.e., Cartesian) coordinates
- HEX stands for hexadecimal numbers as the default for binary operations
- R means Real, as opposite to Complex, CAS mode
- The equal sign (=) stands for CAS Exact mode, as opposite to ~, which means Approx mode
- 'X' means that the default CAS independent variable (stored in VX, is the upper case X)

Thus, before starting any operation involving algebraic or calculus manipulations (i.e., most operations in the calculator) make sure that the icon R= is present in the upper part of the display.

Preface to Volume I

The first three chapters of Volume I will help you become familiar with the general operation of the calculator. The volume contains 12 chapters that include the following subjects:

Introduction. The first chapter describes the calculator's features, how to update the ROM, Web sites that provide additional information, calculator operating modes, setting of flags, basic memory operations, etc.

The HP 49 G keyboard. The second chapter describes the calculator's keyboard in detail and can be used as a quick reference resource when using the calculator.

Basic calculator operation. The third chapter describes the general calculator operation, communication between calculators, special characters, powers of ten, angle modes and coordinate systems, mathematical constants included in the calculator ROM, use of the command catalog, and using units in arithmetic operations.

Calculations with real numbers. The fourth chapter describes the CAS and provides examples of calculations with real numbers using standard functions (exponential, logarithmic, powers, trigonometric functions), display options, hyperbolic functions, operations with non-decimal numbers, probability functions, etc.

Calculation with complex numbers. The fifth chapter describes complex numbers and their operation in the calculator, as well as functions of complex variables.

Introduction to lists, functions, and programs. Lists are important mathematical objects used by the HP 49 G calculator, and it dedicates a number of functions and menus to their operation. This chapter describes how to handle lists and how to related them to functions defined in the calculator. It also provides a brief introduction to the use of *User RPL* programs.

HP 49 G programming basics. This chapter provides a more detailed introduction to programming the HP 49 G using *User RPL*. Examples of programming and debugging are provided that introduce the user to simple input-output operations, branching, loops, logical expressions, and related subjects.

Algebra and related subjects. This chapter presents detailed explanations on the use of the Computer Algebraic System (CAS) functions for algebraic manipulation of expressions including polynomials, fractions, exponential and logarithmic terms, trigonometric identities, expansion and factorization. A brief presentation of modular arithmetic is presented to introduce the MOD operation.

Vectors. This chapter presents exercises in the use of the matrix writer, creation of one-dimensional arrays (or vectors), manipulation of vectors, operations with vectors, vectors in Cartesian and other coordinate systems, and a plethora of application examples taken from the physical sciences.

Matrices and linear algebra. This chapter covers an extensive amount of material on matrix theory, matrix creation, matrix manipulation, matrix operations, row and column operations, and matrix characterization. It also includes theory and exercises on linear algebra applications such as different methods for the solution of linear systems of equations, calculation of eigenvalues and eigenvectors, matrix factorization, and matrix quadratic forms. The chapter concludes with a number of applications of matrices and linear algebra to practical problems in science and engineering.

Graphics and character strings. This chapter includes a number of exercises that teach the user how to effectively use the graphics capabilities of the HP 49 G calculator. Examples are included of all possible types of graphs in the calculator: functions, polar coordinates, parametric plots, conics, differential equations, truth plots, bar plots, scatter plots, slope fields, wireframe and Fast3D plots, contour plots, y-slice animations, grid maps, and parametric surface plots. The chapter also introduces the multiple functions that the calculator provides for interactive and programmable use of graphics. Examples of interactive graphics functions and their equivalent programs are presented. This chapter briefly covers concepts related to binary number operations because they are commonly used in programming graphics. The chapter concludes with the incorporation of character strings into graphics.

Solution to equations. Chapter 12, the last in this volume, presents an exhaustive set of examples for the solution of single and multiple equations using symbolic and numerical functions in the calculator. The Chapter includes a review of solution of polynomial equations and introduces financial calculations typical of engineering economics. The chapter describes the use of the calculator's numerical solver for single equations, and of graphical and numerical methods for solving systems of non-linear equations. The use of the calculator Multiple Equation Solver (MES) is included for the solution of triangles (a typical trigonometric application) and motion in polar coordinates (engineering mechanics).

In summary, the contents of this volume cover all the calculator features involving pre-calculus mathematics, graphics, programming, and linear algebra. Chapters 1 through 4 cover the basics for simple calculator operations, i.e., the basic type of operations that calculator users at all levels should be able to master. Complex numbers, lists, and programming constitute the next level of mastery of the calculator that all users should be familiar with. These subjects are covered in Chapters 5 through 7. Algebraic operations, covered in Chapter 8, and graphics, covered in Chapter 11, are useful for users taking pre-calculus mathematics in High-School or College levels. Chapters 9 and 10 cover the fundamental operations necessary for a linear algebra course at the College level. Elementary concepts from these chapters can be used in a High-School-level linear algebra course.

1 Introduction

This book is intended to be a survey of applications of the HP 49 G programmable/graphics calculator in science and engineering mathematics. The presentation of the calculator features and the exercises presented herein emphasize engineering applications in RPN mode. However, a few exercises in algebraic mode are also presented.

Notation used in this book

Keystrokes for the HP49 G keys are shown between brackets. For example, [ENTER] indicates the use of the ENTER key located at the lower right corner of the keyboard. Keystrokes that require either [←], [→], or [ALPHA], are followed by the appropriate instruction, which may not necessarily be the main key label. For example, the sequence [←][UPDIR] indicates pressing the [←] key, followed by the [VAR] key, which has the instruction UPDIR as a secondary label.

The keys at the top row of the keyboard, labeled F1 through F6, are referred to as *soft menu keys*. The operation of these keys will depend on the current content of the keys shown at the bottom of the display. Also, when using the soft menu keys, we will indicate the appropriate instruction between brackets. For example, to get to a specific subdirectory from the HOME directory, say E202, we will indicate that the key [E202] must be pressed. To find out which one of the white keys is the appropriate one, look at the labels associated with those keys at the bottom of the display.

For entering numbers or variable names, we could, for example, indicate the keystroke sequence, as in the instruction: ['][ALPHA][ALPHA][E][2][0][2][ALPHA], the keystroke sequence to enter in the display the variable name 'C303'. Or, we could simply indicate: press ['], and type C303.

The HP 49 G operating system or ROM

The HP 49 G is not like any other calculator. Think of it as a pocketsize computer. As with any computer, there is a Central Processing Unit (CPU) that is managed by an operating system. Unlike the HP 48 G, whose operating system is permanently stored in conventional ROM (read only memory), and cannot be changed by the user, the operating system of your HP 49G can be upgraded by downloading new versions from the internet, or copying it from another HP 49 G. To obtain improved versions of the operating system check out the HP website:

http://www.hp.com/calculators/graphing/rom/

In that website, you will find instructions such as:

> Update your HP 49G
> Download a new ROM version to your calculator using the new
> _PC Connectivity Kit_. Or if you already have it get the ROM update (version 1.16). To view your ROM version type VERSION on the HP 49G.
>
> Beta ROM:
> _Version 1.17-5_

By clicking in the corresponding highlighted text, you can download the software required for your HP 49 G to communicate with your computer, or the most current version of the operating system (ROM). The information above corresponds to version _1.17-5 beta_ of the HP 49 G ROM. _Beta_ means simply that the version is still under development, and that should be used with caution. If you want to use a tested version of the ROM, click on the _PC Connectivity Kit_ text to get the current version. To communicate with a PC or Macintosh computer you need to have an appropriate cable, which can be purchased at many electronic stores directly or through the Internet, or constructed yourself using instructions available from many Internet Websites. The software help feature has instructions for upgrading your ROM. Be warned, however, that upgrading the ROM requires about 30 minutes, so reserve plenty of time for such operation.

The ROM can also be loaded into a calculator from another calculator by using the command ROMUPLOAD in the receiving calculator. Instructions for ROM transfers between calculators are also available when you download the PC Connectivity Kit software. ROM transfer between calculators also takes about 30 minutes. Also, fresh batteries should be installed before beginning transfer since the Flashing process requires more than the usual operating current.

Keeping up to date with the HP 49 G

To keep up to date with the HP 49 G visit the following website regularly:

http://www.hp.com/calculators/graphing/49g_info.html

In that website you will find technical information about the calculator and other items of interest. Technical manuals for the calculator are available at the following website:

http://www.hp.com/calculators/graphing/49g_tech.html

Another way to keep up to date with new developments with the HP 49 G calculator, is to subscribe to, and check regularly, the discussion group _comp.sys.hp48_. Although it is called _hp48_, it is the forum for discussion, report of bugs, questions and answers about both the HP 48 G and the HP 49 G calculators. It may occur that, in the future, a new group (perhaps called comp.sys.hp49) will be created, but _comp.sys.hp48_ is the group to visit at this time.

Programs for the HP 49 G

As you know, your calculator is not only a traditional scientific/graphic calculator, it is also programmable. That means that you can develop your own routines that perform certain numerical and logical operations of your choice. Many people around the world develop programs for the HP 49 G and make them available, most of them for free, to users with access to the Internet. There is a number of websites were such programs are available. You can find them by doing an Internet search with your web browser. One of the best collection of programs for the HP 48 G and HP 49 G is, however, found under the URL:

http://www.hpcalc.org

(If that URL doesn't work, try the alternative address: http://165.234.32.14)

This website was developed for the HP 48 G and earlier HP calculators. Therefore, many of the links you will find are related to the HP 48 G. There are a number of them related to the HP 49 G that permits you to download programs in a variety of categories such as math, statistics, engineering, etc.

There are two programs that I have found to be very useful for my applications (mainly civil and environmental engineering, mathematics, and statistics): *Solvesys*, a library that allows you to solve a number of non-linear equations simultaneously (also available at: http://solvesys.cjb.net/); and, *LibMaker*, a program that allows you to create libraries out of your own directories in the calculator.

Some HP 49 G programming concepts

This section is intended to clarify some concepts about programs in the HP 49 G calculators. A *program* in the HP 49 G calculator, is simply a collection of commands stored in a variable name that are performed whenever that variable name is invoked. A *programmed command* can be invoked directly from the keyboard, or by listing it in another program.

Programming Languages

There are several ways that a program can be created. The simplest is to use the calculator keyboard to create the program in User RPL language. *RPL* is the name of the programming language native to the HP 48/49 G calculators. The *User RPL* language is a subset of the RPL language which is relatively easy to access from the keyboard, and, with some experience, relatively easy to program. There exists another subset of RPL, called *System RPL*, which is more cryptic and requires a longer learning curve to master. System RPL produces programs that optimize use of memory and time resources in the calculator. Programs can also be created for the HP 49 G calculator by using *Assembler* languages. Assembler language is just one level above the binary language (machine language) used by computers. Programming in Assembler language requires developing the program in a computer and then downloading it into the calculator. Software exists that lets the user program an HP 48 G or HP 49 G calculator in User RPL or System RPL in a computer. In algebraic mode it is possible to program the HP 49 G using a language called HP Basic.

Hewlett Packard provides an emulator program for the HP 49 G that you can use on a PC with a UserRPL editor to debug programs prior to downloading to the HP 49 G as an alternative to keying them all on in on keypad.

Libraries

There is a type of program called a *library* that can be loaded and installed in your calculator and is accessible by using the [➝][LIB] keystroke sequence. Libraries reside in *memory ports*, of which your HP 49 G calculator has three: ports 0, 1, and 2. Ports are visible by using the same keystroke sequence used to access libraries (You may need to press [NXT] to get to the listing of ports if you have three or more libraries active in your calculator). Programs defined in libraries are available to the user regardless of the current working directory. That makes library programs very accessible and convenient. Also, library programs cannot be modified, so there is very little risk that you will accidentally erase or modify a library. The program *LibMaker*, mentioned above, allows you to create your own libraries after you have created User RPL programs in a given sub-directory within your calculator. More details about creating and using libraries will be provided later.

Calculator operating modes

Most HP calculators, particularly those for engineering applications, use what is referred to as *the Reverse Polish Notation* (*RPN*) mode. Most other calculators use an *algebraic entry mode*, which mimics the way that we write arithmetic or algebraic expressions in paper. The RPN mode is more efficient in the use of calculator memory and processing time, and became very well known and widely used in the engineering community when adopted by HP back in the 1970's. The HP 49 G allows you to choose between both RPN and algebraic entry modes. I prefer to use the RPN mode not only because I am more familiar with it from my experience with the HP 48 G, but also because, by using the RPN mode, one can use much of the existing expertise from the HP 48 G.

Changing the calculator mode

To change the calculator mode, press the [MODE] key in the calculator. If you haven't changed your default, you will have in the calculator an input screen, labeled CALCULATOR MODES, with the first line indicating the Operating Mode as *Algebraic*. To change mode to *RPN*, toggle the [+/-] button until RPN is shown, or use the [CHOOS] soft key and select RPN. Press [OK] to return to normal calculator operation. A similar process is used to change the operating mode to algebraic. For the time being, keep the operating mode in the *algebraic* option.

Comparing algebraic mode with RPN mode

The following program illustrates a calculation performed using both the algebraic and RPN modes. The arithmetic expression to calculate is:

$$\sqrt{\frac{3 \cdot (5 - \frac{1}{3 \cdot 3})}{23^3}} + \exp(2.5)$$

To enter this expression in the calculator we can use the *equation writer* as follows:

[EQW] [√x] [3] [×] [↰][( )] [5] [-] [1] [÷] [3] [×][3] [▲][▲][▲][▲][▲][▲][▲]

[÷] [2] [3] [y^x] [3] [▶][▶] [+] [↰] [e^x] [2] [.] [5] [ENTER]

After pressing [ENTER] the calculator displays the expression:

$$\sqrt{(3*(5-1/(3*3))/(23^3+EXP(2.5))}$$

Pressing [ENTER] again will provide the following value (accept Approx. mode on, if asked, by pressing [OK]):

$$3.49051563628$$

You could also type the expression directly into the display without using the equation writer as follows:

[√x] [↰][( )] [3] [×] [↰][( )] [5] [-] [1] [÷] [3] [×][3] [▶][÷] [2] [3] [y^x] [+] [↰] [e^x] [2] [.] [5]
[ENTER]

to obtain the same result.

Let's try now the same operation in reverse polish notation (RPN). First, clear the screen by using

[↪] [CLEAR].

Then, change the operating mode to RPN by following the procedure indicated above. Notice that in RPN operating mode, the display shows several levels of output labeled, from bottom to top, as 1, 2, 3, etc. This is referred to as the *stack* of the calculator. The different levels are referred to as the *stack levels*, i.e., stack level 1, stack level 2, etc.

Note: Years ago when memory was expensive and therefore very limited, there were only three or four stack levels or *registers*. These were referred to as *register x* (for stack level 1), *register y* (for stack level 2), *register z* (for stack level 3), and *register t* (for stack level 4). This notation is still used particularly when describing calculator functions that use only four levels of the stack. Informally, these four registers are referred to simply as *x*, *y*, *z*, and *t*. Today we have an stack size limited only by available memory, with far many more levels than there are letters in the alphabet.

Basically, what RPN means is that, instead of writing an operation such as

$$3 + 2,$$

as

[3][+][2][ENTER],

we write first the operands, in the proper order, and then the operator, i.e.,

[3] [ENTER] [2] [ENTER] [+].

As you enter the operands, they occupy different stack levels. Entering [3][ENTER] puts the number 3 in stack level 1 (register x). Next, entering [2][ENTER] pushes the 3 upwards to occupy stack level 2 (register y). Finally, by pressing [+], we are telling the calculator to apply the operator, or program, [+] to the objects occupying registers y and x. The result, 5, is then placed in register x.

Let's try some other simple operations before trying the more complicated expression used earlier for the algebraic operating mode:

35×2	[3][5] [ENTER] [2] [×]
$123/32$	[1][2][3] [ENTER] [3][2] [÷]
4^2	[2] [ENTER] [2] [y^x]
$\sqrt[3]{27}$	[2][7] [ENTER] [3] [↵][$^x\sqrt{y}$]

Notice the position of the y and the x in the last two operations. The base in the exponential operation is y (stack level 2) while the exponent is x (stack level 1) before the key [y^x] is pressed. Similarly, in the cubic root operation, y (stack level 2) is the quantity under the root sign, and x (stack level 1) is the root.

Try the following exercises:

$(5 + 3) \times 2$	[5][ENTER] [3][ENTER]	Calculates (5 +3) first.
	[2] [×]	Completes the calculation.

Let's try now,

$$\sqrt{\frac{3 \cdot (5 - \frac{1}{3 \cdot 3})}{23^3}} + \exp(2.5)$$

[3][ENTER]	Enter 3 in x
[5][ENTER]	Enter 5 in x, 3 moves to y
[3][ENTER]	Enter 3 in x, 5 moves to y, 3 moves to z
[3][×]	Place 3 and multiply, 9 appears in x
[1/x]	Calculate 1/(3×3) = 1/9 = .111111, last value placed in x; 5 in y; 3 in z
[-]	Calculate 5 - 1/(3×3) = 4.888888, which occupies x now; 3 in y
[×]	Calculate 3× (5 - 1/(3×3)) = 14.66666, this value appears in x now.
[2][3][ENTER]	Enter 23 in x, 14.66666 moves to y.
[3] [y^x]	Enter 3, calculate 233 = 12167, this value appears in x now. 14.66666 in y.
[÷]	Calculate (3× (5 - 1/(3×3)))/23³ = 1.205×10⁻³, this value now appears in x (*).
[2][.][5] [↵] [e^x]	Calculate $e^{2.5}$ = 12.18249, this value appears in x, y shows previous value.
[+]	Calculate (3× (5 - 1/(3×3)))/23³ + $e^{2.5}$ = 12.18369, this value appears in x.
[√x]	Calculate √ ((3× (5 - 1/(3×3)))/23³ + $e^{2.5}$) = 3.4905156, this value appears in x.

Obviously, the algebraic mode is easier to use for completing this calculation. However, even when in the RPN mode you can still use the Equation Writer to write algebraic expressions. Thus, the same steps used earlier to write the expression with the Equation Writer in algebraic mode, can be used in the RPN mode as follows:

[EQW] [√x] [3] [×] [↵][( )] [5] [-] [1] [÷] [3] [×][3] [▲][▲][▲][▲][▲][▲][▲]
[÷] [2] [3] [y^x] [3] [▶][▶] [+] [↵] [e^x] [2] [.] [5] [ENTER]

The resulting expression is shown in stack level 1 (register x). Pressing [ENTER] again, however, will not calculate the numerical result of the expression. It will simply enter a copy of the expression in register x, pushing the original expression to register y. This operation is useful if you want to preserve an intact copy of any expression for future use. To obtain a numeric value press [→][EVAL] ([SYMB] key) or [→][→NUM] ([ENTER] key).

> Note: My personal bias towards the RPN mode will require you to use that calculator mode in most of the applications that follow. Therefore, make sure that your calculator is set to RPN mode by pressing [MODE] and toggling the [+/-] key until RPN shows in the *Operating Mode..* field. Press [OK] to return to normal calculator display.

Flags

A flag is a variable or object that specifies a given setting of the calculator or an option in a program. Flags in the HP 49 G calculator are labeled with numbers. Flags that are used to control the process in a user-defined program are called *user flags*, and referred to by using positive integer numbers. Flags used to control calculator settings, i.e., *system flags*, are labeled with negative integer numbers.

To see the current flag setting press the [MODE] button, and then the [FLAGS] soft key (i.e., F1). You will get a screen labeled *SYSTEM FLAGS* listing flag numbers and the corresponding setting. (Note: in this screen the system flag labels use positive integer numbers. Negative integer numbers are used in the stack or in programs to set or clear system flags). A flag is said to be *set* if you see a check mark (✓) in front of the flag number. Otherwise, the flag is *not set* or *cleared*. To change the status of a system flag press the soft key [✓CHECK] while the flag you want to change is highlighted. You can use the vertical arrow keys to move about the list of system flags.

Example of flag setting: general solutions vs. principal value

For example, the default value for system flag 01 is *General solutions*. What this means is that if an equation has multiple solutions, all the solutions will be returned by the calculator, most likely in a list. By pressing the [✓CHECK] soft key you can change system flag 01 to *Principal value*. This setting will force the calculator to provide a single value known as the principal value of the solution.

To see this at work, first clear system flag 01 (i.e., set it to *General solutions*). Press [OK] twice to return to normal calculator display. We will try a quadratic equation solution. Say, solve

$$t^2 - 3t + 5 = 0.$$

Use the following keystrokes:

[EQW] [ALPHA][↵][T] [y^x][2] [▶] [-] [3] [×] [ALPHA][↵][T] [+] [5] [▲][▲] [→][=] [0]
[ENTER] [ENTER] (To keep a second copy of the equation)
[→] ['] [ALPHA][↵][T] [ENTER]
[ALPHA][ALPHA] [Q][U][A][D] [ENTER]

The solution is given as a list:

{t=(1.5,1.65831239▶

To see the two solutions press [▼]. The complete list of solutions is:

$$\{\text{`t = (1.5,-1.65831239518)' `t = (1.5,1.65831239518)'}\}$$

Press [ENTER] to return to normal calculator display. Press [⇐] to clear the stack.

Now, let's change the setting of system flag 01 to *Principal value* by using:

[0][1][+/-] [ENTER] [alpha][alpha] [S][F] [ENTER] (This produces no output)

Solve for the equation once more using:

[↦] ['] [ALPHA][↩][T]   [ENTER]
[ALPHA][ALPHA] [Q][U][A][D] [ENTER]

The solution is now:

$$\{t=(1.5,1.65831239▶$$

Press [▼] to see a unique solution (the principal value):

$$t = (1.5,1.65831239518)$$

Press [ENTER] to return to normal calculator display. Press [⇐] to clear the stack.

Notes:
[1] For a quadratic equation of the form $ax^2+bx+c = 0$, the two general solutions are given by

$$x = (-b/2a)\pm [(b/2a)^2+c]^{1/2}.$$

while the principal value would use only the positive sign in front of the square root.

[2] A result which include an ordered or the form (*a,b*), such as those shown above for *t*, represents a *complex number*. Typically, complex numbers are represented as *a+bi*, where *i* is the unit imaginary number defined as $i^2 = -1$.

Other flags of interest

Bring up once more the current flag setting by pressing the [MODE] button, and then the [FLAGS] soft key. Make sure to clear system flag 01 if it was left set from the previous exercise. Use the up and down arrow keys to move about the system flag list.

Some flags of interest and their preferred value for the purpose of the exercises that follow in this manual are:

02 *Constant → symb*: Constant values are kept as symbols
03 *Function → symb*: Functions are not automatically evaluated, instead they are loaded as symbolic
  expressions.
27 '*X+Y*i*' → *(X,Y)*: Complex numbers are represented as ordered pairs
60 [α][α] *locks*: The sequence [alpha][alpha] locks the alphabetic keyboard
91 *MTRW: Matrix*: The matrix writer produces a matrix, i.e., [[a_{11} a_{12} a_{13}]...], rather than a list of
  lists, {{a_{11}, a_{12}, a_{13}},...}.

Press [OK] twice to return to normal calculator display.

Objects and their types in the HP 49 G calculator

Your calculator is accompanied by a small booklet called *The HP 49 G Pocket Guide*. It can be used as a quick reference for figuring out the operation of calculator commands. It will be very handy once you have enough experience to understand the operation of the most commonly used commands. At this point I want to use it to show you all the types of objects that you can operate with in your calculator. Open your Pocket Guide to the very last page, page 80, to see a list of object types. The table shows 31 different types of objects, with their corresponding type number, name, and an example to illustrate the concept. Some of the most useful type of objects are listed below.

⁂ For most real-number arithmetic operations you need only object type 0, *real numbers*.

⁂ *Complex numbers*, object type 1, are an extension of real numbers that include the unit imaginary number, $i = \sqrt{(-1)}$. A complex number, e.g., $3 + 2i$, is written as *(3, 2)* in the HP 49 G calculator.

⁂ Vector and matrix operations utilize objects of type 3, *real arrays*, and, if needed, type 4, *complex arrays*. Objects type 2, *strings*, are simply lines of text (enclosed between quotes) produced with the alphanumeric keyboard.

⁂ A *list* is just a collection of objects enclosed between curly brackets and separated by spaces in RPN mode (the space key is labeled [SPC]), or by commas in algebraic mode. Lists, objects of type 5, can be very useful when processing collections of numbers. For example, the columns of a table can be entered as lists. If preferred, a table can be entered as a matrix or array.

⁂ Objects type 8 are *programs in User RPL language*. These are simply sets of instructions enclosed between the symbols << >>.

⁂ Associated with programs are objects types 6 and 7, *Global* and *Local Names*, respectively. These names, or <u>variables</u>, are used to store any type of objects. The concept of *global* or *local* names is related to the scope or reach of the variable in a given program.

⁂ An *algebraic object*, or simply, an *algebraic* (object of type 9), is a valid algebraic expression enclosed between apostrophes.

⁂ *Binary integers*, objects of type 10, are used in some computer science applications.

⁂ *Graphics objects*, objects of type 11, store graphics produced by the calculator.

⁂ *Tagged objects*, objects of type 12, are used in the output of many programs to identify results. For example, in the tagged object: Mean: 23.2, the word Mean: is the tag used to identify the number 23.2 as the mean of a sample, for example.

⁂ *Unit objects*, objects of type 13, are numerical values with a physical unit attached to them.

⁂ *Directories*, objects of type 15, are memory locations used to organize your variables in a similar fashion as folders are used in a personal computer.

⁂ *Libraries*, objects of type 16, as mentioned earlier, are programs residing in memory ports that are accessible within any directory (or sub-directory) in your calculator. They resemble *built-in functions*, objects of type 18, and *built-in commands*, objects of type 19, in the way they are used.

Organizing data in your calculator

Data can be organized in directories and sub-directories in a similar fashion as folders are used to organize data in a computer. When you turn on your calculator, you will see the characters {HOME} listed above the horizontal line at the top of the display. The name between curly brackets is the *path* of the working directory. When you use your calculator for the very first time there will be no sub-directories in the HOME directory. The labels above the soft menu keys will most likely list some variables corresponding to the Calculator Algebraic System or CAS, such as the default unknown variable VX, and others called REALA, PERIO, etc., if anything at all. To see the contents of any variable listed in the soft menu keys, press the key, or press the red [↦] key followed by the corresponding soft menu key.

To create directories, you will need to type the directory name in stack level 1, and use the command CRDIR. For example, suppose you want to create a directory named DIR1 under your HOME directory. Use the following:

[ALPHA][ALPHA]   This locks the alphabetic keyboard
[D][I][R][1] [ENTER]   Enter name. The [ENTER] key disengages the
                                        alphabetic keyboard in this instance.
[↤][PRG][MEM][DIR][CRDIR]   Navigates through menus to activate the CRDIR
                                        (Create DIRectory) command
[VAR]   Recovers the variables menu.

You should have at this point the label [DIR1] associated with the first of your soft menu keys. Notice that the label has the form of a folder, indicating that DIR1 is a directory within the HOME directory.

To get inside the DIR1 directory, press the corresponding soft menu key. The path specification at the top of the display will now show the path {HOME DIR1}. (This is similar to having the path C:\HOME\DIR1 in a PC computer.) Of course, there will be no variables stored within this directory and all the soft menu keys will be empty.

You can store any kind of objects in variables within the HOME directory or any subdirectory you create. For example, within the directory DIR1 you can create a couple of variables as follows:

1) Store the value of 324.5 into a variable that we will call A:
   [3][2][4][.][5][ENTER] [ALPHA][A] [STO>]

2) Store the string "MY DIRECTORY" in variable TITLE:
   [↦]["][ALPHA][ALPHA] [M][Y][SPC][D][I][R][E][C][T][O][R][Y][ENTER]
   [ALPHA][ALPHA] [T][I][T][L][E][ENTER] [STO>]

The soft menu keys will show the labels [TITLE] and [A] corresponding to the variables of the same name.

By creating directories and sub-directories within the HOME directory, you can develop your own directory tree to store any number of variables, algebraic objects, programs, etc. To move down the directory tree just press the soft menu key corresponding to the directory you want to access. To move up the directory tree use the command UPDIR, accessible through [↤][UPDIR].

You can visualize this tree structure by using: [↤][FILES]. This will show a window titled FILE MANAGER showing the following information:

```
0: IRAM          234KB
1: ERAM          254KB
2: FLASH         1080KB
Home             234KB
Etc.
```

The first three lines represent the memory ports where program libraries are stored. Line 4 and subsequent is the picture of your directory tree, with the HOME directory as the trunk of the tree. You can move through the directory tree by using the arrow keys. To access any particular directory or sub-directory, just highlight its name and press [OK]. The resulting screen will list all the variables in your directory or sub-directory indicating their type and size in KB. To access the normal calculator display within the selected directory or sub-directory, press [NXT][NXT][HALT]. To recover the soft menu keys, press [VAR].

More details on the use of variables and the FILES command can be found in the calculator's manual.

Note: Use the [VAR] key anytime to recover the listing of your variables in the soft key menu at the bottom of the calculator display.

2 The HP 49 G keyboard

In the previous examples we have already used many of the keys in the calculator's keyboard. In this chapter we explore the keyboard in more detail.

The HP 49 G keyboard consists of 51keys, most of them organized in five columns, except for the four arrow keys near the upper right corner of the keyboard, and the six soft menu keys (or soft keys) at the top of the keyboard. Each key in the keyboard can access more than one object or operation by combining it with the alphabetic key [ALPHA], the blue left-shift key [←], and the red right-shift key [→]. Additional characters are accessible by combining [ALPHA][←] and [ALPHA][←] with some of the keys.

Primary function of each key

The primary, or main, function of each key is shown as a white label on the key. For example, the main function of the green [ALPHA] key is to access the alphabetical keyboard. Similarly, the main function of the [√x] key is to extract the square root of the value in register x.

Alphabetic characters

The gray keys also show a white, uppercase letter in green background, indicating the upper-case alphabetic character that can be placed in the display by pressing that key after the [ALPHA] key. <u>Alphabetic characters</u> are also available in the six soft menu keys at the top of the keyboard.

Notice that the key corresponding to the letter X, also has a main function to enter the letter X. The letter X occupies a special place in the algebraic, calculus, and graphical operations of the calculator, being the default name of the variable used for those operations. In other words, as in mathematical textbooks *x* is the default unknown variable, *X is the default unknown variable in the HP 49 G calculator*. To verify this, check that there is a variable in your calculator called VX in your HOME directory. Press the corresponding soft menu key to verify that it contains the value 'X'.

Left-shift and right-shift functions

Most keys have a third function associated with them and activated by pressing the blue left-shift key [←]. The left-shift function associated with each key is shown above the key towards the left in blue color. For example, the left-shift function associated with the [SIN] key is the ASIN function (arcsine). The fourth function of a key is associated with the red right-shift key [→]. This function is shown above and to the right of each key in red color. The right-shift function associated with the [SIN] key, for example, is summation symbol Σ.

Secondary functions of the soft menu (F1-F6) keys

As indicated earlier, the soft menu keys are used to access the objects corresponding to the current menu as displayed below stack level 1 in RPN mode. They also have the alphabetic characters A, B, C, D, E, and F, associated with them. Notice that right above the soft menu keys there are blue labels corresponding to the *graphic functions* Y=, WIN, GRAPH, 2D/3D, TBLSET, and TABLE. Warning: *These functions are only accessible, by pressing [↤] followed by the corresponding soft menu key, when the calculator is operating in algebraic mode.* In RPN mode, pressing [↤] followed by any particular soft menu key, will store the contents of register *x* into the variable corresponding to the soft menu key. Therefore, make sure not to use the left-shift (blue label) functions associated with the soft menu keys (F1-F6) when in RPN mode. You can still access the graphic commands in the soft keys *in RPN mode* by holding down the left-shift key [↤] while pressing the desired soft menu key.

The arrow keys

The arrow keys allow the user to move the cursor within the display in the four main directions: up, down, left, and right. Combining these keys with the blue left-shift key will move the cursor to the first or last line (up or down) in a list, or to the first or last character in a line (left or right) within the visible window of the display. Using the red right-shift key before pressing any of the arrow keys will move the cursor to the absolute first or absolute last line in a list, or to the absolute first or absolute last character of a line, even if those positions are not visible in the display.

Utility keys

The six keys located in two rows to the left of the oddly shaped arrow keys, and the row immediately below, can be referred to as utility keys. They allow the user access to many basic utilities of the calculator. Let's discuss them in detail:

[APPS]	Produces a CHOOSE box listing a number of function menus including plot, input/output, library of constants, numeric solver, time & date, equation writer, and file manager. Many of these menus were accessible through keys in the HP 48 G calculator.

[↤][FILES]	Provides access to the file manager (see above for some more information).

[↦][BEGIN] of an object

 	that will be highlighted or selected for copying or replacing. This is therefore, an editing function.

This right-shift function is used to mark the beginning character of a segment

---

[MODE]	Used to modify different settings in the calculator. We have used this key to change the operating mode from algebraic to RPN, and vice versa. Other options that can be changed include: number format, angle measure, and coordinate system. Using the [CHOOS] soft menu key when the appropriate option is highlighted, will show the different settings available for that option (e.g., Degrees, Radians, Grads, for the angle measure). Other options, which can be changed by using the [CHK] soft menu key, are: sound a beep when mistakes are made, provide a 'key click' sound, and recover last stack.

Within this screen you can also access other options for which different settings can be selected. These options are accessed by pressing the soft menu keys [FLAGS], [CAS], or [DISP]. The [CANCL] soft menu key cancels the last selection and returns you to normal calculator display. The [OK] key, saves any change made to the MODE options, and returns you to normal calculator display.

⁜ The [FLAGS] command permits the user access to a list of 120 System Flags. These "flags" are system variables that determine options such as whether an underflow calculation should be force to become zero, or should show an error message. Use the up and down arrow keys to navigate through the list of system flags. Any system flag showing a check mark to the left of its ID number has had its default changed to the alternative setting. You can change a system flag setting by using the [CHK] soft menu key. Press [OK] to leave the FLAGS environment.

⁜ The [CAS] command accesses the Calculator Algebraic System settings. Some of these settings include the default independent variable ('X' is the original setting), a modulo parameter (see manual), and options for numeric values versus constant names (e.g., e, i, π), approximated (e.g., 0.5) versus exact (e.g. 1/2) values, complex results allowed in some functions, verbose messages for calculus information, step-by-step algebraic and calculus manipulation of some expressions, increasing versus decreasing power for polynomial ordering, rigorous results (whether to simplify |x| to x), simplify or not non-rational expressions. Use the [CHK] key to change any of these settings, [CANCL] to cancel any changes in settings and return to the MODE environment, [OK] to save changes and return to the MODE environment.

⁜ The [DISP] command access a screen where you can make changes to settings controlling the calculator display. You can change the system font, change options for editing, for the display of expressions in the stack, and for the equation writer (EQW). You can change the number of lines in the header in the display, select whether to show the current time in the display, and whether you want a digital or analog clock for the latter. [CANCL] and [OK] soft menu keys are available in this screen to cancel or accept any change in the display settings.

[←][CUSTOM]   Provides access to a customized keyboard. In other words, you can re-define the keyboard operation and access the customized keyboard through this operation. For more details consult the calculator manual.

[→][END]   This right-shift function is used to mark the ending character of a segment of an object

that will be highlighted or selected for copying or replacing. This is also an editing function.

[TOOL]   Provides a soft menu with a number of editing operations and operations on variables. The commands available are:

⁜ [EDIT] Used to edit the contents of register x.
⁜ [VIEW] Displays contents of register x in full screen.
⁜ [RCL] Recalls contents of variable the variable whose name is in register x.

- [PURGE] Deletes the variable whose name is listed in register x. If a list of variables is placed in register x, the entire list of variables is purged with this command. The [PURGE] command does not work with directory names.
- [CLEAR] Clears the stack.
- [STACK] Provides a menu of stack operations:

>[DUP][SWAP][OVER][ROT][UNROT].
>
>Pressing [NXT] access the following operations within the STACK menu:
>
>[ROLL][ROLLD][PICK][UNPIC][PICK3][DEPTH].
>
>Pressing [NXT] once more produces:
>
>[DUP2][DUPN][DROP2][DROPN][DUPDU][NIP].
>
>Pressing [NXT] once more produces the menu:
>
>[NDUPN][ ][ ][ ][ ][PRG].
>
>Of all the STACK operations listed, you most likely be using regularly the operations [DUP] and [SWAP]:

- [DUP]: Duplicates the content of register x into register y.

- [SWAP]: Swaps the contents of registers x and y.

>The [PRG] command is the same as the keystroke combination [↩][PRG].

[■][i]	Places the complex unit, i, in the display. If the complex mode is not activated, you will be asked if you want it activated. If you say NO, then no action takes place. If you say YES, either the symbol i or the value (0., 1.) is placed in register x.
[■][	] this function	The vertical bar	is used to indicate an evaluation, as in $(x^2 + 1)|_{x=2}$. To use you need to have an algebraic expression in register y, and a list of variables and values in register x. Try this example: [EQW][X][y^x][2][▶][+][1][ENTER] [↩][{][X] [SPC] [2] [ENTER] [→][	][ENTER] The result is $\qquad 2.^2 + 1.$
[VAR]	Shows the names of variables in your working directory as labels associated with the soft menu keys.
[■][UPDIR]	Moves to the directory immediately above the working directory. Repeated applications this command will eventually land you in the HOME directory.
[■][COPY]	Copies highlighted characters within an object into a temporary storage (the clipboard). This is another editing command.

[STO▶]   Stores object in register y into variable whose name is listed in register x.

[←][RCL]   Recalls contents of variable whose name is listed in register x.

[→][CUT]   Cuts highlighted characters within an object. This is yet another editing command.

[NXT]   Because there are only six soft menu keys, only six commands or variable names can be seen at a time. The [NXT] command allows the user to move to the next set of six commands or variable names, and so on, until all commands or variable names have been shown.

[←] [PRV]   Shows the previous set of six commands or variable names, if any, in the current menu or directory.

[→] [PASTE]   Pastes characters in the clipboard into the location in an object indicated by the cursor's position. This is yet another editing command.

[HIST]   Provides a list of the most recent commands, i.e., a history of calculator usage.

[←][CMD]   Shows a list of the last four commands used in the calculator. You can re-use any of those four commands by highlighting it and pressing [OK].

[→][UNDO]   Cancels last stack operation.

[CAT]   Provides a list, or catalog, of all commands available in the calculator. Command names shown in italics corresponds to user-loaded libraries. To access a particular command, press [ALPHA] followed by the first letter of the command. Then, use the up and down arrow keys to highlight the desired command. Press [OK] to activate the command. You can also enter more than the first letter in a command by using [ALPHA][ALPHA] followed by the few first letters of the command. Press [ALPHA] when you're done, before pressing [OK] to activate the command.

[←][PRG]   Use it to access programming menus.

[→][CHARS]   Provides access to all the alphanumeric characters available in the HP 49 G. Use the arrow keys to navigate across and down the list of characters. As you highlight any particular character, the keystroke sequence that you can use to generate such character directly from the keyboard will be shown in the lower left corner of the display. If no keystroke sequence is available, the lower left corner of the display will be blank. To copy a given character directly from the list, highlight the desired character and press the soft menu key labeled [ECHO1]. To copy more than one character at a time, select the characters desired one by one, pressing [ECHO] for each one of them. When finished selecting characters, press [ENTER]. The character or set of characters "echoed" to the display will show up at the bottom of the display as a character string. Pressing [ENTER] will convert the characters to an algebraic expression that will be placed in register x.

[EQW]   Starts the equation writer.

[⇦][MTWR]   Starts the matrix writer.

[⇨][']   Places a set of apostrophes in the display leaving the cursor ready for typing an algebraic expression between them.

[SYMB]   Lets the user access a number of menus for symbolic operations:

        [ALG]   for algebraic manipulation.
        [ARITH]   for arithmetic manipulation.
        [CALC]   for calculus operations.
        [GRAPH]   for graphical operations.
        [SOLVE] for solution of equations and differential equations.
        [TRIG]   for manipulating trigonometric expressions.
        [NXT][EXPLN]   for manipulating expressions involving natural logarithms and exponential functions.

These menus and the operations they contain are discussed elsewhere.

[⇦][MTH]   Lets the user access a number of menus for mathematics operations (same as the [MTH] key in the HP 48 G):

        [VECTR]   for vector operations.
        [MATRX]   for matrix operations.
        [LIST]   for operations with lists.
        [HYP]   for hyperbolic functions.
        [REAL]   standard functions that apply to real numbers (e.g., absolute value).
        [BASE]   for conversions between decimal, binary, octal, and hexadecimal number bases.
        [NXT][PROB]   for functions involving probability calculations.
        [FFT]   for Fast Fourier Transform and inverse applications.
        [CMPLX]   standard functions that apply to complex numbers (e.g., conjugate).
        [CONST]   access to the list of constants, such as π, i, e, available in the calculator.

[⇦][EVAL]   Evaluates expression in register x. The result of the evaluation depends on the type of expression in stack level 1. For example, pressing [EVAL] when a list of objects is in register x will decompose the list showing the objects in the different levels of the display, with the first object in the topmost level of display necessary. Most algebraic expressions will be simplified, and numerical results provided when pressing [EVAL].

[⇦]   Drops contents of register x, letting every display level move one level downwards.

[⇦][DEL] Deletes contents from all levels of the display (practically the same as [↪][CLEAR]).

[⇨][CLEAR] Clears display (practically the same as [⇦][DEL]).

Mathematical operations keys

The operation of the next two rows of keys should be obvious from the labeling of the keys. These could qualified as mathematical operations keys. Here is a quick run down of their operation:

[y^x] Calculates the x power (stack level 1) of the value in register y.

[⇦][e^x] Calculates the exponential function of the value in register x.

[⇨][LN] Calculates the natural logarithm (logarithm base e) of the value in register x.

[$\sqrt{x}$] Calculates the square root of the value in register x.

[⇦][x^2] Calculates the square of the value in register x.

[⇨][$^y\sqrt{x}$] Calculates the y-th root of the value in register x.

[SIN] Calculates the sine of the value in register x.

[⇦][ASIN] Calculates the sine inverse function of the value in register x.

[⇨][Σ] In the Equation Writer or in an algebraic expression in stack level 1, this keystroke combination provides the summation symbol.

[COS] Calculates the cosine of the value in register x.

[⇦][ACOS] Calculates the cosine inverse function of the value in register x.

[⇨][∂] In the Equation Writer or in an algebraic expression in stack level 1, this keystroke combination provides the derivative symbol. If an algebraic expression is provided in register y, and a variable in register x, this keystroke combination will calculate the derivative of the expression in y with respect to the variable in x.

[TAN] Calculates the tangent of the value in register x.

[⇦][ATAN] Calculates the tangent inverse function of the value in register x.

[⇨][∫] In the Equation Writer or in an algebraic expression in stack level 1, this keystroke combination provides the integration symbol. You can calculate an integral by using this keystroke sequence provided you have the variable of integration in register x, the expression to be integrated in register y, the upper limit of integration in register z, and the lower limit of integration in register t.

[EEX]	this key is used to enter powers of ten in the calculator. A number such as -32.345×10^{-23}, is entered into the HP 49 G calculator as
	[3][2][.][3][4][5][+/-][EEX][2][3][+/-] [ENTER],
	and displayed as -3.2345E-22
[←][10^x]	Calculates the antilogarithm of the value in register x.
[→][LOG]	Calculates the base-10 logarithm of the value in register x.

[+/-]	Entered after a number will change the sign of that number. Used while in a choose box, will toggle through the different options.
[←][≠]	Enters the "not equal" sign.
[→][=]	Enters the "equal" sign.

[X]	Enters the letter X.
[←][≤]	Enters the "less than or equal" sign.
[→][<]	Enters the "less than" sign.

[1/x]	Calculates the inverse of the value in register x.
[←][≥]	Enters the "greater than or equal" sign.
[→][>]	Enters the "greater than" sign.

[÷]	Divides the contents of the y register by those of the x register. Enters the fraction symbol (/) in programs and algebraic expressions.
[←][ABS]	Calculates the absolute value of vectors or the modulus of a complex number.
[→][ARG]	Calculates the argument, or angle, formed by the vector representing a complex number and the x (or real) axis.

The ALPHA key

[ALPHA]	Press it once followed by any alphabetic key to enter that letter in upper case.
[ALPHA][ALPHA]	Locks the alphabetic keyboard in upper case (default flag setting). Pressing [←] before typing any letter when the alphabetic keyboard is locked, will enter that letter in lower case.
[ALPHA][←]	Press this keystroke combination followed by any alphabetic key to enter that letter in lower case.
[ALPHA][←][ALPHA][ALPHA]	This keystroke sequence will lock the alphabetic keyboard in lower case for the default flag setting. Pressing [←] before typing any letter when the alphabetic keyboard is locked in lower case, will enter that letter in upper case.

Numeric keypad

The next four rows of keys include the [ALPHA], [←], and [→], in the first column, the numeric pad in columns two through four, and basic arithmetic operations in the last column. A description of the keys follows:

The <u>numeric keypad</u> consists of the keys [7][8][9][4][5][6][1][2][3][0], listed by rows first, then by columns. Their main function is to enter the corresponding digit. The numeric keypad is completed with the keys for the decimal point [.] and the space [SPC]. All of the numeric keypad keys have left-shift and right-shift functions as listed here:

[←][S.SLV]	Symbolic SoLVer: access menu for symbolic solver operations.
[→][NUM.SLV]	NUMeric SoLVer: access menu for numeric solver operations ([→][SOLVE] in HP 48 G).
[←][EXP&LN]	access menu for algebraic manipulations using exponential functions and natural logarithms.
[→][TRIG]	access menu for algebraic manipulations using trigonometric functions.
[←][FINANCE]	access menu for financial calculations.
[→][TIME]	access menu for time and alarm setting and operations.

[⇦][CALC]	access menu for calculus operations.
[⇨][ALG]	access menu for algebraic operations.

[⇦][MATRICES]	access menu for matrix operations (new for the HP 49 G, [MTH][MATR] used in the HP 48 G).
[⇨][STAT]	access choose list for statistical data analysis.

[⇦][CONVERT]	access menu for unit conversion tools.
[⇨][UNITS]	access unit menus.

[⇦][ARITH]	access menus for arithmetic operations.
[⇨][CMPLX]	access menu for complex number operations.

[⇦][DEF] used to define a function given an expression of the form 'F(X) = ...' in register x. For example, to define the function

$$F(X) = X^2,$$

type:

[EQW][ALPHA][F][⇦][( )][X][▶] [⇨][=] [X][y^x][2] [ENTER] [⇦][DEF] [VAR]

A variable F is available in your soft key menu. Enter [⇨][F] to see the contents of that variable. The result is a program:

$$<< \rightarrow X \ 'X^2.' \ >>$$

This is interpreted in the following way:

→ X means "enter the value in stack level 1 as the local variable X in this program',
'X^2.' means "take the value in local variable X and raised to the second power. The result shows up in register x".

Basically, the program evaluates the function F(X) for the current value of the x register, when you press the soft menu key [F]. As an example, try

[2][F]

The result is, of course, 4.

[⇨][LIB] access libraries and memory ports.

[⬅][#]   use it to enter numbers in binary, octal, or hexadecimal bases.

[➡][BASE]   access menu for conversions between different numerical bases.

[⬅][∞]   enters the symbol for infinity.

[➡][→]   enters the slender arrow character. This character is used to indicate entering local variables in a program (see example for F(X) above). It can also be used as a letter in labeling variables.

[⬅][::]   the double colon is used to specify location of objects stored in port memory.

[➡][↵]   enters the feed line character for text output.

[⬅][π]   enters the constant π.

[➡][,]   enters a comma.

Arithmetic operation keys

The following three keys correspond to the <u>arithmetic operations</u> of multiplication, subtraction, and addition, as well as to <u>the ON and ENTER keys</u>.

[×]   multiplies contents of registers x and y. In programs and algebraic expressions it enters the multiplication symbol *

[⬅][[ ]]   enters a set of square brackets leaving the cursor ready to fill up its interior. Square brackets are used to create vectors and arrays.

[➡][" "]   enters a set of double quotes leaving the cursor ready to fill up the space in between with characters. Double quotes are used to enter text strings.

[−]   subtract contents of register x from those of register y. In programs and algebraic expressions, it enters the subtraction symbol -

[⬅][()]   enters a set of parentheses leaving the cursor ready to fill up its interior. Parentheses are used to group terms in algebraic expressions, and to write complex numbers.

[➡][_]   enters the an underline character. The underline character is used mainly with unit objects.

[+]	adds contents of registers x and y. In programs and algebraic expressions, it enters the addition symbol +
[←][{}]	enters a set of curly brackets leaving the cursor ready to fill up its interior. Curly brackets are used to create lists of objects.
[→][" "]	enters a set of double quotes leaving the cursor ready to fill up the space in between with characters. Double quotes are used to enter text strings.

[ON]	used to turn the calculator on, or to cancel any operation when pressed by itself while the calculator is already on. (Notice the white label CANCEL below the key).
[←][CONT]	CONTinue. Use to continue certain type of calculator operations.
[→][OFF]	turns the calculator off.

[ENTER]	used to enter input into stack level 1 (register x).
[←][ANS]	ANSWER. Recalls the last answer obtained.
[→][→NUM]	obtains the numerical value of the expression in register x.

3 Basic calculator operation

The previous two chapters were aimed at getting you acquainted with your calculator keyboard, as well as introducing some basic concepts related to its operation. In this section we present several basic calculator functions useful in numerical computations.

Undo, Arg, and Cmd

- If you make a mistake in your calculations you may be able to recover your stack by pressing [↦][UNDO] ([HIST] key).
- If you want to re-use the arguments in the latest calculation, try [↦][ARG] ([÷] key).
- To re-use a previous command you can use the keystroke sequence [↩][CMD]. This will show a list of the last four commands used in the calculator. Highlight the command you want to re-use, and press [OK].

Deleting variables

Sometimes you will need to delete variables in the directories and subdirectories to free up memory space. Here are some hints on how to perform such operations:

- Press [VAR] to show all variables in the current directory. The variables will be shown at the bottom of the display, corresponding to the white buttons in the top row of the calculator. If there are more than six variables in any given directory, use the [NXT] or [↩][PREV] keys to view all of them.

- To delete a single variable or subdirectory in the current directory: place the name of the variable in display's level 1 by pressing [ALPHA][ALPHA] and typing the variable name, followed by [ENTER], or, simply, by entering [↦] ['] and pressing the soft menu key corresponding to the variable of interest. Then, press [TOOL][PURG]. Press [VAR] to recover the variables menu. The variable name corresponding to the variable that you just purged will no longer be available.

- To delete several variables at once, create a list of the variables to be purged, by using [↩][{}]; then, press the soft menu keys corresponding to each of the variables to be purged. When the list is complete, press [ENTER]. To purge the list, press [TOOL][PURG]. The names of the purged variables will disappear from the labels of the soft menu keys.

Transferring data between two HP 49 G calculators

Suppose you want to use the HP 49 G cable to transfer a variable called MYDAT to another calculator. The procedure for transferring directories or variables is as follows:

- Receiver calculator:

1. Press [APPS][▼][OK] to select the Input/Output (I/O) functions.
2. Press [▼][OK] to select the option Get from HP 49 from the choose box.

- Sender calculator:

1. Press [APPS][▼][OK] to select the Input/Output (I/O) functions.
2. Press [OK] to select the option Send to HP 49 from the choose box.
3. Press [CHOOS] and select the name of the object to be transferred (MYDAT, in this case).
4. Press [SEND]

Note: in step 3 for the sender calculator you can choose a list of variables by pressing [←][{}] while the *Name:* field is selected. Then, press [VAR], and press the soft menu keys corresponding to all the variables you want to transfer at once. Each variable name will be placed in the list, separated by spaces. Press [ENTER] when done selecting variables. To send the list of variables, press [SEND].

Transferring data from the HP48 to the HP 49

The HP 49 G can be operated as a receiver calculator according to the instructions presented immediately above. For the HP 48 G/G+/GX calculator on the other end of the wire use the following commands:

- HP 48 G/G+/GX as sender calculator:

1. Press [↱][I/O] to select the Input/Output (I/O) functions.
2. Press [▼] four times to highlight the option Transfer... This will produce a screen called TRANSFER, which will show the current I/O parameters for the HP 48 G/G+/GX calculator. Modify them, if needed, so that the following values are set:

   ```
   PORT: Wire      TYPE: Kermit
   NAME:
   FMT: ASC        XLAT: Newl   CHK:3
   BAUD:9600       PARITY:None  _OVRW
   ```

3. Select the name or list of names to be sent to the HP 49 G calculator. These will show in front of the *NAME:* field in the TRANSFER screen.
4. Press [SEND] when ready to send the variable(s) to the HP 49 G calculator.

Transferring data from the HP49 to the HP 48

- HP 48 G/G+/GX as receiver calculator:

1. Press [↱][I/O] to select the Input/Output (I/O) functions.

2. Press [▲] to highlight the option Start Server. This will produce a blank screen with the message *Awaiting Server Cmd.*

   ⚜ HP 49 as sender calculator:

1. Press [APPS][▼][OK] to select the Input/Output (I/O) functions.
2. Press [▼] four times to select the option highlight the option Transfer... This will produce a screen called TRANSFER, which will show the current I/O parameters for the HP 49 G calculator. Modify them, if needed, so that the following values are set:

   ```
   PORT: Wire      TYPE: Kermit
   NAME:
   FMT: ASC        XLAT: Newl   CHK:3
   BAUD:9600       PARITY:None  _OVRW
   ```

3. Select the name or list of names to be sent to the HP 49 G calculator. These will show in front of the *NAME:* field in the TRANSFER screen.
4. Press [SEND] when ready to send the variable(s) to the HP 49 G calculator.

Note: not all variables can be transferred between the HP 48 G/G+/GX and the HP 49 G calculators. In such cases you will get a *Syntax Error* message and the transfer will be aborted.

How to type Greek letters and other characters

To type characters now shown in the keyboard, the general procedure is to use [→][CHARS] to access the table of text characters, as indicated in the keyboard description above. You can use the arrow keys ([◄],[►],[▲], and [▼]) to move to any desired character. If the character is not available in the current screen, you can scroll the character list up or down using the appropriate key until the desired character appears in the display.

Once the desired character is highlighted, press [ECHO1] to copy it to the stack and return to normal calculator display. Press [ENTER] to enter the character, as an algebraic expression, in register x.

To copy more than one character, select the characters, one by one, and press [ECHO] after selecting each of them. When done, press [ENTER]. Press [ENTER] again to enter the string of characters, as an algebraic expression, in register x.

Keyboard shortcuts for special characters

Many of the Greek letters and mathematical characters can be typed directly from the keyboard by preceding the letter keys with the keystroke combination [ALPHA][→]. The characters corresponding to each of the letters of the alphabet combined with [ALPHA][→] are as follows:

A:	α	B:	β	C:	Δ	D:	δ	E:	ε	F:	ρ
G:	(none)	H:	(none)	I:	|	J:	(none)	K:	(none)	L:	(none)
M:	μ	N:	λ	O:	'	P:	Π	Q:	^	R:	√
S:	σ	T:	θ	U:	τ	V:	ω	W:	=	X:	<
Y:	>	Z:	/								

Some special characters are also available by combining the keystroke sequence [ALPHA][↱] with the numeric pad, as follows:

1: ~	2: !	3: ?	4: (see calculator)	5: \
6: ∠	7: (none)	8: (none)	9: (none)	0: →

Changing the display format

Typically, the display format used is the *standard format* (STD), which adjust the number of decimals shown according to whether or not the number displayed is an integer. For example, if you enter the number 2.5 ([2][.][5]) when the standard format is active, the display will show just 2.5. Press now, [↱][LN], and you get .916290731874. The standard format uses up to 12 decimal places for non-integer results.

- Use the following keystroke sequence to access the format change environment:

    [↰][PRG][NXT][MODES][FMT].

- To fix the number of decimals places used, say to three decimal places, use the following keystroke sequence: [3][FIX]. The displayed result will now be 0.916.

- To return to the standard display format you may press [STD]. The current display format will be marked by a dot in the corresponding white key at the top of the keyboard.

Note that in the format change environment there are also soft menu keys labeled [SCI][ENG][FM,][ML]. The [SCI] and [ENG] keys refer to what calculator manufacturers call *Scientific* and *Engineering* notations, respectively, which provides results using powers of ten.

The [FM,] button, when marked by a dot, changes the decimal point to a coma. Press [FM,] and the displayed number will read 0,916. Press [FM,] again to select the decimal point.

Press [VAR] to return to your variable menu.

Note: An alternative way to change the format is to press the [MODE] key, then press [▼] to access the Number Format... field. Use the CHOOSE box to change the number format, or toggle the [+/-] button until the proper selection is displayed in the number format field. If setting the number format to fix, make sure to indicate the number of significative figures in your result in the appropriate field.

Exercises using different number formats

Try the following exercises:

Using the standard format, enter the number 2.5689, press [ENTER] four times:

   [2][.][5][6][8][9] [ENTER][ENTER][ENTER][ENTER]

Calculate:

[×]	Result: 6.59924721
[×]	Result: 16.9528061578
[×]	Result: 43.5500637388
[↰][x^2]	Result: 1896.60805165
[↰][x^2]	Result: 3597122.10158
[↰][x^2]	Result: 1.29392874137E13 (power-of-ten notation)
	i.e., $1.2939...\times 10^{13}$.

Notice that all the results show 12 significative figures when using the standard number format.

Change the mode to FIX with 3 decimals:

[MODE][▼][CHOOS][▼][OK][▶][3][OK]

and repeat the same calculations as before:

[2][.][5][6][8][9] [ENTER][ENTER][ENTER][ENTER]

Calculate:

[×]	Result: 6.599
[×]	Result: 16.952
[×]	Result: 43.550
[↰][x^2]	Result: 1896.608
[↰][x^2]	Result: 3597122.101
[↰][x^2]	Result: 1.293E13

Change the mode to SCIENTIFIC with 5 significative figures:

[MODE][▼][CHOOS][▼] [▼] [OK][▶][5][OK]

and repeat the same calculations as before:

[2][.][5][6][8][9] [ENTER][ENTER][ENTER][ENTER]

Calculate:

[×]	Result: 6.59925E0
[×]	Result: 1.69528E1
[×]	Result: 4.35501E1
[↰][x^2]	Result: 1.89661E3
[↰][x^2]	Result: 3.59712E6
[↰][x^2]	Result: 1.29393E13

Note: the main characteristic of the *scientific* number format is that there is always one integer digit.

Change the mode to ENGINEERING with 5 significative figures:

[MODE][▼][CHOOS][▼][OK] [OK][▶][5][OK]

and repeat the same calculations as before:

[2][.][5][6][8][9] [ENTER][ENTER][ENTER][ENTER]

Calculate:
[×]	Result: 6.59925E0
[×]	Result: 16.9528E0
[×]	Result: 43.5501E0
[↩][x^2]	Result: 1.89661E3
[↩][x^2]	Result: 3.59712E6
[↩][x^2]	Result: 12.9393E12

> Note: the main characteristic of the *engineering* number format is that the power of ten is a multiple of 3.

Entering numbers as powers of ten

Typically, powers of ten are used to write numbers whose absolute values are relatively large (e.g., 1.2×10^{13}) or very small (e.g. 2.3×10^{-16}). The HP 49 G, as most calculators and computer software, uses the notation $E \pm n$, where n stands for an integer number ($n \leq 499$), to indicate a power of ten. The character sequence $E \pm n$ is used to replace the $\times 10^{\pm n}$ component of a number, if needed, when entering the number into the calculator display.

For example, the number 1.2×10^{13} is entered by using [1][.][2][EEX][1][3][ENTER]. This number is displayed as 1.2E13 if using the standard calculator display. A number with a negative power of ten, e.g., 2.3×10^{-16}, is entered as [2][.][3][EEX][1][6][+/-][ENTER]. This number will be displayed as 2.3E-16.

Changing the angle mode and coordinate system

Using the [MODE] key also provides access to the Angle Measure... option, whose setting can be changed to Degrees, Radians, or Grads (π radians = 180 degrees = 200 grads). As with the number format, angle units can be changed by pressing [CHOOS] and selecting the appropriate units, or by toggling the [+/-] sign until the desired units appear in the proper field.

You can access the angle/coordinate change environment directly from the keyboard by pressing:

[↩][PRG][NXT][MODES][ANGLE].

In this environment, you will see the following soft menu key labels:

[DEG][RAD][GRAD][RECT][CYLIN][SPHER],

referring to the angle mode as in (sexagesimal) DEGrees, RADians, or (decimal) GRADes; and to the RECTangular, CYLINdrical (polar), and SPHERical coordinate systems. For most applications we use the rectangular coordinate system, and the angles in degrees or radians.

As an example, press the keys labeled [RECT] and [DEG] to set rectangular coordinates and angle in degrees. Then, enter a vector of three components, say [2 3 5]. Use the following keystroke sequence:

[↩][[ ]][2][SPC][3][SPC][5][SPC][ENTER]

The calculator assumes that the three components of the vector correspond to the x,y,z, components of Cartesian or rectangular coordinates. The current selection of angle and coordinates is shown in the top line of the display as DEG XYZ.

If you now press [CYLIN], the three-dimensional vector gets transformed to

$$[3.606 \angle 56.310\ 5.00],$$

where the symbol $\angle$ indicates an angle. (The character can be typed in by using the sequence [ALPHA][→][6]). Notice that the characters R∠Z appear instead of XYZ in the upper left corner of the display. This change indicates that the components of the vector are now the polar cylindrical coordinates r = 3.606, θ = 56.310°, and, z = 5.00.

Pressing the key [SPHER] will produce the following vector:

$$[6.165 \angle 56.310 \angle 35.796],$$

which correspond to the spherical coordinates, ρ = 6.165, θ = 56.310°, φ = 35.796°. The upper left corner of the display will show the characters: R∠∠ instead of R∠Z.

We could change our angle units to radians by pressing [RAD]. Notice that the vector in the display now reads:

$$[6.164 \angle 0.983 \angle 0.625],$$

and the upper left corner of the display shows the characters RAD in front of the cylindrical coordinate descriptor R∠∠. If you press the key [GRAD], the vector will be displayed as

$$[6.164 \angle 62.567 \angle 39.773],$$

while the upper left corner of the display shows the characters GRAD in front of the cylindrical coordinates descriptor R∠∠. The decimal GRADes are not commonly used in practice.

Try the following exercise:

[MTH][VECTR][NXT][RECT] [←][[ ]][2][SPC][3][SPC][ENTER] [CYLIN]

If the RAD indicator is shown, you will get

$$[3.606 \angle 0.983].$$

Press [DEG], to get

$$[3.606 \angle 56.310].$$

Important relationships between angle units

As a reminder, recall that the basic transformation between angle units is as follows:

$\theta^r / \theta^\circ = \pi/180,$
$\theta^r / \theta^d = \pi/200,$
$\theta^\circ / \theta^d = 90/100 = 9/10.$

Quick conversions from degrees to radians and vice versa

Quick conversions from degrees to radians, and vice versa, can be accomplished by using the sequence:

[↰] [MTH][REAL][NXT][NXT][D→R], and [R→D], respectively.

For example, try:

[3][7] [↰] [MTH][REAL][NXT][NXT][D→R]   to convert 37° to 0.646 rad.

Also, try

[3][.][1][4] [↰] [MTH][REAL][NXT][NXT][R→D]   to convert 3.14 rad to 179.909°.

Important relationships between coordinate systems

Coordinate transformations are given by the following expressions:

- Rectangular & cylindrical coordinates:

$$r = (x^2 + y^2)^{1/2}, \qquad \tan\theta = y/x,$$
$$x = r\cos\theta, \qquad y = r\sin\theta$$

- Rectangular & spherical coordinates:

$$\rho = (x^2 + y^2 + z^2)^{1/2}, \quad \tan\theta = y/x, \quad \tan\phi = z/(x^2+y^2)^{1/2},$$
$$x = \rho\sin\phi\cos\theta, \quad y = \rho\sin\phi\sin\theta, \quad z = \rho\cos\phi.$$

Make sure that you reset your coordinate system to rectangular before continuing with this tutorial.

Additional examples on angle measure and coordinate system conversions

Assuming that the coordinate system is set to Cartesian, the angle measure to degrees, and the number format to standard. Enter a vector [1 2 5], i.e., the Cartesian components of the vector are
$$x = 1, y = 2, z = 5:$$

[↰][[ ]] [1][SPC][2][SPC][5][ENTER]   Result: [1. 2. 5.]

Change to cylindrical coordinates:

[↰][MTH][VECTR][NXT][CYLIN]   Result: [2.236 ∠63.43....]

Change number format:

[MODE][▼][CHOOS][▼][OK][▶][2][OK]   Result: [2.236 ∠63.43 5.00]

In cylindrical coordinates, therefore, the components of the vector are r = 2.24, θ = 64.43°, and z = 5.00. Now, let's change coordinates to spherical:

[SPHER]   Result: [5.48 ∠63.43 ∠24.09].

The components of this vector in spherical coordinates are ρ = 5.48, θ = 64.43°, and φ = 24.09°.

Keeping the vector currently in register x, enter the following vector in spherical coordinates:

[3.2 ∠27.5 ∠16]:

[←][[ ]] [3][.][2][SPC] [ALPHA][↱][6] [2][7][.][5][SPC] [ALPHA][↱][6] [1][6][ENTER]

Resulting in: [3.20 ∠27.50 ∠16.00]:

Add the two vectors:

[+]   Result: [8.61 ∠53.48 ∠20.35].

Convert to different coordinate systems:

[RECT]   Result: [1.78 2.41 8.08].

[CYLIN]   Result: [3.00 ∠53.48 8.08]:

HP 49 G standard mathematical constants

The following are the mathematical constants used by your calculator:

	e:	the base of natural logarithms.
	i:	the imaginary unit, $i^2 = -1$.
	π	the ratio of the length of the circle to its diameter.
	MINR:	the minimum real number available to the calculator.
	MAXR:	the maximum real number available to the calculator.

To have access to these constants, use the combination:

[←][MTH][NXT][CONST].

The calculator display will show buttons corresponding to the following variables:

[E][2.718][I][(0.00)][π][3.142].

Press [NXT] to get

[MINR][1.000][MAXR][1.000].

Press [2.718] to get the value of *e*, 2.718 in the display. If you press [E], you will get the variable name in the display, namely, 'e'. To get the numerical value, press [↵][→NUM].

Similar results are obtained by using the other built-in constants or their values from the soft menu keys.

Physical constants available in the HP49 G calculator

There is a library of physical constants built in into the calculator. This library of physical constants is associated with the HP 48 G menu for the equation library (EQ LIB). The equation library is a set of equation sets commonly used in physics and engineering that is built in into the HP 48 G. The developers of the HP 49 G decided not to include the equation library in the new calculator. However, the other menus associated with the equation library are still available in the HP 49 G calculator, although the way to access them is partially hidden.

Utility menus adapted from the HP 48 G

After many years of experience with the HP 48 G I have grown attached to certain utility menus that I have found to be very useful in calculator operations. I have created a directory, called UTL48 (for UTiLities 48), that contains the following menus:

[STATm]	to access the soft menu for statistical applications.
[PLOtm]	to access the soft menu for plotting.
[SYMBm]	to access the old soft menu for symbolic operations.
[EQLIBm]	to access the equation library menu. Although there is not an equation library in the HP 49 G, as there is in the HP 48 G, the utilities listed under this menu are available in the HP 49 G for ROM version 1.16 on.
[LIBm]	to access the old soft menu for library operations.
[SOLVEm]	to access the old SOLVER menu. This is actually a left over from the HP 48 S series, but I still find it very useful for some applications.

You can create this directory in your calculator too within the HOME directory by following these instructions:

[ALPHA][ALPHA][U][T][L][4][8][ENTER]   Enter the name 'UTL48' in register x
[↵][PRG][MEM][DIR][CRDIR]   Create directory UTL48
[VAR][UTL48]   Enter directory UTL48

The keystrokes shown below will create the programs corresponding to the different menus listed above:

[↵][<< >>] [7][4][.][0][1] [SPC] [ALPHA][ALPHA] [M][E][N][U] [ENTER]
[ALPHA][ALPHA] [S][O][L][V][E][↵][M]   [STO▶]

[↵][<< >>] [1][1][0][.][0][1] [SPC] [ALPHA][ALPHA] [M][E][N][U] [ENTER]
[ALPHA][ALPHA] [L][I][B][↵][M] [STO▶]

[↵][<< >>] [1][1][3][.][0][1] [SPC] [ALPHA][ALPHA] [M][E][N][U] [ENTER]

[ALPHA][ALPHA] [E][Q][L][I][B][↵][M] [STO▶]

[→][<< >>] [9][3][.][0][1] [SPC] [ALPHA][ALPHA] [M][E][N][U] [ENTER]
[ALPHA][ALPHA] [S][Y][M][B][↵][M] [STO▶]

[→][<< >>] [8][1][.][0][2] [SPC] [ALPHA][ALPHA] [M][E][N][U] [ENTER]
[ALPHA][ALPHA] [P][L][O][T][↵][M] [STO▶]

[→][<< >>] [9][6][.][0][1] [SPC] [ALPHA][ALPHA] [M][E][N][U] [ENTER]
[ALPHA][ALPHA] [S][T][A][T][↵][M] [STO▶]

Using the *LibMakr* library, you can create a library out of this directory, that you can then load onto your port memory, and have these soft key menus readily accessible in any of your directories.

Accessing the EQ LIB menus

One way to access the EQ LIB menus is to use the program [EQLIBm] proposed above. If you have that one ready, go ahead and press [EQLIBm]. If you don't have it ready, or do not want to move to that directory at this point, simply enter the following in your stack to activate that menu:

[1][1][3][.][0][1] [ENTER]
[ALPHA][ALPHA] [M][E][N][U] [ENTER]

As a result, you get the following soft menu keys:

[EQLIB][COLIB][MES][UTILS]

The [EQLIB] key has no definition in the HP 49 G, as mentioned before. [MES] provides access to the Multiple Equation Solver (MES), which will be discussed later. [UTILS] provides some utility functions used in selected engineering disciplines such as fluid mechanics. At this point, however, we are interested only in using the [COLIB] (COnstants LIBrary) command.

Accessing the constants library

Press [COLIB] to access the soft menu keys:

[CONLI][CONST][   ][   ][   ][EQLIB]

Pressing [CONLI] will generate a table of physical constants. The window is labeled CONSTANTS LIBRARY, and it lists the symbol and meaning of several physical constants starting with Avogadro's number. By using the up and down arrow keys you can navigate through the list of constants.

The soft menu keys available for this option include:

[SI][ENGL][UNIT][VALUE][→STK][QUIT]

By pressing [SI] or [ENGL] you select the system of units for the constants, SI = Systeme International (the International system of units), and ENGL = English, British, or Imperial system of units (ES).

Press the [VALUE] key to see the values of the constants listed in the list. If the [UNIT] key is activated (a dark square is shown), the units in the selected system are shown. Toggle the [UNIT] key to deselect unit display. The values of the constants are shown without units. Press [UNIT] once more to recover the units display.

If you want to copy the value of a particular constant to the stack, select the constant and press [→STK].
Press [QUIT] to leave the Constant Library environment.

The function CONST

The soft menu that is now available includes a function, [CONST], and a the [EQ LIB] folder. The CONST function is used to recover the value of a constant if the name of the constant is shown as an algebraic expression in register x. The value reported will correspond to the system of units selected last within the library constant environment. Try the following exercise: g is the name of the constant corresponding to the acceleration of gravity. Type:.

[ALPHA][←][G][ENTER] [CONST]

The value of g will be shown in the display in the selected system of units.

Accessing the constant library through the command catalog

The constant library can be accessed directly from the keyboard by using the [CAT] key. The command to access the library is [CONLI] (actually, the full name is CONLIB), therefore, you should search for that command in the catalog. Use the following keystrokes:

[CAT]
[ALPHA][ALPHA][C][O][ALPHA],

then, use the down arrow key to scroll down the command catalog until the CONLIB command is highlighted. When ready, press [OK]. The list of physical constants will be available for you .

Using the command catalog

The command catalog is accessible by pressing the [CAT] key. It produces an ordered list of all of the calculator's ROM commands as well as library functions loaded by the user. Library function names are shown in *italics*. The catalog lists commands starting with a few names that use non-alphabetical characters (i.e., !, %, %CH, etc.), then listing in alphabetical order those commands whose names start with upper-case letters of the English alphabet. After exhausting the letters of the alphabet, the catalog lists commands that start with other non-alphabetical characters, lower-case letters of the English alphabet, and some that start with Greek letters and arrows.

To see the start of the catalog use [←][▲]. Then use the down arrow key, [▼], to see the few commands that start with non-alphabetical characters before the command ABCUV (which starts the alphabetical listing of command names) shows up in the catalog. To see the end of the catalog use [←][▼]. The very last command is the symbol >> (which signals the end of a program). Use the upper arrow key, [▲], to see the commands at the bottom of the catalog. There are quite a few commands that start with non-alphabetic characters before reaching the last command whose name uses an upper-case letter (ZVOL).

To access a particular command, press [ALPHA] followed by the first letter of the command. Then, use the up and down arrow keys to highlight the desired command. Press [OK] to activate the command. You can also enter more than the first letter in a command name by using [ALPHA][ALPHA] followed by two or more of the first letters of the command. Press [ALPHA] when you're done, before pressing [OK] to activate the command.

Working with units in the HP 49 G calculator

Working with units entails searching through a number of menus in the calculator where the units are defined. Due to the lengthy process involved in assigning units when using the HP4G or GX calculator, my personal preference is to work without units as much as I can. When you do that, however, make sure you are using a consistent system of units (i.e., all standard base unit in the SI, or all in the English system).

To illustrate the use of units, let's calculate a force, F, given the mass,

$$m = 3.5 \text{ kg},$$

and the acceleration,

$$a = 2.3 \text{ cm/s}^2.$$

From Newton's second law,
$F = m\ a$,

we have,
$F = (3.5 \text{ kg})(2.3 \text{ cm/s}^2)$.

To perform this calculation use the following:

[3][.][5]	Enter the numeric value of m
[↱][UNITS][NXT][MASS][KG]	Select kg units in the "mass" menu.
[2][.][3] [↱][UNITS][SPEED][CM/S] acceleration are	Select units of velocity since units of not available.
[1][↱][UNITS][TIME][S]	Select units of time (s), to calculate a = v/t.
[÷]	Gives units of acceleration (cm/s^2)
[×]	To calculate F = m·a (`8.050_Kg*cm/s^2`)
[ENTER][ENTER]	Make 2 more copies of the result for future use.

Next, we demonstrate some operations that can be performed using the UNITS menu:

[↱][UNITS][TOOLS][UBASE]	Converts to basic units of the SI system.
[⇐]	Drop contents of level 1.
[UVAL]	Eliminates units, retains numeric value.
[⇐]	Drop contents of level 1.
[1][↱][UNITS][NXT][FORCE][LBF] in	Enters 1 lbf in level 1. We'll convert the value

[→][UNITS][TOOLS][UBASE][CONVE] level 2, which is in Kg cm/s2, into lbf (pound force), a unit in the English System.
Converts value in level 2 to units in level 1.

The latest operation, i.e., *unit conversions*, is about the most useful I can think of in the UNITS menu.

This operation can also be accessed by pressing [↩][CONVERT] (the key for [6]), followed by soft menu keys [UNITS][TOOL][CONVE].
A complete list of units is presented in pages 12 and 13 of the HP 49 POCKET GUIDE.

A household practical problem using units

Here is a practical problem from a household kitchen:
- How many pints in a gallon?

[→][UNITS][VOL][1][NXT][GAL][1][PT]
[↩][CONVERT][UNITS][TOOLS][CONV] (Result: 1 gallon = 8 pints);

- How many cups in a pint?:

[→][UNITS][VOL] [1][NXT][PT] [1][NXT][CU]
[↩][CONVERT][UNITS][TOOLS][CONV] (Result: 1 pint = 2 cups).

Universal gas law

The ideal gas law is written as

$$pV = nRT,$$

where:
- p = gas pressure,
- V = gas volume,
- n = number of moles (gmol),
- R = universal gas constant, and
- T = absolute temperature.

We indicated earlier in this chapter that the value of R is readily available from your calculator by using the function CONST(R). To see what is the value of R use:

[→]['][ALPHA][ALPHA][C][O][N][S][T][↩] [( )] [R][ALPHA][→][EVAL]

The value provided is 8.31451_J/(gmol*K). Suppose that for n = 0.5 mol (or gmol), we let T = 280_K, and V = 0.1_m^3, let's calculate the value of p provided with these values of T and V, use:

[.][5][→][UNITS][NXT][MASS][NXT][NXT][mol]   Enter 0.5_mol
[▼][▶][▶][▶][▶][ALPHA][↩][G][ENTER]   Modify it to read 40_gmol
[2][8][0][→][_][ALPHA][K][ENTER]   Enter value of T
[×][×]   Calculate nRT
[.][0][1]][→][UNITS][VOL][m^3] [÷]   Enter V, and calculate nRT/V

43   ©2000 – Gilberto E. Urroz
All rights reserved

The result is 116403.14_J/m^3. This result does not look like units of pressure. Let's do some conversion of units as follows:

[ENTER]   Copy the value in the stack
[↵][CONVERT][UNITS][TOOLS][UBASE]   Converts to basic S.I. units, 116403.14_kg/(m*s^2)

This result still does not look like units of pressure. So, let's try converting to *Pascals*, the preferred unit of pressure in the S.I. system, by using:

[1][↱][UNITS][NXT][PRESS][Pa]   Enter 1_Pa
[][CONVERT][UNITS][TOOLS][CONVE]   Converts quantity in level 2 to units in level 1

The result is 116403.14_Pa, or 116.40kPa.

4 Calculations with real numbers

In this chapter we introduce the Calculator Algebraic System and show simple operations with real and complex numbers.

What is the CAS in the HP 49 G calculator?

CAS stands for *Calculator Algebraic System*. This is the ROM-based software that controls the algebraic (symbolic) operations within the calculator. The CAS presents a variety of options that can be set by the user by pressing the [MODE] key, and then pressing the [CAS] soft key. This generates a screen called *CAS MODES*, which lets you set the CAS options. To see an explanation for the different options use the arrow keys to move to the different fields in the *CAS MODES* screen. The description of a given field is shown at the bottom of the calculator display.

CAS options

The different options available in the *CAS MODES* screen are:

- *Indep var:* 'X' -- Enter independent variable name. By default 'X' is the preferred independent variable. This value is stored in a variable called VX, which appears in the HOME directory of the calculator. Many of the algebraic operations in the HP 49 G use this default independent variable.
- *Modulo*: 3 -- Enter modulo value. The modulo value is used in operations with polynomials. The default value is 3.
- *Numeric* -- replace constants by values? When this option is selected, the results will be evaluated to numerical, rather than symbolic, expressions.
- *Approx* -- Perform approx calculations? Selecting this option will produce numerical results in decimal format rather than as fractions for rational numbers or algebraic expressions for other type of results.
- *Complex* -- Allow complex numbers? This option allows for complex numbers to result from certain operations.
- *Verbose* -- Display calculus information? This option allows the calculator to provide information relevant to some calculus operations (derivatives, integrals, etc.).
- *Step/Step* -- Perform operations step by step? Selecting this option allows the calculator to show intermediate steps in operations such as integration, matrix inversion, solution of linear equations, etc.
- *Incr Pow* -- Increasing polynomial ordering? Selecting this option will show polynomial terms in order of increasing power rather than the default setting of decreasing power order.
- *Rigurous* -- Don't simplify |X| to X If selected, will keep absolute value expressions from defaulting to the positive value (the principal value).
- *Simp* -- Simplify non rational expr? Needs no explanation.

Many of these settings are requested by the calculator when a particular result requires it. For example, if you try to get the square root of a negative number while in REAL (rather than COMPLEX) mode, the calculator will request that you change the mode to COMPLEX. If you refuse, the calculator aborts the operation and provides a square root error: <!> √ Error: Mode switch cancelled

Checking current CAS settings

To check the current CAS settings you need to just look at the top line in the calculator display in normal operation. For example, you may see the following setting:

RAD XYZ DEC R = 'X'

This stands for RADians for angular measurements, XYZ for Rectangular (Cartesian) coordinates, DECimal number base, Real numbers preferred, = means "exact" results, and 'X' is the value of the default independent variable.

Another possible listing of options could be

DEG R∠Z HEX C ~ 't'

This stands for DEGrees as angular measurements, R∠Z for Polar coordinates, HEXagesimal number base, Complex numbers allowed, ~ stands for "approximate" results, and 't' as the default independent variable.

Checking the calculator mode

When in RPN mode the different levels of the stack are listed in the left-hand side of the screen. When the ALGEBRAIC mode is selected there are no numbered stack levels, and the word ALG is listed in the top line of the display towards the right-hand side.

Real number calculations

This section shows some examples of calculations with real numbers. Some of these calculations may, under the right conditions, produce a complex result. We will show you how to deal with such situations, but a complete coverage of complex numbers is not presented until later in this book. Set number format to standard. Also, within CAS select exact, as opposite to approximate, and real as opposite to complex.

[MODE] [▼] (toggle [+/-] key until Std shows in the field)

[CAS] [▼][▼] (toggle [+/-] key, or press the [✓CHK] key, until the option Approx is unchecked.)

[▶] (toggle [+/-] key, or press the [✓CHK] key, until the option Complex is unchecked.)

Press [OK][OK] to return to normal calculator display.

The calculations with real numbers will be presented as exercises for you to try, as follows:

✤ Exponential, e^x, e.g., [1][↰][e^x] results in EXP(1). The reason for this result is that we selected the *exact* option, which results in algebraic formulae for the results. To obtain a numeric result, we can use [↪][→NUM]. The result is 2.7182... When the *approximate* setting is selected, the numerical result is automatic.

✤ Natural logarithm, LN, e.g., [5][↪][LN] results in LN(5). Press [↪][→NUM]., to obtain the numerical value: 1.60943...

Note: the exponential and natural logarithm functions are inverse functions of each other, i.e.,

if $y = \ln x$, then, $x = e^y$,

and vice versa.

✤ Power, y^x, e.g., 5^{-2}, [5][ENTER][2][+/-] [y^x] results in an error message: ^ Error: Negative integer. The reason is that, with the *exact* mode selected for results, all numbers entered without a decimal part are considered integers. When raising a number to a integer power, the calculator will try to multiply the base an integer number of times to calculate the power. This works fine for positive integers, but not for negative integers. To work with negative integer powers, we need to select the *approximate* mode:
[MODE] [CAS] [▼][▼][▼][+/-][OK][OK]

Try the operation again: [5][ENTER][2][+/-] [y^x], resulting in .04.
An alternative sequence for this operation is: [5][SPC][2][+/-][y^x].

The calculation of a power in the approximate method uses logarithms and anti-logarithms to calculate the power, i.e., $a^x = exp(x \cdot \ln a)$.

✤ Square power, x^2, e.g., [5][↰][x^2] results in 25.

✤ Square root, √x, e.g., [8] [√x] results in 2·√2, in the exact mode. The numerical value, obtained with [↪][→NUM], is 2.828427...

✤ x-th root of y, $^x\sqrt{}\,y$, e.g., [2][7][+/-][SPC][3] [↪][$^x\sqrt{}\,y$] results in the value -3.

Let's check out what happened if the exact mode is selected:

[MODE] [CAS] [▼][▼][+/-][OK][OK]
[2][7][+/-][SPC][3] [↪][$^x\sqrt{}\,y$]

In this case, you are asked to select the complex mode before the calculator provides a result. This is so because we now are using the exact mode. In the approximate mode condition, the calculator selects, by default, the principal root. At this point, if you were to select the option NO for the complex mode, then an error will be reported. Try it: [▼][OK]. The result is the message: XROOT Error: Mode switch cancelled.

Repeat the exercise, but this time accept the change of mode to complex:

[2][7][+/-][SPC][3] [↪][$^x\sqrt{}\,y$] [OK]

Now, since the angle measure was set to degrees, you get asked if you want to change that mode to radians. If you refuse the change, you will get the same error obtained above when refusing to change to complex. Thus, press [OK], to obtain the result:

'3*EXP(i*π/3)',

which is a complex number. (More about complex numbers later).

- Powers of 10, 10^x, e.g., [3][.][1][←][10^x] will produce the value 1258.92541179.

- Base-10 logarithm, LOG, e.g., [5][→][LOG] produces the value LOG(5) if the CAS is set to EXACT mode. Press [→][→NUM]., to obtain the numerical value: 0.69897...

These two operations are inverse of each other, i.e., $10^{\log x} = x$, and $\log 10^x = x$.

Modifying the display to obtain a textbook-like appearance

This frame refers to modifications that you can make to the display to change from the one-line algebraic expression, such as the result, '3*EXP(i*π/3)', to a more textbook-like format, such as:

$$3 \cdot EXP\left(\frac{i \cdot \pi}{3}\right)$$

To change to a textbook-like appearance, use:

[MODE][DISP] [▼] [▼][▶], toggle the [+/-] key until the *Textbook* option is checked. Then, press [OK][OK] to return to normal calculator display.

Changing from complex to real mode

Before continuing, I suggest we abandon the complex mode by using:

[MODE][CAS] [▼] [▼] [▶], toggle the [+/-] key until the complex option is unchecked.

- Trigonometric functions, SIN, COS, TAN, can have their arguments in any of the valid angle units: degrees, radians, grads. Inverse trigonometric functions, ASIN, ACOS, ATAN, will return the resulting angular measure in the units indicated in the upper left corner of the display. Try the following examples:

First, change mode to degrees: [MODE][▼][▼] , toggle the [+/-] key until Degrees is selected. Then, press [OK]. The characters DEG are shown in the upper left corner of the display. Try these exercises:

- [4][5] [SIN] results in SIN(45), a symbolic result. To get the numerical result, use [→][→NUM].
- [2][6][5] [COS], results in COS(265). Numerical result, [→][→NUM] results in -8.7155..E-2
- [2][5][+/-][TAN], results in -TAN(25). Numerical result is -0.466307

More complicated expressions involving trigonometric functions can be built by using the Equation Writer, for example:

$$\frac{1+\tan\left(\frac{\pi}{3}\right)}{1+\tan\left(\frac{\pi}{4}\right)}$$

The angles in this expressions are given in radians. To change the angle mode to radians we can access the CALCULATOR MODES screen and change the Angle Measure feature, or we can just simply type RAD into register x and press [ENTER], i.e., [ALPHA][ALPHA][R][A][D][ENTER]. Then, enter the expression to be evaluated:

[EQW] [1][+][TAN] [←][π] [÷] [3] [▶][▶][▶] [÷] [1][-][TAN] [←][π][▶] [÷] [4] [ENTER]

Using [→][EVAL] you get (1+√3)/2. If you have used instead [→][→NUM], the result would have been 1.36602...

Some functions accessed through the MTH menu

The MTH (MaTH) menu permits access to a number of functions as described below. Press [←][MTH] to activate this menu. The following folders (function menus) are available:

[VECTR][MATRX][LIST][HYP][REAL][BASE]

Pressing next reveals the menus:

[PROB][FFT][CMPLX][CONST]

Obviously, the folders [VECTR] and [MATRX] correspond to operations with vectors and matrices. These two topics will be discussed separately in later chapters. The folder [LIST] deals with lists of objects. The functions contained in the LIST folder are discussed in a later chapter in relation to some simple programming with User RPL. The [CMPLX] menu relates to complex number calculations which is the subject of the next chapter. The remaining MTH folders (i.e., HYP, REAL, BASE, PROB, FFT, and CONST) are described below.

HYP menu: hyperbolic functions, including SINH, COSH, TANH, and the inverse hyperbolic functions, namely, ASINH, ACOSH, ATANH. The [NXT] menu includes the functions EXPM and LNP1, with

$$EXPM(x) = exp(x) - 1,$$

and

$$LNP1(x) = ln(x+1).$$

Press [MTH] to return to the MTH menu.

Examples of the hyperbolic functions menu

[2][.][5][SINH] : sinh 2.5 = 6.05020..   [2][ASINH][→][→NUM] : $\sinh^{-1}(2)$ = 1.4436...
[2][.][5][COSH] : cosh 2.5 = 6.13228..   [2][ACOSH][→][→NUM] : $\cos^{-1}(2)$ = 1.3169...
[2][.][5][TANH] : tanh 2.5 = 0.98661..   [.][2][ATANH]   : $\tanh^{-1}(0.2)$ = 0.2027...
[NXT] provides the second menu:
[2][EXPM] [→][→NUM]   => 6.38905....   [1][LNP1][→][→NUM]   => 0.69314....

REAL menu: functions applicable to real numbers mainly (some functions also applicable to complex numbers), such as:

- [%]   : calculates the x percentage of y
- [%CH] : calculates 100(y-x)/x, i.e., the percentage change
- [%T] : calculates 100 x/y
- [MIN] : minimum value of x and y
- [MAX] : maximum value of x and y
- [MOD] : y mod x = residual of y/x

Press [NXT] to get to the next menu set:

- [ABS] : calculates the absolute value, |x|
- [SIGN] : determines the sign of x, i.e., -1, 0, or 1.
- [MANT]   : determines the mantissa of a number based on $\log_{10}$.
- [XPON] : determines the power of 10 in the number
- [IP]   : determines the integer part of a real number
- [FP]   : determines the fractional part of a real number

Press [NXT] once more to get the final menu set:

- [RND] : rounds up y to x decimal places
- [TRNC] : truncate y to x decimal places
- [FLOOR] : closest integer that is less than or equal to x
- [CEIL] : closest integer that is greater than or equal to x
- [D→R] : converts degrees to radians
- [R→D] : converts radians to degrees.

Press [NXT] [MTH] to recover the MTH menu.

Examples of functions from the REAL menu

[2][0][ENTER][5][.][2][%]   : 5.2 % of 20 is 1.04
[2][2][ENTER][2][5][%CH]   : percentage change from 22 to 25 is 13.6363...
[5][0][0][ENTER][2][0][%T]   : 20 is the 4% of 500
[2][+/-][ENTER][2][MIN]   : min(-2,2) = -2
[2][+/-][ENTER][2][MAX]   : max(-2,2) = 2
[1][5][MOD][4]   : 15 mod 4 = residual of 15/4 = 3

Press [NXT] to get to the next menu set:

[3][+/-][ABS]   : |-3| = 3
[5][+/-][SIGN]   : sign(-5) = -1
[2][5][4][0][MANT]   : mantissa of 2540 = 2.540
[2][5][4][0][XPON]   : exponent of 2540 = 3
[2][.][3][5][IP]   : integer part of 2.35 is 2
[2][.][3][5][FP]   : fractional part of 2.35 is 0.35

Press [NXT] once more to get the final menu set:

```
[1][.][4][5][6][7][ENTER][2][ RND  ]     : rounds up 1.4567 to 1.46
[1][.][4][5][6][7][ENTER][2][ TRNC ]     : truncates 1.4567 to 1.45
[2][.][3][FLOOR]                              : floor(2.3) = 2
[2][.][3][ CEIL  ]                              : ceil(2.3) = 3
[4][5][ D→R ]                            : 45° = 0.78539 rad
[1][.][5][ R→D ]                              : 1.5 rad = 85.943669... °.
```

Press [NXT] [MTH] to recover the MTH menu.

BASE menu: conversion between various numerical base representations, i.e., hexadecimal, binary, decimal, as well as bit, byte, and logical functions. These functions are mainly of interest to computer scientists. Please consult the calculator documentation for further instructions on how to use these functions. A brief introduction to these operations is presented in Chapter 11.

PROB menu: (Press [NXT] to reach this and other menus). Probability and statistic functions, including:

- [COMB] : combinations of y objects taken x at a time (in a combination the order of selection is irrelevant).

- [PERM] : permutations of y objects taken x at a time (in a combination the order of selection is important to consider).

- [!] : for an integer number n, this is the factorial of the number, namely n(n-1)(n-2)...3 2 1 = n! For a real number x, this is the Gamma function, Γ(x-1).

- [RAND] : random number generator. It generates a random number uniformly distributed in [0,1].

- [RDZ] : sets the seed for the random number generator.

Press [NXT] to obtain the following menu:

- [UTPC] : Upper-tail probability for the Chi-squared (χ^2) probability distribution.
- [UTPF] : Upper-tail probability for the F probability distribution.
- [UTPN] : Upper-tail probability for the Normal probability distribution.
- [UTPT] : Upper-tail probability for the Student-t probability distribution.
- [NDIST] : Probability density function for the Normal distribution.

Press [MTH] to recover the MTH menu.

Combinatorics, random numbers, and probability functions

Factorials, permutations, and combinations

The *factorial* of an integer n is defined as: $n! = n \cdot (n-1) \cdot (n-2)...3 \cdot 2 \cdot 1$. By definition, $0! = 1$.

Factorials are used in the calculation of the number of permutations and combinations of objects. For example, the number of *permutations* of r objects from a set of n distinct objects is

$$_nP_r = n(n-1)(n-1)...(n-r+1) = n!/(n-r)!$$

Also, the number of *combinations* of n objects taken r at a time is

$$\binom{n}{r} = \frac{n(n-1)(n-2)...(n-r+1)}{r!} = \frac{n!}{r!(n-r)!}$$

To simplify notation, I'll use $P(n,r)$ for permutations, and $C(n,r)$ for combinations.

The HP48 series calculators provide functions to calculate factorials, permutations and combinations by entering [MTH][NXT][PROB]. The soft key displays now the functions:

[COMB][PERM][!][RAND][RDZ]

The operation of the first three functions is described below:

[COMB]: Calculates the number of combinations of n (in level 2) items taken r (in level 1) at a time;
[PERM]: Calculates the number of permutations of n (in level 2) items taken r (in level 1) at a time;
[!]: Factorial of a positive integer (in level 1). For a non-integer, x [!], returns $\Gamma(x+1)$, where $\Gamma(x)$ is known as the *Gamma* function.

For example, to calculate 8!, enter: [8][MTH][NXT][PROB][!].

The display shows a value of 40320.

To calculate P(15,3), enter

[1][5][ENTER][3][ENTER][↵][MTH][NXT][PROB][PERM],

or, if you are already in the MTH\PROB subdirectory, enter

[1][5][ENTER][3][ENTER][PERM].

The result is 2730.

To calculate C(15,3) enter

[1][5][ENTER][3][ENTER][↵][MTH][NXT][PROB][COMB],

or, if you are already in the MTH\PROB subdirectory, enter

[1][5][ENTER][3][ENTER][COMB].

The result is 455.

The Gamma function

The *Gamma function* is defined by

$$\Gamma(\alpha) = \int_0^\infty x^{\alpha-1} e^{-x} dx$$

This function has the property that ,

$\Gamma(\alpha) = (\alpha-1) \Gamma(\alpha-1)$, for $\alpha > 1$,

therefore, it can be related to the factorial of a number, i.e.,

$\Gamma(\alpha) = (\alpha-1)!$,

when α is a positive integer.

In the HP 49 G, the factorial function [!] is extended to any real number to define the Gamma function as

$\Gamma(x) = (x-1)!$

Therefore, to obtain $\Gamma(2.5)$, for example, use

[2][.][5][ENTER][1][-][←][MTH][NXT][PROB][!].

The result is $\Gamma(2.5)$ = 3.32335097045.

Generating random numbers

The HP48G series calculator has a random number generator that returns a real number between 0 and 1. The generator is able to produce sequences of random numbers. However, after a certain number of times (a very large number indeed), the sequence tends to repeat itself. For that reason, the HP48G generator is referred to as a *pseudo-random* number generator. To generate a random number with your calculator, press:

[←][MTH][NXT][PROB][RAND].

To generate a sequence of numbers just keep pressing the [RAND] soft key.

If you want to generate a sequence of number and be able to repeat the same sequence later, you can change the "seed" of the generator by entering a given number in level 1 and pressing [RDZ] before generating the sequence. Random number generators operate by starting with a "seed" number that is transformed into the first random number of the series. The current number then serves as the "seed" for the next number and so on. By "re-seeding" the sequence with the same number you can reproduce the same sequence more than once. For example, try the following:

[.][2][5][RDZ]   Use 0.25 as the "seed."
[RAND]   First random number = 0.75285...
[RAND]   Second random number = 0.51109...
[RAND]   Third random number = 8.5429...E-2 = 0.085429....

Re-start the sequence:

[.][2][5][RDZ]   Use 0.25 once more as the "seed."
[RAND]   First random number = 0.75285...
[RAND]   Second random number = 0.51109...
[RAND]   Third random number = 8.5429...E-2 = 0.085429........

If you press [RDZ] with no value in the display, the generator will take a number based on the calculator's clock time and use it as the seed.

The pseudo-random number generator provided in your calculator produces random numbers with a *uniform distribution* in the interval [0,1]. To learn more about uniform distributions check out section 5.5 in your textbook.

Examples of probability calculations for continuous random variables

The probability distribution for a continuous random variable, X, is characterized by a function f(x) known as the *probability density function* (pdf). The pdf has the following properties: f(x) > 0, for all x, and

$$\int_{-\infty}^{+\infty} f(x)dx = 1.$$

Probabilities area calculated using the *Cumulative Distribution Function* (cdf), F(x), defined by

$$P[X < x] = F(x) = \int_{-\infty}^{x} f(\xi)d\xi.$$

Where P[X<x] stands for "the probability that the random variable X is less than the value x".

Normal distribution pdf

The expression for the normal distribution pdf is:

$$f(x) = \frac{1}{\sigma\sqrt{2\pi}} \exp[-\frac{(x-\mu)^2}{2\sigma^2}],$$

where μ is the mean, and σ^2 the variance of the distribution.

To calculate the value of f(μ,σ²,x) for the normal distribution, enter the following values: the mean, μ, in level 3; the variance, σ², in level 2; and, the value x in level 1, then enter

[MTH][NXT][PROB][NXT][NDIST].

For example, check that for a normal distribution, f(1.0,0.5,2.0) = 0.20755374. Use the following sequence:
[1][ENTER][.][5][ENTER][2][ENTER][MTH][NXT][*PROB*][NXT][*NDIST*]

Normal distribution cdf

The HP48G series calculator has a function UTPN that calculates the upper-tail normal distribution, i.e.,

$$UTPN(x) = P(X>x) = 1 - P(X<x).$$

To obtain the value of the upper-tail normal distribution UTPN we need to enter the following values: the mean, μ, in level 3; the variance, σ², in level 2; and, the value x in level 1, then enter

[MTH][NXT][*PROB*][NXT][*UTPN*].

For example, check that for a normal distribution, with μ = 1.0, σ² = 0.5, UTPN(0.75) = 0.638163. Use the following sequence:

[1]ENTER][.][5][ENTER][.][7][5][ENTER][MTH][NXT][*PROB*][NXT][*UTPN*]

Different probability calculations for normal distributions [X is N(μ,σ²)] can be defined using the function UTPN, as follows:

P(X<a) = 1 - UTPN(μ, σ²,a)

P(a<X<b) = P(X<b) - P(X<a) = 1 - UTPN(μ, σ²,b) - (1 - UTPN(μ, σ²,a))
= UTPN(μ, σ²,a) - UTPN(μ, σ²,b)

P(X>c) = UTPN(μ, σ²,c)

Example: Using μ = 1.5, and σ² = 0.5, find (a) P(X<1.0); (b) P(X>2.0); (c) P(1.0<X<2.0).

(a) P(X<1.0) = 1 - P(X>1.0) = 1 - UTPN(1.5, 0.5, 1.0). Enter:

[1][ENTER][1][.][5][SPC][.][5][SPC][1][MTH][NXT][*PROB*][NXT][*UTPN*][-].

The result is P(X<1.0) = 0.239750.

(b) P(X>2.0) = UTPN(1.5, 0.5, 2.0). Enter:

[1][.][5][SPC][.][5][SPC][2] ([MTH][NXT][*PROB*][NXT]) [*UTPN*].

The result is P(X<2.0) = 0.239750.

(c) P(1.0<X<2.0) = F(1.0) - F(2.0) = UTPN(1.5,0.5,1.0) - UTPN(1.5,0.5,2.0). Enter

[1][.][5][SPC][.][5][SPC][1][UTPN] [1][.][5][SPC][.][5][SPC][2][*UTPN*] [-]

The result is P(1.0<X<2.0) = 0.7602499 - 0.2397500 = 0.524998.

The Student-t distribution

The Student-t, or simply, the t-, distribution has one parameter ν, known as the degrees of freedom. The probability distribution function (pdf) is given by

$$f(t) = \frac{\Gamma(\frac{\nu+1}{2})}{\Gamma(\frac{\nu}{2}) \cdot \sqrt{\pi\nu}} \cdot (1+\frac{t^2}{\nu})^{-\frac{\nu+1}{2}}, -\infty < t < \infty$$

where $\Gamma(\alpha) = (\alpha-1)!$ is the gamma function defined above.

The HP48G/GX provides for values of the upper-tail (cumulative) distribution function for the t-distribution using [UTPT] given the value of t and the parameter ν. The definition of this function is, therefore,

$$UTPT(\nu,t) = \int_t^\infty f(t)dt = 1 - \int_{-\infty}^t f(t)dt = 1 - P(T \leq t)$$

To use this function, enter ν in level 2 and t in level 1, then press [UTPT]. Recall that to get to the probability functions you need to use the keystroke sequence:

[MTH][NXT][*PROB*][NXT].

For example, to calculate UTPT(5, 2.5), use the following:

[5][ENTER][2][.][5][ENTER][*UTPT*]

The result is: UTPT(5,2.5) = 2.7245...E-2
Alternatively, you can use:

[5][SPC][2][.][5][*UTPT*].

Different probability calculations for the t-distribution can be defined using the function UTPT, as follows:

P(T<a) = 1 - UTPT(ν,a)

P(a<T<b) = P(T<b) - P(T<a) = 1 - UTPT(ν,b) - (1 - UTPT(ν,a)) = UTPT(ν,a) - UTPT(ν,b)

P(T>c) = UTPT(ν,c)

The Chi-squared (χ^2) distribution

The Chi-squared (χ^2) distribution has one parameter v, known as the degrees of freedom. The probability distribution function (pdf) is given by

$$f(x) = \frac{1}{2^{\frac{v}{2}} \cdot \Gamma(\frac{v}{2})} \cdot x^{\frac{v}{2}-1} \cdot e^{-\frac{x}{2}}, v>0, x>0$$

The HP48G/GX provides for values of the upper-tail (cumulative) distribution function for the χ^2-distribution using [UTPC] given the value of t and the parameter v. The definition of this function is, therefore,

$$UTPC(v,x) = \int_t^\infty f(x)dx = 1 - \int_{-\infty}^t f(x)dx = 1 - P(X \leq x)$$

To use this function, enter v in level 2 and x in level 1, then press [UTPC]. For example, to calculate UTPC(5, 2.5), use the following:

[5][ENTER][2][.][5][ENTER][UTPC]

The result is: UTPC(5,2.5) = 0.776495...

Alternatively, you can use:

[5][SPC][2][.][5][UTPC].

Different probability calculations for the Chi-squared distribution can be defined using the function UTPC, as follows:

P(X<a) = 1 - UTPC(v,a)

P(a<X<b) = P(X<b) - P(X<a) = 1 - UTPC(v,b) - (1 - UTPC(v,a)) = UTPC(v,a) - UTPC(v,b)

P(X>c) = UTPC(v,c)

The F distribution

The F distribution has two parameters vN = numerator degrees of freedom, and vD = denominator degrees of freedom. The probability distribution function (pdf) is given by

$$f(x) = \frac{\Gamma(\frac{vN+vD}{2}) \cdot (\frac{vN}{vD})^{\frac{vN}{2}} \cdot F^{\frac{vN}{2}-1}}{\Gamma(\frac{vN}{2}) \cdot \Gamma(\frac{vD}{2}) \cdot (1 - \frac{vN \cdot F}{vD})^{(\frac{vN+vD}{2})}}$$

The HP48G/GX provides for values of the upper-tail (cumulative) distribution function for the F distribution using [UTPF] given the value of F and the parameters vN and vD. The definition of this function is, therefore,

$$UTPF(vN, vD, F) = \int_{t}^{\infty} f(F)dF = 1 - \int_{-\infty}^{t} f(F)dF = 1 - P(\Im \le F)$$

To use this function, enter vN in level 3, vD in leve2, and F in level 1, then press [UTPF]. For example, to calculate UTPF(10,5, 2.5), use the following:

[1][0][ENTER][5][ENTER][2][.][5][ENTER][UTPF]

The result is: UTPF(5,2.5) = 0.776495...
Alternatively, you can use:

[1][0][SPC][5][SPC][2][.][5][UTPF].

Different probability calculations for the F distribution can be defined using the function UTPF, as follows:

P(F<a) = 1 - UTPF(vN, vD,a)

P(a<F<b) = P(F<b) - P(F<a) = 1 -UTPF(vN, vD,b)- (1 - UTPF(vN, vD,a))
         = UTPF(vN, vD,a) - UTPF(vN, vD,b)

P(F>c) = UTPF(vN, vD,a)

FFT menu: Fast Fourier Transform - brief introduction

This menu is useful in transforming a signal (time domain) into the frequency domain, through the function [FFT] (Fast Fourier Transform), or a spectrum in the frequency domain into a signal in the time domain through the function [IFFT] (Inverse Fast Fourier Transform). The signal or spectrum used as input for these functions is provided as a vector with an even number of elements.

Use [MTH][NXT][FFT] to access the FFT menu.

Examples:

[↵][[ ]][1][SPC][2][SPC][3][SPC][4][ENTER][FFT]

Gives the following spectrum:
[(10,0), (-2,2),(-2,0),(-2,-2)]

To see the full result, press [▼], which starts the matrix editor. Use the arrow keys to move from element to element. Press [ENTER] to return to normal calculator display.

With the spectrum in level 1 of the display, press [IFFT], to recover the signal [1 2 3 4].
Press [MTH] to recover the MTH menu.
More details on the calculation of FFTs and IFFTs is presented in Volume 2 of this book.

5 Calculations with complex numbers

Introduction to complex numbers

A complex number z is a number written as

$$z = x + iy,$$

where x and y are real numbers, and i is the imaginary unit defined by $i^2 = -1$.

The complex number x+iy has a real part,
$$x = Re(z),$$
and an imaginary part,
$$y = Im(z).$$

We can think of a complex number as a point P(x,y) in the x-y plane, with the x-axis referred to as the real axis, and the y-axis referred to as the imaginary axis. Thus, a complex number represented in the form x+iy is said to be in its Cartesian representation.

A complex number can also be represented in polar coordinates (polar representation) as

$$z = re^{i\theta} = r \cdot \cos\theta + i\, r \cdot \sin\theta$$

where
$$r = |z| = (x^2+y^2)^{1/2}$$

is the magnitude of the complex number z, and

$$\theta = Arg(z) = \arctan(y/x)$$

is the argument of the complex number z.

The relationship between the Cartesian and polar representation of complex numbers is given by the Euler formula:
$$re^{i\theta} = \cos\theta + i\sin\theta$$

The complex conjugate of a complex number $z = x + iy = re^{i\theta}$, is

$$\bar{z} = x - iy = re^{-i\theta}.$$

The complex conjugate of z can be thought of as the reflection of z about the real (x-) axis. Similarly, the negative of z,

$$-z = -x - iy = -re^{i\theta},$$

can be thought of as the reflection of z about the origin.

© 2000 – Gilberto E. Urroz
All rights reserved

The CMPLX menus - basic complex number applications

There are two CMPLX (complex numbers) menu available in the calculator.

CMPLX menu through MTH

The first CoMPLeX menu presented herein is accessed by using [←][MTH][NXT][CMPLX]. The first menu shows the following functions:

[RE]: Real part of a complex number
[IM]: Imaginary part of a complex number
[C→R]: Takes a complex number (x,y) and separates it into its real and imaginary parts.
[R→C]: Takes the numbers in registers x and y and forms the complex number (y,x)
[ABS]: Calculates the magnitude of a complex number or the absolute value of a real number.
[ARG]: Calculates the argument of a complex number.

Press [NXT] to see the second menu:

[SIGN]: Calculates a complex number of unit magnitude as $z/|z|$.
[NEG]: Changes the sign of the complex number in register x of the display.
[CONJ]: Produces the complex conjugate of the complex number in register x.

CMPLX menu defined in the keyboard

A second CMPLX menu is accessible by using the right-shift option associated with the [1] key, i.e., [→][CMPLX]. The resulting menu include the following functions:

[ARG]: Calculates the argument of a complex number.
[ABS]: Calculates the magnitude of a complex number or the absolute value of a real number.
[CONJ]: Produces the complex conjugate of the complex number in register x.
[i]: Enters the imaginary unit, $i = (-1)^{1/2}$.
[IM]: Imaginary part of a complex number
[NEG]: Changes the sign of the complex number in register x of the display.

Press [NXT] to see the second menu:

[RE]: Real part of a complex number
[SIGN]: Calculates a complex number of unit magnitude as $z/|z|$.

Polar representation of a complex number

To obtain the polar representation of a complex number change the coordinate system to cylindrical. You can perform that change by using: [←][PRG][MXT][MODES][ANGLE]. This menu will show the current settings for angular measurement and coordinate system highlighted by a rectangle in the corresponding labels. Press [CYLIN] to change to cylindrical (or polar) coordinates.

Examples of basic complex number operations

Make sure your calculator is in RPN mode, angle units set to radians, coordinates set to rectangular, and the CAS is set to Complex. Enter the complex number

$$z = -3+5i,$$

as

[←][( )][3][+/-][→][,][5][ENTER].

Press [ENTER] eight times to place eight copies of the number z in the stack. Then, activate the CMPLX menu from the keyboard, i.e., [→][CMPLX]. Try the following exercises:

- Find the argument: [ARG], returns θ = 2.111...rad. Use [←] to drop this result from the stack.

- Find the magnitude: [ABS], returns r = 5.83095.... Use [←] to drop this result from the stack.

- Find the conjugate: [CONJ], returns $\bar{z}$ = (-3.,-5.). Use [←] to drop this result from the stack.

- Find imaginary part: [IM], returns Im(z) = 5. Use [←] to drop this result from the stack.

- Find the negative of z: [NEG], returns -z = (3., -5.).

- Find real part: [NXT][RE], returns Re(z) = -3. Use [←] to drop this result from the stack.

- Find a unit normal vector corresponding to z: [SIGN], returns u = (-0.514495..., 0.8574...). Press [ENTER] to return to normal calculator display.

- To check that the last result is indeed of length 1, use [NXT][ABS]. Use [←] to drop result from stack.

Note: the functions ABS and ARG can be accessed directly by combining the left- and right-shifts, respectively, with the [÷ $_z$] key.

Activate the alternate CMPLX menu by using [↩][MTH][NXT][CMPLX]

- Separate the complex number into is real and imaginary parts: [C→R], resulting in register y = -3., and register x = 5.
- To put the complex number back together use [R→C].

Polar representation

- To see the polar representation of the number use: [CAT][ALPHA][C], use the down arrow key to highlight CYLIN, then press [OK]. This changes the coordinate system to cylindrical (or polar) coordinates, and results in the number being displayed as (5.8309, ∠ 2.11), or $z = 5.83095 \cdot e^{2.111i}$.
- Notice that if the coordinates are set to cylindrical any complex number will automatically be represented in its polar form even if you enter its Cartesian representation. For example, try entering:

$$[↩][()][1][→][,][5][ENTER].$$

The result returned by the calculator is (5.09901, ∠ 1.373).

- To return to the Cartesian representation change the coordinates back to rectangular by using [CAT][ALPHA][R], highlighting RECT, and pressing [OK]. The number will be displayed now as (1.,5.).

Complex number calculations

Complex numbers can be added, subtracted, multiplied, and divided. The rules for these operations are shown below:

Let

$$z = x + i \cdot y = r \cdot e^{i\theta},$$

$$z_1 = x_1 + i \cdot y_1 = r_1 \cdot e^{i\theta_1},$$

and

$$z_2 = x_2 + i \cdot y_2 = r_2 \cdot e^{i\theta_2},$$

be complex numbers. In these definitions the numbers x, y, x_1, x_2, y_1, and y_2 are real numbers.

Addition: $\qquad z_1 + z_2 = (x_1 + x_2) + i \cdot (y_1 + y_2)$

Subtraction: $\qquad z_1 - z_2 = (x_1 - x_2) + i \cdot (y_1 - y_2)$

Multiplication: $\qquad z_1 \cdot z_2 = (x_1 \cdot x_2 - y_1 \cdot y_2) + i \cdot (x_1 \cdot y_2 + x_2 \cdot y_1) = r_1 \cdot r_2 \cdot e^{i(\theta_1 + \theta_2)}$

Multiplication of a number by its conjugate results in the square of the number's magnitude, i.e.:

$$z \cdot \bar{z} = (x + i \cdot y) \cdot (x - i \cdot y) = x^2 + y^2 = r^2 = |z|^2$$

Division:

$$\frac{z_1}{z_2} = \left(\frac{z_1}{z_2}\right) \cdot \left(\frac{\bar{z}_2}{\bar{z}_2}\right) = \frac{z_1 \cdot \bar{z}_2}{|z_2|^2} = \frac{x_1 \cdot x_2 + y_1 \cdot y_2}{x_2^2 + y_2^2} + i \frac{y_1 \cdot x_2 - x_1 \cdot y_2}{x_2^2 + y_2^2} = \frac{r_1}{r_2} \cdot e^{i(\theta_1 - \theta_2)}.$$

Powers:

$$z^n = (r \cdot e^{i \cdot \theta})^n = r^n \cdot e^{i \cdot n\theta}$$

Roots: because the argument θ of a complex number z has a periodicity of 2π, we can write

$$z = r \cdot e^{i \cdot (\theta + 2k\pi)}, \quad \text{for } k = 0, 1, 2, \ldots$$

There are n n-th roots of z calculated as

$$\sqrt[n]{z} = z^{1/n} = r^{1/n} \cdot e^{i\frac{(\theta + 2k\pi)}{n}}, \quad k = 0, 1, 2, \Lambda \ (n-1).$$

Examples of operations with complex numbers

To enter complex numbers simply enter them as ordered pairs, as shown earlier. Try the following exercises. Remember to press [ENTER] after entering a complex number:

(-2.,3.) + 5 = (3.,3.)   (4.,2.)+(0,-1.) = (4.,1.)   (-3.2, -5.2) - (6.6, -2.) = (-9.8,-3.2)

(2.,3.)·(-1.,4.) = (-14.,5.)   (2.,3.)/(5.,-2.) = (0.137,0.655)   (2.,3.)³ = (-46., 9.)

√(-1.,5.) = (1.431,1.746)   (2.,2.)⁻² = (0.,-0.125)

Note: roots, other than the square, can not be calculated directly in the HP 49 G. You will have to find the magnitude, r, and argument, θ, of the number, and then use the expression shown above for $z^{1/n}$. For example, to find the three cubic roots of 1 = (1.,0.). First find r = |(1.,0.)| = 1.0, θ = Arg(1.,0.) = 0., thus,

$$\sqrt[3]{z} = z^{1/3} = 1^{1/3} \cdot e^{\frac{2k\pi}{3}i}, \quad k = 0, 1, 2.$$

To find the roots use the following:

[EQW][↰][e^x][2][×][ALPHA][↰][k][×][↰][π][×] [↰][i] [▲][▲] [÷][3][ENTER][ENTER][ENTER]

This will keep three copies of the expression EXP(2kiπ/3). To evaluate the three roots you need to replace values of k = 0, 1, and 2. You can do that replacement by using the evaluation symbol (|) available in the [TOOL] key when combined with the right-shift key, i.e., [→][|]. The replacement is defined by using lists of the form {k 0}, {k 1}, and {k 2}.

For k = 0, use:

[←][{}] [ALPHA][←][K] [SPC] [0] [ENTER] [→][|] [ENTER] [→][EVAL]

resulting in the value of 1.0. Use [⇦] to drop this result from the stack.

Try the replacement {k 1}, to get -(1-i$\sqrt{3}$)/2. With [→][→NUM], the result is (-0.5,0.8660). Use [⇦] to drop this result from the stack.

Finally, try the replacement {k 2}, to get -(1+i$\sqrt{3}$)/2. With [→][→NUM], the result is (-0.5,-0.8660). Use [⇦] to drop this result from the stack.

Functions of a complex variable

We defined a complex variable z as $z = x + iy$, where x and y are real variables, and $i = (-1)^{1/2}$. We can also define another complex variable

$$w = F(z) = \Phi + i\Psi,$$

where, in general,

$$\Phi = \Phi(x,y), \text{ and } \Psi = \Psi(x,y),$$

are two real functions of (x, y). These real functions can also be given in terms of the polar coordinates (r, θ) if we use the polar representation for z, i.e.,

$$z = r \cdot e^{i\theta} = r(\cos\theta + i\cdot\sin\theta).$$

In such case,

$$\Phi = \Phi(r,\theta), \text{ and } \Psi = \Psi(r,\theta).$$

Recall that the coordinate transformations between Cartesian and polar coordinates are:

$r = (x^2 + y^2)^{1/2}$, $\tan\theta = y/x$,
$x = r\cos\theta$, $y = r\sin\theta$

The complex variable w is also known as a complex function. Another name for a complex function is "mapping." Thus, we say F(z) is a mapping of z. In geometric terms, this means that any figure in the x-y plane gets "mapped" onto a different figure on the Φ–Ψ plane by the complex function F(z).

As an example, take the function

$$w = F(z) = \ln(z) = \ln(r \cdot e^{i\theta}) = \ln(r) + i\theta.$$

We can identify the functions

$$\Phi = \Phi(r,\theta) = \ln(r), \text{ and } \Psi = \Psi(r,\theta) = \theta,$$

as the real and imaginary components, respectively, of the function ln(z). Using the transformations indicated above we can also write,

$$\Phi = \Phi(x,y) = \ln[(x^2+y^2)^{1/2}] = (1/2) \ln(x^2+y^2), \text{ and } \Psi = \Psi(x,y) = \tan^{-1}(y/x).$$

Example: Expanding ln(z) using the HP 49 G calculator

The following keystrokes can be used to obtain the real and imaginary parts of the complex function w = ln(z). We will take advantage of this exercise to show you a few of the algebraic functions of the HP 49 G calculator, although we will not discuss such functions in depth until later.
First, type in the function LN(z), then create the equality z = r·EXP(i·θ), as follows:

[→]['][→][LN][ALPHA][↵][Z][ENTER]
[→]['][ALPHA][↵][Z][→][=] [ALPHA][↵][R] [×][↵][e^x] [↵][i] [×] [ALPHA][→][T] [ENTER]

Then, use the SUBSTitute command in the [→][ALG] menu to replace z with its polar representation:

[→][ALG][SUBST]

This results in LN(r·EXP(i·θ)). To expand this logarithmic expression, we use the command TEXPA in the [↵][EXP&LN] menu:

[↵][EXP&LN] [TEXPA]

which results in LN(r) + LN(EXP(i·θ)). We know that LN and EXP are inverse functions, but apparently the HP 49 G doesn't, therefore, we help a little by simplifying the result to LN(r) + i·θ. This can be performed by editing the latter result by pressing [▼][EDIT]. Press the right-arrow key [▶] until the cursor is on top of the i in the expression. Use the backspace key [⇐] to clear up the characters LN(EXP(. Then, use the right-arrow key, [▶], and the backspace key, [⇐], to clear up the two parentheses left out to the right of θ in the expression. Press [ENTER] twice to return to normal calculator display. We have thus separated the function LN(z) into LN(r) + i·θ. (The moral of this story is: do not expect the calculator to do everything algebraic for you. You need to understand the operation you are performing in the calculator to be able to interpret, and, perhaps, modify, the results given by the machine.)

Example: Expanding z^2 using the HP 49 G calculator

Let's try another example by expanding the function z^2, with z = x + i·y. Use the equation writer to enter the following expression:

[EQW] [↵][( )] [ALPHA][↵][X] [+] [↵][i] [×] [↵][ALPHAX][Y] [▶][▶][▶] [y^x][2] [ENTER]

This produces the result (x + i·y)^2. To expand the expression use: [→][EVAL].

The result is: $x^2+2iyx-y^2 = (x^2-y^2)+i(2xy)$.

Functions of complex numbers in the HP 49 G calculator

Exponential, logarithmic, trigonometric, hyperbolic, and other functions can be applied directly to complex numbers by entering the numbers as ordered pairs, and then entering the appropriate function name. In the following exercises the complex numbers are shown as they will look like when entered in the screen. For example, we would show the value (3., -2.), which is entered as: [↵][( )][3][→][,][2][+/-][ENTER], followed by, say, [↵][e^x], to indicate the operation exp((3.,-2.)) = (-8.35...,-18.26...). In traditional Cartesian representation of complex numbers this operation would be written as exp(3-2i) = -8.35-18.26i.

Try the following exercises:

(5.,2.) [↵][e^x] : exp(5+2i) = -61.76+134.95i.   (-1.,2.)[→][LN]: ln(-1+2i) = 0.80+2.03i

(3.,4.) [↵][x^2]: $(3+4i)^2$ = -7+24i   (5.,5.) [√x]: √(5+5i) = 2.46+1.02i

(-2.,-2) [SIN]: sin(-2-2i) = -3.42+1.51i   (0.,-3.) [COS]: cos(-3i) = 10.07

(7.5,2.2)[TAN]: tan(7.5+2.2i) = 0.02+1.02i   3.5 [ASIN]: sin^{-1}(3.5) = 1.57-1.92i

6 Lists, functions, and programs

In this chapter we show how to define functions of one or more variables in the HP 49 G calculator. Functions are converted into User RPL programs, therefore, some basic concepts about programs are also introduced. Functions can be applied to lists of numbers, therefore, we introduce in this chapter basic notions of lists and their operations.

Lists

A list is just a collection of objects enclosed between curly brackets and separated by spaces in RPN mode (the space key is labeled [SPC]), or by commas in algebraic mode. Objects that can be included in a list are numbers, letters, character strings, variable names, and/or operators contained between curly brackets. Lists are useful for manipulating data sets and in some programming applications.

Some examples of lists are: { t 1 }, {"BETA" h2 4}, {1 1.5 2.0}, {a a a a}, { {1 2 3} {3 2 1} {1 2 3}}, etc. In the examples shown below we will limit ourselves to numerical lists.

Entering a list of objects

To enter a <u>list of numbers</u> in RPN mode, simply activate the curly brackets by pressing [←][{}], then enter the list elements pressing [SPC] to include a space between them. For example, to enter the list {0 -1 3 5} use:

[←][{}] [0] [SPC] [1][+/-] [SPC] [3] [SPC] [5] [ENTER].

When entering a <u>list that includes algebraic objects</u> make sure that the algebraic objects are enclosed between single quotes. An algebraic object consisting of a single character can be entered without the single quotes. For example, to enter the list {t -1 t2+1 5}, which combines numbers with algebraic objects, use:

[←][{}] [ALPHA][←][T] [SPC] [1][+/-] [SPC] [→]['][ALPHA][←][T] [y^x] [2] [+] [1] [▶] [SPC] [5] [ENTER].

Operations with lists of numbers

Simple operations with lists are accessible by using the keystroke sequence: [←][MTH][LIST].

The operations thus available are:

[ΔLIST][ΣLIST][ΠLIST][SORT][REVLI][ADD]

The other operations in this menu are defined as follows:

[ΔLIST]: produces a list of increments between consecutive elements in the original list, e.g.,
{1 2.1 3.5 4.2 3.8} [ΔLIST] produces {1.1 1.4 0.7 -0.4}.

[ΣLIST]: calculates the sum of the elements in the list. E.g., {1 2.1 3.5 4.2 3.8} [ΣLIST] produces 14.6.

[ΠLIST]: calculates the product of the elements in the list. E.g., {1 2.1 3.5 4.2 3.8} [ΠLIST] produces 117.306.

[SORT]: sorts elements in increasing order. E.g., {1 2.1 3.5 4.2 3.8} [SORT] produces {1 2.1 3.5 3.8 4.2}.

[REVLI]: reverses order of elements in list. E.g., {1 2.1 3.5 4.2 3.8} [REVLI] produces {3.8 4.2 3.5 2.1 1}

[ADD]: this function is used to add a constant to all elements of a list, as in:

{1 2} [ENTER] 2 [ENTER][ADD], which produces {3 4},

or to add two lists of the same size, as in:

{1 2} [ENTER] {3 4} [ENTER][ADD], which produces the list {4 6}.

Note: the [+] key is used to <u>concatenate</u> two lists or a list and a single object. For example, the concatenation of the lists {1 2} and {3 4} is accomplished by using:

{1 2} [ENTER] {3 4} [ENTER][+]

which produces the list {1 2 3 4}.

The following operation adds the number 5 to this list: [5][+], producing the list {1 2 3 4 5}.

The remaining <u>arithmetic operations</u>, namely [-], [×], and [÷], can be used with lists of the same size to operate on them term by term. You can also use them with a list and a constant to perform the required operation on each term of the list and the constant. Try the following examples:

{5 2 3}[ENTER]{4 2 -1}[ENTER] [-], you will get the list {1 0 4} (term-by-term operation)

{5 2}[ENTER]{2}[ENTER][-] produces an error message because the two lists are not of the same size.

{5 2}[ENTER] 2 [ENTER][-], subtracts 2 from each element in the list producing the list {3 0}.

{-1 0 1}[ENTER]{2 3 4}[×] produces the list {-2 0 4} (term-by-term operation)

{3 4 5}[ENTER] 4 [×] produces the list {12 16 20}

{25 35 12}[ENTER]{5 7 3}[÷] produces the list {5 5 4} (term-by-term operation)

{20 40 15} [ENTER] 5 [÷] produces the list {4 8 3}

Composing and decomposing lists

Additional operations with lists are available by pressing [←][PRG][LIST]. In particular, we will describe the use of the operations [OBJ→], [→LIST].

[OBJ→] is used to decompose a list into its elements. By using this operation the elements of the list are placed in different levels of the display, with level 1 displaying the number of elements, level 2 displaying the last element on the list, level 3 displaying the second-to-last element on the list, and so on. For example, {0.5 3.2 6.5} [←][PRG][LIST][OBJ→] produces the following display:

```
4:                0.5
3:                3.2
2:                6.5
1:                  3   (← number of elements)
```

[→LIST] is used to create a list using the values available in the different display levels. The number of elements in the list must be placed in level 1. The value in level 2 will become the last element in the list, the value in level 3 will become the second-to-last element in the list, and so on. For example, the following display:

```
4:                0.5
3:                4.2
2:                8.5
1:                  2   (← number of elements)
```

will produce this list: {4.2 8.5} when the keystroke sequence [←] [PRG][LIST][→LIST] is used.

Please notice that [OBJ→] and [→LIST] are inverse operations.

Functions

Defining a function of one variable

Defining a function consists in creating a variable that contains the definition of the function. For example, suppose we want to define the function

$$f(x) = x^3$$

There are two ways to enter the expression that defines the function:

- Using the equation editor:

    [EQW] [ALPHA][←][F] [←][( )] [ALPHA][←][X] [▶] [→][=] [ALPHA][←][X] [y^x][3] [ENTER]

- Entering the expression directly into level 1 of the display:

    ['] [ALPHA][←][F] [←][( )] [ALPHA][←][X] [▶] [→][=] [ALPHA][←][X] [y^x][3] [ENTER]

To define the function, use the keystroke sequence: [←][DEF] (use the key for [2]).

Press [VAR] if needed. There will be a new soft key named f, i.e., [f]. To see how the function is stored into the variable f, press [→][f]. Level 1 of the display now shows the following:

$$<< \rightarrow x \ 'x^3 ' \ >>$$

This expression represents a program written in User RPL language that can be interpreted as follows: take the value in level 1 and assign it to x (→ x), then, calculate the expression between quotes ($'x^3'$). The result is then placed in level 1 of the display (register x).

Notice that the function definition procedure used above (through [←][DEF]), translates the function definition (in general, 'f(x) = expression containing x') into a program of the form

$$<< \rightarrow x \ 'expression \ containing \ x' >>.$$

Note: RPL programs are always enclosed within the double quotes << >>.

Evaluating the function

Let's now evaluate our function f(x) for a given value of x, say, x = 5. First, you need to enter the value of x in level 1 of the display by pressing [5][ENTER]. Next, press the soft key for f, i.e., [f]. The result is 125, or f(5) = 125.

We could evaluate the function using a list as the argument. For example, let's find the value of f({1 2 3}). First, enter the list in level 1 of the display:

[←][{}] [1] [SPC] [2] [SPC] [3] [ENTER]

Then, press the button [f]. The result is {1 8 27}, i.e., f[{1 2 3}] = {1 8 27}. Notice that the program defining f(x) applied the function definition (namely, x^3) to each element of the argument list and produced a new list with the same number of elements.

When list arguments fail

Let's now try defining a different function,

$$g(x) = \frac{\sinh(x)}{1 + x^2}.$$

Use the following keystroke sequence to create the variable g:

[→]['] [ALPHA][←][G] [←][( )] [ALPHA][←][X] [▶] [→][=] [←][MTH][HYP][SINH] [ALPHA][←][X]
[▶] [÷][←][( )] [1] [+][ALPHA][←][X] [y^x][2] [ENTER] [←][DEF].

Press [VAR] to recover your variable menu. Now, evaluate g(3.5), by entering the value of the argument in level 1 ([3][.][5][ENTER]) and then pressing [g]. The result is 1.2485..., i.e., g(3.5) = 1.2485... Try also obtaining g[{1 2 3}], by entering the list in level 1 of the display ([↰][{}] [1] [SPC] [2] [SPC] [3] [ENTER]), and pressing [G]. In this case we get a division error message. The calculator display looks as follows:

```
<!>/ Error:
Invalid Dimension
```

Press [ON] to recover the normal calculator display. You will see the following output:

```
4:
3:
2:    { SINH(1)  SINH(2)  SIN
1:                 {1 1 4 9}
```

Notice that the list in level 1 has four elements. Now, drop that list by pressing [⌫]. The list now in level 1 has only three elements. Apparently, the program defined by [g] tried to perform a division between two lists that have different number of elements, and that is what causes the division error reported.   (Most arithmetic operations between lists require lists with the same number of elements).

How did we end up with lists of different sizes if we started with a single list of size 3? The explanation lies in the way that the operator [+] applies to lists. The operator [+], when applied to lists, does not act as a simple addition, instead it is known as the concatenation operator and its function is to attach (or concatenate) the two lists. For our particular situation, the list x = {1 2 3} is squared in the denominator (1+x^2) of the expression that defines g(x) producing the list {1 4 9}. Then, this list is concatenated to the number 1 by way of the operator [+] in the denominator (1+x^2). This result in the list {1 1 4 9} = 1 + {1 4 9} shown in level 1 of the display (see above). The list in level 2 of the display represents SINH(x). In order to calculate the function g(x), the calculator tries to divide SINH(x), which is a list of three elements, by (1+x^2), which is a list of four elements, thus producing a division error.

The moral of this exercise is the following: when trying to evaluate a function whose definition includes the plus (+) sign, if the function argument is a list the result most likely will produce an algebraic error. For example, when we evaluated f[{1 2 3}], we obtained a reasonable result (namely, {1 8 27}), because the definition of the function (f(x) = x^3) did not include any plus signs.

An example of programming

To get around using the [+] sign, the HP48G/GX calculator offers the function ADD that applies particularly to lists, but that can be used to represent addition of other numbers. Instead of using the definition currently stored for the variable [G], namely, << → x 'SINH(x)/(1+x^2) >>, we'll replace it with the following program: << 'x' STO x SINH 1 x SQ ADD / 'x' PURGE >>. To key in the program follow these instructions:

Keystroke sequence:	Produces:	Interpreted as:
[↱][<< >>]	<<	Start an RPL program
[↱]['][ALPHA][↰][X][▶][STO]	'x' STO	Store the contents of level 1 into variable x
[ALPHA][↰][X]	x	Place x in level 1
[↰][MTH][HYP][SINH]	SINH	Calculate sin of level 1

[1][SPC][ALPHA][↰][X][↰][x²]	1 x SQ	Enter 1 and calculate x^2
[↰][MTH][LIST][ADD]	ADD	Calculate (1+x2),
[÷]	/	then divide
[↱]['][ALPHA][↰][X][▶]	'x'	
[ALPHA][ALPHA][P][U][R][G][E][ALPHA]	PURGE	Purge variable x
[ENTER]		Program displayed in level 1

To save the program use:

[↱]['][ALPHA][↰][G][STO]

Press [VAR] to recover your variable menu, and evaluate g(3.5) by entering the value of the argument in level 1 ([3][.][5][ENTER]) and then pressing [g]. The result is 1.2485..., i.e., g(3.5) = 1.2485...Same as before. Try also obtaining g[{1 2 3}], by entering the list in level 1 of the display ([↰][{}] [1] [SPC] [2] [SPC] [3] [ENTER]), and pressing [g]. The result now is {SINH(1)/2 SINH(2)/5 SINH(3)/10}, if your CAS is set to EXACT mode. If your CAS is set to APPROXIMATE mode, the result will be {0.5876.. 0.7253... 1.0017...}.

Global and local variables

The program [g], defined above, can be displayed as

<< 'x' STO x SINH 1 x SQ ADD / 'x' PURGE >>

by using [↱][g].

Note: [↱][g] is equivalent to recalling the contents of variable g by using: [↱]['][' g][↰][RCL].

Notice that the program uses the variable name x to store the value placed in register x (level 1 of stack) through the programming steps 'x' STO. The variable x, while the program is executing, is stored in your variable menu as any other variable you had previously stored. After calculating the function, the program purges (erases) the variable x so it will not show in your variable menu after finishing evaluating the program. If we were not to purge the variable x within the program its value would be available to us after program execution. For that reason, the variable x, as used in this program, is referred to as a global variable. One implication of the use of x as a global variable is that, if we had a previously defined a variable with the name x, its value would be replaced by the value that the program uses and then completely removed from your variable menu after program execution.

From the point of view of programming, therefore, a global variable is a variable that is accessible to the user after program execution. It is possible to use a local variable within the program that is only defined for that program and will not be available for use after program execution. The previous program could be modified to read:

<< → x << x SINH 1 x SQ ADD / >> >>

The arrow symbol (→) is obtained by combining the right-shift key with the [0] key, i.e., [↱][→]. Also, notice that there is an additional set of programming symbols (<< >>) indicating the existence of a sub-program, namely << x SINH 1 x SQ ADD / >>, within the main program. The main program starts with the combination → x, which represents assigning the value in register x (level 1 of stack) to a local variable x. Then, programming flow continues within the sub-program by placing x in the stack, evaluating SINH(x), placing 1 in the stack, placing x in the stack, squaring x, adding 1 to x, and dividing stack level 2 (SINH(x)) by stack level 1 (1+x^2). The program control is then passed back to the main program, but there are no more commands between the first set of closing programming symbols (>>) and the second one, therefore, the program terminates. The last value in the stack, i.e., SINH(x)/ (1+x^2), is returned as the program output.

The variable x in the last version of the program never occupies a place among the variables in your variable menu. It is operated upon within the calculator memory without affecting any similarly named variable in your variable menu. For that reason, the variable x in this case is referred to as a variable local to the program, i.e., a local variable.

Note: to modify program [g], place its contents in the stack by using [→][g]. Then activate the stack editor by pressing [▼]. Use the arrow keys ([◄] [►] [▲] [▼]) to move about the program. Use the backspace/delete key, [←], to erase any unwanted characters. To add program symbols (i.e., << >>), use   [↵][<< >>], since these symbols come in pairs you will have to enter them at the start and end of the sub-program and erase one of its components with the delete key to produce the required program, namely:

<< → x << x SINH 1 x SQ ADD / >> >>.

When done editing the program press [ENTER] and store the result in variable [g] by using [↵][g].

Global Variable Scope

Any variable that you define in the HOME directory or any other directory or sub-directory will be considered a global variable from the point of view of program development. However, the scope of such variable, i.e., the location in the directory tree where the variable is accessible, will depend on the location of the variable within the tree. The directory tree is basically the schematic representation of the memory locations within the HOME directory. To see your current directory tree, move to the HOME directory, if you are not already there ([CAT][ALPHA][ALPHA][H][O][ALPHA][OK]), then activate the FILES command: [↵][FILES] (the APPS key). The resulting screen shows your directory tree with the root in the HOME directory and branching out into your other directories and sub-directories. You can use the arrow keys to move about the directory tree. Press [OK][ON] to return to normal calculator display.

The rule to determine a variable's scope is the following: a global variable is accessible to the directory where it is defined and to any sub-directory attached to that directory, unless a variable with the same name exists in the sub-directory under consideration. Consequences of this rule are the following:

⁂ A global variable defined in the HOME directory will be accessible from any directory within HOME, unless redefined within a directory or sub-directory.

⁂ If you re-define the variable within a directory or sub-directory this definition takes precedence over any other definition in directories above the current one.

⁂ When running a program that references a given global variable, the program will use the value of the global variable in the directory from which the program is invoked. If no variable with that name exist in the invoking directory, the program will search the directories above the current one, up to the HOME directory, and use the value corresponding to the variable name under consideration in the closest directory above the current one.

⬇ A program defined in a given directory can be accessed from that directory or any of its sub-directories.

> All these rule may sound confusing for a new calculator user. They all can be simplified to the following suggestion: Create directories and sub-directories with meaningful names to organize your data, and make sure you have all the global variables you need within the proper sub-directory.

Functions defined by more than one expression

In this section we discuss the treatment of functions that are defined by two or more expressions. An example of such functions would be

$$f(x) = \begin{Bmatrix} 2 \cdot x - 1, & x < 0 \\ x^2 - 1, & x > 0 \end{Bmatrix}$$

The HP 49 G provides the function IFTE (IF-Then-Else) to describe such functions.

The IFTE function

The IFT function is written as `IFT(condition, operation_if_true, operation_if_false)`. If condition is true then operation_if_true is performed, else operation_if_false is performed. For example, we can write 'IFTE(x>0, x^2-1, 2*x-1)', to describe the function listed above. One way to define the function is to use 'f(x) = IFTE(x>0,x^2-1, 2*x-1)' and then entering [←][DEF]. For this case use:

[EQW] [ALPHA][←][F] [←][( )]   [ALPHA][←][X] [▶] [→][=] [←][PRG][BRCH][NXT][IFTE]
[ALPHA][←][X] [→][>] [0] [▶] [ALPHA][←][X] [y^x] [2] [▲][▲] [-] [1]
[▶] [2][×] [ALPHA][←][X] [-] [1] [ENTER]

To define the function, use [←][DEF]. Press [VAR] to recover your variable menu. The function [f] should be available in your soft key menu. Press [→][f] to see the resulting program:

<< → x 'IFTE(x>0, x^2-1, 2*x-1)' >>

To evaluate the function enter a number and press [f], for example:[2][f]   produces f(2) = 3, while [2][+/-] [f] produces f(-2) = -5.

Combined IFTE functions

To program a more complicated function such as

$$g(x) = \begin{Bmatrix} -x, & x < -2 \\ x+1, & -2 \leq x < 0 \\ x-1, & 0 \leq x < 2 \\ x^2, & x \geq 2 \end{Bmatrix}$$

you can combine several levels of the IFTE function, for example, type the function

    `'g(x) = IFTE(x<-2, -x, IFTE(x<0, x+1, IFTE(x<2, x-1, x^2)))'`,

then press [←][DEF]. Check that g(-3) = 3, g(-1) = 0, g(1) = 0, g(3) = 9.

Generating a table of values for a function

The soft key functions (F1 through F6) can be combined with the left-shift key, [←], to access the calculator's graphing functions. While detailed instructions for using graphs are presented in a later chapter, in this section we will introduce their use with the purpose of generating tables of values for a function.

> Note: to access the graphing functions in RPN mode hold the left-shift key and the selected soft key simultaneously. In algebraic mode you can press the left-shift key first and then the desired soft key.

✦ We will generate values of the function f(x), defined above, for values of x from -5 to 5, in increments of 0.5. First, we need to ensure that the graph type is set to FUNCTION. To achieve this press [←][2D/3D] (simultaneously, if in RPN mode). The field in front of the Type option will be highlighted. If this field is not already set to FUNCTION, press the soft key [CHOOS] and use the up and down keys, [▲][▼], to select the FUNCTION option, then press [OK].

✦ Next, press [▼] to highlight the field in front of the option EQ. Press [VAR] to show your variable menu. Find the soft key corresponding to the function f, and press [→][f], then press [OK]. This will place a copy of the program that generates values of the function f(x) in the field corresponding to the option EQ. However, in order to evaluate the desired function table we need to edit this program so that only the function expression, namely, `'IFTE(x>0, x^2-1, 2*x-1)'` remains.

✦ To edit the contents of EQ, first press [▲] to highlight the corresponding input field, then press [EDIT]. Use the right-arrow key [▶] to move to the position immediately to the left of the first single quote (press [▶] five times), then use the backspace/delete key, [⇦], 5 times to delete the characters << →x . Next, press the combination [→][▶] and press [▶], if needed, to reach the space past the program closing character, >>. Finally, use the backspace/delete key, [⇦], to delete the character >>. When done, press [ENTER].

⬇ The field in front of the option Indep (for independent variable) will be highlighted now. Since the independent variable we are using is x (lowercase), while the calculator's default independent variable is X (uppercase), you may need to edit this field by pressing [EDIT]. Using the left-arrow key, [◄], move to the left of the closing quote, then use the backspace/delete key, [⌫], to erase the X, and enter [↤][ALPHA][X]. When done, press [OK].

⬇ To accept the changes made to the PLOT SETUP screen press [NXT][OK]. You will be returned to normal calculator display.

The EQ variable

If the EQ field in the PLOT SETUP was empty before you chose the function f above, then you did not have a variable EQ in your directory previous to the exercise above. After you return to normal calculator display, you will notice that there is a new variable called EQ. This variable is generated automatically by the calculator when doing a plot setup (if no EQ variable exists, otherwise the calculator replaces its value with that provided in the PLOT SETUP screen). The EQ variable is used in many other calculator functions always to store an algebraic expression or a program to be plotted or solved for.

⬇ The next step is to access the Table Set-up screen by using the keystroke combination [↤][TBLSET] (i.e., soft key F5) - simultaneously if in RPN mode. This will produce a screen where you can select the starting value (Start) and the increment (Step). Enter the following: [5][+/-][OK] [0][.][5][OK] [0][.][5][OK] (i.e., Zoom factor = 0.5). Toggle the [✓CHK] soft key until a check mark appears in front of the option Small Font if you so desire. Then press [OK]. This will return you to normal calculator display.

The TPAR variable

After finishing the table set up, your calculator will create a variable called TPAR (Table PARameters) that store information relevant to the table that is to be generated. To see the contents of this variable, press [↦][TPAR].

⬇ To see the table, press [↤][TABLE] (i.e., soft key F6) - simultaneously if in RPN mode. This will produce a table of values of x = -5, -4.5, ..., and the corresponding values of f(x), listed as Y1 by default. You can use the up and down arrow keys, [▲][▼], to move about in the table. You will notice that we did not have to indicate an ending value for the independent variable x. Thus, the table continues beyond the maximum value for x suggested early, namely x = 5.

Some options available while the table is visible are [ZOOM][BIG] and [DEFN]:

⬇ The [BIG] key simply changes the font in the table from small to big, and vice versa. Try it.

⬇ The [ZOOM] key, when pressed, produces a menu with the options: In, Out, Decimal, Integer, and Trig. Try the following exercises:

⬥ With the option In highlighted, press [OK]. The table is expanded so that the x-increment is now 0.25 rather than 0.5. Simply, what the calculator does is to multiply the original increment, 0.5, by the zoom factor, 0.5, to produce the new increment of 0.25. Thus, the zoom in option is useful when you want more resolution for the values of x in your table.

⬥ To increase the resolution by an additional factor of 0.5 press [ZOOM], select In once more, and press [OK]. The x-increment is now 0.0125.

⬥ To recover the previous x-increment, press [ZOOM][▲] to select the option Un-zoom. The x-increment is increased to 0.25.

⬥ To recover the original x-increment of 0.5 you can do an un-zoom again, or use the option zoom out by pressing [ZOOM][▼][OK].

⬥ The option Decimal in [ZOOM] produces x-increments of 0.10.

⬥ The option Integer in [ZOOM] produces x-increments of 1.

⬥ The option Trig in [ZOOM] produces increments related to fractions of π, thus being useful when plotting trigonometric functions.

⬥ To return to normal calculator display press [ENTER] or [ON].

As an exercise, try generating a table of values for g(x) for x starting at -3 with increments of 0.5, then try the different zoom options.

Functions of more than one variable

Functions of more than one variable can be defined in a similar manner as functions of one variable, i.e., enter the function expression of the form `function_name($x_1, x_2, ..., x_n$) = expression containing $x_1, x_2, ..., x_n$` in level 1 of the stack. Then, use the command [↵][DEF]. Press [VAR], if needed. A program that evaluates the function will be store in a variable called `function_name`. To evaluate the function for a set of values of $x_1, x_2, ..., x_n$, enter the values of the independent variables in the stack in the order $x_1, x_2, ..., x_n$, then press the soft key button corresponding to the function name.

For example, let's define a simple function, say h(x,y,z) = x+y+z, by using:

[EQW] [ALPHA][↵][H] [↵][( )] [ALPHA][↵][X] [SPC] [ALPHA][↵][Y] [SPC] [ALPHA][↵][Z] [▶] [→][=] [ALPHA][↵][X] [+] [ALPHA][↵][Y] [+] [ALPHA][↵][Z] [ENTER]

Note: it is not necessary to type the commas between the independent variables x, y, z in the function call h(x,y,z). It is only necessary to use the space key. The commas are generated automatically.

To create the program that defines the function use: [↵][DEF]. This creates a variable [h] in the variable menu. To see the contents of this variable, press [→][h]. The program stored in h is:

$$<< \to x\ y\ z\ 'x+y+z' >>$$

The ordering of the variables to the right of the input arrow in the program indicates the order in which the corresponding input values must be entered into the calculator display, namely, x in level 3, y in level 2, and z in level 1, before pressing [h]. For example, to evaluate h(-3, 5, 0.5) use:

[3][+/-][ENTER] [5][ENTER] [0][.][5] [ENTER] [h],

which results in h(-3, 5, 0.5) = 2.5. Alternatively, you can enter the values of x,y,z separated by spaces, i.e.,
[3][+/-] [SPC] [5] [SPC] [0][.][5] [ENTER] [h],
or,
[3][+/-] [SPC] [5] [SPC] [0][.][5] [h].

Applications of function definitions – probability distributions

We indicated in an earlier chapter that the HP 49 G calculator includes pre-defined functions that allow the user to calculate the upper-tail probability distribution (or cumulative distribution function, cdf) for the normal, Student-t, Chi-square (), and F distributions. In this section, we use what we have learnt about function definitions to create other probability distribution functions.

Discrete probability distributions

A random variable is said to be discrete when it can only take a finite number of values. For example, the number of rainy days in a given location can be considered a discrete random variable because we count them as integer numbers only. Let X represent a discrete random variable, its mass distribution function (mdf) is represented by

$$f(x) = P[X=x],$$

and

i.e., the probability that the random variable X takes the value x.

The mass distribution function must satisfy the conditions that

$$f(x) > 0, \text{ for all } x,$$

and

$$\sum_{all\ x} f(x) = 1.0$$

A cumulative distribution function (cdf) is defined as

$$F(x) = P[X \leq x] = \sum_{k \leq x} f(k).$$

Defining summations in the HP 49 G

To define a summation in the HP 49 G use the combination [→][Σ] (SIN key). When used in the equation editor, spaces will be provided for the summation parameters. For example, the sum

$$\sum_{n=1}^{\infty} \frac{1}{n^2}$$

Can be typed <u>using the equation writer</u> as follows:

[EQW] [→][Σ] [ALPHA][↰][N] [▶] [1] [▶] [↰][∞] [▶] [1] [÷] [ALPHA][↰][N] [y^x] [2] [ENTER]

If you have your display set to textbook, you will get the exact expression shown above. If you change your display to the default one-line display, you will get the expression:

'Σ(n=1,∞,1/n^2)'

which suggest the general form of the summation expression for line input, namely:

'Σ(control_variable=starting_value,ending_value,expression)'

To <u>evaluate</u> the current expression, if it is a convergent series, first make sure that the CAS is set to EXACT. Then, use [→][EVAL]. The calculator's CAS, when set to the EXACT mode, will seek a symbolic result, which in this case is given as 'π^2/6'.

To <u>enter</u> a summation expression <u>directly into display level 1</u>, without using the equation writer, you need to open a set of single quotes and then type the expression as indicated in the example below. Here we are typing again the summation used above:

[→]['] [→][Σ] [ALPHA][↰][N] [→][=] [1] [→][,] [↰][∞] [→][,] [1] [÷] [ALPHA][↰][N] [y^x] [2] [ENTER]

Binomial probability distribution function

Create a directory HOME\STATS\DFUN (Discrete FUNctions). We will define some probability distributions and distributions functions in that directory.

To define the binomial distribution function, enter the following keystrokes:

['][α][α][↰][B][↰][P][↰][D] [↰] [( )] [↰] [X] [↰] [,] [↰] [N] [↰] [,] [↰] [P][α][▶][↰][=]
[MTH][NXT][PROB][COMB][α][α][↰][N][↰][,][↰][X] [α] [▶]
[×] [α][↰] [P][y^x] [α][↰] [X]
[×] [↰] [( )][1][-][α][↰] [P] [▶] [y^x] [↰] [( )] [α][↰] [N][-] [α][↰] [X]
[ENTER]

The calculator display will show:

1:    'bpd(x,n,p)=COMB(n,
      x)*p^x*(1-p)^(n-x)'

This statement defines a function bpd (binomial probability distribution) of a discrete random variable x = 0, 1, ..., n, with parameters n and p. We'll now define the function into a variable 'bpd'. Enter:

$$[\leftarrow][DEF].$$

Press [VAR], if needed. To see the contents of the variable 'bdp', enter:

$$[\rightarrow][\ BPD\ ].$$

The display will show a program:

```
1:    << → x n p 'COMB(n,x
      )*p^(1-p)^(n-x) '
      >>
```

The sequence [←][DEF] converts the function definition, 'bpd(x,n,p)=COMB(n,x)*p^x*(1-p)^(n-x)', into the program listed above.

To use the function we just defined, we need to enter three arguments, namely, x, n, and p, into display levels 3, 2, and 1, respectively. To do that, simply use the keystroke sequence:

$$x\ [ENTER]\ n\ [ENTER]\ p\ [ENTER].$$

To determine the value of bpd(x,n,p), press the soft key corresponding to the label [BPD].
For example, to calculate bpd(2,5,0.5), enter:

[2][ENTER][5][ENTER][0][.][5][ENTER][BPD]   The display shows a value of 0.3125.

Binomial cumulative distribution function

To define the binomial distribution function, type in the following function definition:

$$\text{'BDF(x,n,p)} = \Sigma(k=0,x,bdp(k,n,p))\text{'}$$

Then, press [←][DEF]. The variable BDF is now defined. If you press [→][BDF], you'll get the following program:
$$<< \to x\ n\ p\ '\ \Sigma(k=0,x,bdp(k,n,p))\ ' >>$$

Calculate the value of BDF(3,5,0.5), by pressing:

 [3][ENTER][5][ENTER][0][.][5][ENTER][BDF]   The display shows a value of 0.8125.

Other discrete probability distributions and distribution functions

The following function definitions will provide you with other probability distributions and distribution functions for discrete random variables:

Poisson Probability Distribution: 'PoPD(x,λ) = λ^x*EXP(-λ)/x!'
Poisson Distribution Function: 'PoDF(x,λ) = Σ(k=0,x,'PoPD(k,λ))'
Hypergeometric Prob. Distr.:   'hpd(x,n,a,N)=COMB(a,x)*COMB(N-a,n-x)/COMB(N,n)'
Geometric Distribution Function: 'gpd(x,p) = p*(1-p)^(x-1)'

Remember to press [↵][DEF] after entering the function definition in level 1 of the display. To facilitate locating the functions, keep them all under the same directory DFUN. If you define all the functions suggested in this section, your white keys should have the following labels (they may be in different order):

[BPD][BDF][POPD][PODF][HPD][GPD]

Also, BDF uses bpd, and PoDF uses PoPD, therefore, you must define bpd and PoDF before using BDF or PoDF.

> Note: If the functions in your directory are not in the order shown above and you want to place them in that order, try the following:
>
> (1) Create a list that looks like this: { bpd BDF PoPD PoDF hpd gpd} and place it in level 1. To create the list use these keystrokes:
>
> [↵][{}] [BPD][BDF][POPD][PODF][HPD][GPD][ENTER]
>
> (2) Use the following keystroke sequence to re-order the variables:
>
> [↵][MEMORY][DIR][ORDER].

Examples:

Calculate the following:	Keystrokes:	Result:
PoPD(x=3, λ=2.2)	[3][SPC][2][.][2][POPD]	0.19663867...
PoPF(x=5, λ = 1.5)	[5][SPC][1][.][5][POPF]	0.99554401...
hpd(x=2,n=10,a=20,N=100)	[2][SPC][1][0][SPC][2][0][SPC][1][0][0][HPD]	0.31817062...
gpd(x=5,p=0.25)	[5][SPC][.][2][5][GPD]	0.07910156...

Continuous probability functions

In this section we describe several continuous probability distributions including the gamma, exponential, beta, and Weibull distributions. These distributions are described in any statistics textbook. Some of these distributions make use of a the Gamma function defined earlier, which is calculated in the HP 49 G calculator by using the factorial function as $\Gamma(x) = (x-1)!$, for any real number x.

The gamma distribution

The probability distribution function (pdf) for the gamma distribution is given by

$$f(x) = \frac{1}{\beta^\alpha \Gamma(\alpha)} \cdot x^{\alpha-1} \cdot \exp(-\frac{x}{\beta}), \text{ for } x > 0, \alpha > 0, \beta > 0;$$

The corresponding (cumulative) distribution function (cdf) would be given by an integral that has no closed-form solution.

The exponential distribution

The exponential distribution is the gamma distribution with a = 1. Its pdf is given by

$$f(x) = \frac{1}{\beta} \cdot \exp(-\frac{x}{\beta}), \ for \quad x > 0, \beta > 0;$$

While its cdf is given by

$$F(x) = 1 - \exp(-x/\beta), \ for \ x>0, \ \beta >0.$$

The beta distribution

The pdf for the gamma distribution is given by

$$f(x) = \frac{\Gamma(\alpha+\beta)}{\Gamma(\alpha) \cdot \Gamma(\beta)} \cdot x^{\alpha-1} \cdot (1-x)^{\beta-1}, for \ 0<x<1, \alpha>0, \beta>0;$$

As in the case of the gamma distribution, the corresponding cdf for the beta distribution is also given by an integral with no closed-form solution.

The Weibull distribution

The pdf for the Weibull distribution is given by

$$f(x) = \alpha \cdot \beta \cdot x^{\beta-1} \cdot \exp(-\alpha \cdot x^{\beta}), \quad for \ x>0, \alpha>0, \beta>0$$

While the corresponding cdf is given by

$$F(x) = 1 - \exp(-\alpha \cdot x^{\beta}), \quad for \ x>0, \alpha>0, \beta>0$$

Defining integrals in the HP 49 G - brief introduction

A more detailed presentation of integral calculus is presented in a subsequent chapter. This is just a brief introduction to defining integrals using the equation writer. The integral symbol is obtained by using the right-shift key combined with the TAN key:[↦][∫]. For example, to evaluate the integral

$$\int_0^t \sqrt{x^2+1} \ dx \ ,$$

use:

[EQW] [↦][∫] [0] [▶] [ALPHA][↤][T] [√x] [ALPHA][↤][X] [y^x] [2] [▲][▲] [+] [1] [▶] [ALPHA][↤][X] [ENTER]

To evaluate this integral, with the CAS in EXACT mode, use:

[→][EVAL]

which results in `'-((LN(-t+√(t^2+1))-t* √(t^2+1))/2)'`, or

$$-\frac{LN(-t+\sqrt{t^2+1})-t\cdot\sqrt{t^2+1}}{2}$$

if the calculator is in the textbook display mode.

To define a collection of functions corresponding to the gamma, exponential, beta, and Weibull distributions, first create a sub-directory called CFUN (Continuous FUNctions) and define the following functions:

Gamma function:	`'fgamma(x) = (x-1)!'`
Gamma pdf:	`'gapd(x) = x^(α-1)*EXP(-x/β)/(β^α*fgamma(α))'`
Gamma cdf:	`'gadf(x) = ∫(0,x,gapd(t),t)'`
Beta pdf:	`'βpd(x)= fgamma(α+β)*x^(α-1)*(1-x)^(β-1)/(fgamma(α)*fgamma(β))'`
Beta cdf:	`'βdf(x) = ∫(0,x,βpd(t),t)'`
Exponential pdf:	`'expd(x) = EXP(-x/β)/β'`
Exponential cdf:	`'exdf(x) = 1 - EXP(-x/β)'`
Weibull pdf:	`'Wepd(x) = α*β*x^(β-1)*EXP(-α*x^β)'`
Weibull cdf:	`'Wedf(x) = 1 - EXP(-α*x^β)'`

Remember to press [←][DEF] after entering each function definition in level 1. You also need to create a couple of variables, α and β, and load them with some values, say α = 1.0, β = 2.0. To do this use the following keystrokes:

[1]['][ALPHA][→] [A][STO] [2]['][ALPHA][→] [B][STO]

Finally, for the cdf for Gamma and Beta cdf's, you need to edit the program definitions to add →NUM to the programs produced by [DEF]. For example, for the Gamma cdf, use the following keystroke sequence:

[→][GADF][▼][▼][▼][←][→NUM][ENTER].

The program should look like this: `<< → x '∫(0,x,gapd(t),t)' →NUM >>`

Store the new program into gadf, using:

[←][GADF].

Repeat the procedure for βdf. If you want to order the variables in your directory, use the procedure shown in section 7.2, to get your white buttons to look like this:

[α][β][FGAM][GAPD][GADF][βPD][βDF][EXPD][EXDF][WEPD][WEPF]

Examples:
Unlike the discrete functions defined earlier, the continuous functions defined in this section do not include their parameters (α and/or β) in their definitions. Therefore, you don't need to enter them in the display to calculate the functions. However, those parameters must be previously

defined by entering the corresponding values in the variables α and β. To enter these values use a procedure similar to that suggested above when the variables α and β were first defined.

Practice the following exercises:

⚜ For α = 3, β = 2, use the Gamma distribution to obtain f(12), and P(X>12).
First, store α and β:

$$[3][\hookleftarrow]\ [\ \alpha\ ]\ [2][\hookleftarrow][\ \beta\ ].$$

Then, for f(12), enter

$$[1][2][GAPD],$$

thus f(12) = 0.02230. Also, P(X>12) = 1 - P(X<12) = 1 - F(12), enter

$$[1][ENTER][1][2][GADF][-],$$

thus P(X>12) = 0.05196. (The evaluation of the integral for GADF takes about twenty seconds)

⚜ For α = 2, β = 9, use the Beta distribution to obtain f(0.10) and P(X<0.10).
First, store α and β:

$$[2][\hookleftarrow]\ [\ \alpha\ ]\ [9][\hookleftarrow][\ \beta\ ].$$

Then, for f(0.10), enter

$$[NXT][.][1][\ \beta PD\ ],$$

thus f(0.10) = 3.87420. Also, P(X<0.10) = F(0.10), enter

$$[.][1][\ \beta DF\ ],$$

thus P(X>12) = 0.2639010. (The evaluation of the integral for GADF takes from 5 to 10 seconds).

⚜ For α = 0.2, β = 0.333, use the Weibull distribution to obtain f(5) and P(X<5).
First, store α and β:

$$[VAR][.][2][\hookleftarrow]\ [\alpha]\ [.][3][3][3][\hookleftarrow][\ \beta\ ].$$

Then, for f(5), enter

$$[NXT][5][\ WEPD],$$

thus f(5) = 0.0161738. Also, P(X<5) = F(5), enter

$$[5][WEDF],$$

thus P(X<5) = 0.289518.

⚜ For β = 0.2, use the Exponential distribution to obtain f(2.5) and P(X>2.5).
First, store only β=0.2:

[VAR] [2][┐][β].

Then, for f(2.5), enter

[NXT][2][.][5][EXPD],

thus f(2.5) = 1.8633×10^{-5}. Also, P(X>2.5) = 1 - P(X<2.5) = 1 - F(2.5), enter

[1][ENTER][2][.][5][EXDF][-],

thus P(X>12) = 0.00000372.

Manipulating elements of lists

Elements of a list can be manipulated by using the functions available in the menu the results from [┐][PRG][LIST][ELEM]. The functions thus available are:

[GET][GETI][PUT][PUTI][SIZE][POS]

in the first menu, and, (press [NXT])

[HEAD][TAIL][   ][   ][   ][   ]

Press [NXT] again, to recover the first set of functions.

To show examples of these functions, let's prepare a simple list, say, {2 3 5 7 1 2 9 0 1}, place it in level 1 of the display and press eight times to make eight copies of the list available in the stack. We'll use this list to illustrate the use of the functions listed above.

Determining list size

With the list of interest in level 1, press the soft key [SIZE] to obtain the size of the list. This operation wipes out the list, therefore, if you want to operate more on a given list that is not saved to a variable, make sure to make at least one more copy available in the stack (by pressing [ENTER]). For the list under consideration the size is 9. Use [⇦] to drop this result from the stack.

Extracting and inserting elements in a list

To extract an element of the list, simply enter its position in stack level 1 and press [GET]. For example, to get the third element of the list use [3][GET]. The result is 5. This operation will also wipe out the list in level 2. Use [⇦] to drop this result from the stack.

The function [GETI] also will extract the element indicated (position = I) and place it in level 1 of the stack, but it will also calculate (I+1) and place it in level 2 of the stack, while preserving the original list in level 3 of the stack. Thus, using [3][GETI] in the list above will produce:

3:    {2 3 5 7 1 2 9 0 1}

```
2:                              4
1:                              5
```

The function GETI can be used in programs when extracting elements in sequence. You will operate on the element extracted (level 1) while increasing by one the index controlling your program. At the same time, the list is still available for extracting more elements if needed. Use [◁] twice to drop the latter results from the stack.

Suppose that we want to insert the value 1.2 in position 6 of the list. With the list in level 1, type [6][ENTER] [1][.][2] [ENTER] [PUT]. Thus, the position is entered first and the element to be inserted next, before pressing [PUT]. The list will now look like {2 3 5 7 1 1.2 2 9 0 1}. Use [◁] to drop this result from the stack.

The function [PUTI] will perform the same operation, while increasing the position value by 1 and leaving the list available in level 2. Try the following: [6][ENTER] [1][.][2] [ENTER] [PUTI]. The result is:

```
2:    {2 3 5 7 1 1.2 2 9 0 1}
1:                           7
```

Again, the [PUTI] function is more appropriate for programming purposes. Use [◁], twice, to drop this result from the stack.

Element position in the list

The position of an element in the list can be determined by entering the element value in level 1 while the list is in level 2, then pressing [POS]. Try this exercise, using the list previously defined: [9][POS]. The result is 7. Use [◁], twice, to drop this result from the stack.

Note: If you list an element that is not contained in the list, [POS] will return a value of zero.

HEAD and TAIL functions

Press [NXT] to reach the functions HEAD and TAIL. The [HEAD] function extracts the very first element in the list in level 1. For the case under consideration, pressing [HEAD] will produce the value 2. Use [◁] to drop this result from the stack.

The function [TAIL] returns the list in level 1 without the first element. With the same list in level 1, press [TAIL] to get {3 5 7 1 2 9 0 1}.

Applications of list operations

The following exercises are applications of lists to data processing. The data is assumed to be entered as a list of numbers and available in level 1 of the stack.

Mean, variance, and standard deviation of a sample

In this exercise we calculate the mean value of a small list of numbers. Enter the following list:

{1.0 1.5 2.0 2.0 1.1 0.5 1.2 },

and press [ENTER] three times to keep extra copies of the list available in the stack.

The <u>mean value</u> is defined as the sum of elements divided by the number of elements, i.e.,

$$\bar{x} = \Sigma x_i /n,$$

thus, to calculate this value use:

[↰][MTH][LIST][ΣLIST]	Calculates $\Sigma x_i = 9.3$
[▶]	Swaps levels 1 and 2
[↰][PRG][LIST][ELEM][SIZE]	Produces n = 7
[÷]	Calculates $\boxed{\bar{x} = \Sigma x_i /n = 1.32857...}$

The <u>variance</u> of the list is calculated as

$$s_x^2 = \Sigma(x_i - \bar{x})^2/(n-1).$$

Thus, for the list under consideration, with the list in level 2 and $\bar{x}$ in level 1, calculate the variance as follows.:

[-]	Subtract $\bar{x}$ from every element in the list
[↰][x²]	Squares every element in the resulting list
[↰][MTH][LIST][ΣLIST]	Calculates $\Sigma(x_i - \bar{x})^2 = 1.79428...$
[6][÷]	Divide by (n-1), to calculate $\boxed{s_x^2 = 1.7642...}$
[√x]	This is the <u>standard deviation</u>, $\boxed{s_x = 1.3395...}$

Calculating statistics from grouped data

Grouped data refers to data grouped into classes and presented as a group of class marks (x_i) and their corresponding frequency distribution (f_i). Assuming the we have a total of n data points grouped into k classes, the calculation of the mean and variance proceeds according to the formulas used below.

The HP48 series calculators can be used to calculate the summations in those formulas if the vectors x and f, containing class marks and frequencies, respectively, are entered as lists. To illustrate the calculations, we'll use the following data:
$$x = \{ 2.5\ 3.5\ 4.5\ 5.5\ 6.5 \},\ f = \{3\ 5\ 11\ 7\ 2\}.$$

Separate the two lists and store them in variables x and f, respectively, by using:

{ 2.5 3.5 4.5 5.5 6.5 }	[ENTER]['][ALPHA][↰][X][STO]	Store list into variable x.
{3 5 11 7 2}	[ENTER]['][ALPHA][↰][F][STO]	Store list into variable f.

To calculate $\Sigma x_i f_i$, follow these steps:

[VAR][x][f][×]	Places the x and f lists in display levels 2 and 1, respectively, and multiply their corresponding elements creating a new list.
[↰][MTH][LIST][ΣLIST]	Calculates the sum of all the elements in the list in display level 1. The
	display shows now that $\Sigma x_i f_i = 126$.

To calculate the number of elements in the unordered sample, we use the following keystrokes:

[VAR][F][↰][MTH][LIST][ΣLIST]	The display shows that n = 28.

Since $\Sigma x_i f_i$ is in display level 2 and n is in display level 1, we can calculate the <u>mean value</u> as

$$\bar{x} = \Sigma x_i f_i /n,$$

by pressing [÷], giving as a result $\boxed{\bar{x} = 4.5}$.

Using the same data vectors x and f, we can calculate $\Sigma x_i^2 f_i$ as follows:

[VAR][x][↵][x²] Places the x list in display levels 1, and takes the square of each element in list x.

[f][×] Places the f list in display level 1, while the x list is moved to level 2. We, then, multiply the two lists, element by element.

[↵][MTH][LIST][ΣLIST] Calculates the sum of all the elements in the list in display level 1. The

display shows now that $\Sigma x_i^2 f_i = 599$.

We can use the information obtained so far to calculate the variance, s_x^2, according to the formula shown below. We know that n = 28, $\Sigma x_i f_i = 126$, and $\Sigma x_i^2 f_i = 599$, then,

$$\boxed{s_x^2 = [n\cdot\Sigma x_i^2 f_i - (\Sigma x_i f_i)^2]/[n\cdot(n-1)] = [28\times 599 - 126^2]/[28\times 27] = 1.185185....}$$

The standard deviation is obtained by pressing [√x], the result is:

$$\boxed{s_x^2 = 1.088662...}$$

Mean and variance of a discrete probability distribution

Suppose that we want to calculate the mean, variance and standard deviation for a discrete random variable. We will use the calculator to compute those parameters when the probability distribution is given as a table.
For example, using the data,

$$x = \{0\ 1\ 2\ 3\ 4\ \}, f(x) = \{0.05\ 0.20\ 0.45\ 0.20\ 0.10\}.$$

We begin by creating a subdirectory HOME\STATS\DDIST (for Discrete DISTributions), and moving into that subdirectory for the calculations. Next, enter and store the values of x and f as lists. Recall that to store those any list into a variable you need to type in the list in level 1 of the display, then type the variable name between quotes, and press [STO]. Once both lists have been stored, you will have two soft keys at the top of the keyboard showing the variable names [x] and [f].

First, it is convenient to check if the probability distribution is valid by checking that every element in f is larger than or equal to zero, and by checking that they add up to 1.0. To check that they add up to one use:

[f][↵][MTH][LIST][ΣLIST] If display level 1 shows a 1.0, you can continue with the analysis. If not, the distribution is not valid.

To calculate the mean, μ, we need to multiply the lists x and f and sum the elements of the result, as follows:

[VAR][x][f][×] [↰][MTH][LIST][ΣLIST]   Display level 1 should show 2.1, i.e., $\mu = 2.1$.

To calculate the second moment with respect to the origin, μ'_2, we use:

[VAR][X][↰][x²][F][×]   Calculates the list $x^2 \cdot f$.
[↰][MTH][LIST][ΣLIST]   Calculates $\mu'_2 = 5.4$.

To calculate the variance, with μ in level 2 and μ'_2 in level 1 of the display, try the following:

[▶][↰][x²][-]   Calculates $\sigma^2 = \mu'_2 - \mu^2 = 0.99$

Of course, the standard deviation is calculated using [√x], to get $\sigma = 0.994987...$.

Note: The key [▶], when the edit line (bottom line of the display with no stack number showing) is not active, swaps the contents of lines 1 and 2. This operation has the same effect of the command SWAP, which can be accessed through the command catalog: [CAT][ALPHA][ALPHA][S][W][ALPHA][OK].

Programs for generating lists of numbers

Following we present three programs developed for creating or manipulating lists. These programs are useful in handling numerical lists for some a variety of applications. The program listings are as follows:

LISC:
<< → n x << 1 n FOR j x NEXT n →LIST >> >>

CRLST:
<< → st en df << st en FOR j j df STEP en st - df / FLOOR 1 + →LIST >> >>

CLIST:
<< REVLIST DUP DUP SIZE 'n' STO ΣLIST SWAP TAIL DUP SIZE 1 - 1 SWAP FOR j DUP ΣLIST SWAP TAIL NEXT 1 GET n →LIST REVLIST 'n' PURGE >>

The following are keystroke sequences to get some of the commands in the programs:

SWAP	[ALPHA][ALPHA][S][W][A][P]	STO	[STO]
FOR	[↰][PRG][BRCH][FOR][FOR]	NEXT	[↰][PRG][BRCH][FOR][FOR]
STEP	[↰][PRG][BRCH][FOR][FOR]	→LIST	[↰][PRG][LIST][→LIST]
REVLIST	[↰][MTH][LIST][REVLI]	DUP	[ALPHA][ALPHA][D][U][P]
SIZE	[↰][PRG][LIST][ELEM][SIZE]	ΣLIST	[↰][MTH][LIST][ΣLIST]
TAIL	[↰][PRG][LIST][ELEM][NXT][TAIL]		
GET	[↰][PRG][LIST][ELEM][GET]	FLOOR	[↰][MTH][REAL][NXT][NXT][FLOOR]

The operation of these programs is as follows:

(1) LISC: creates a list of n elements all equals to a constant c.

   Operation: enter n, enter c, press [LISC]

   Example: [5][ENTER][6][.][5][ENTER][LISC] creates the list: {6.5 6.5 6.5 6.5 6.5}

(2) CRLST: creates a list of numbers from n_1 to n_2 with increment Δn, i.e.,
   {n_1, $n_1+\Delta n$, $n1+2\cdot\Delta n$, ... $n_1+N\cdot\Delta n$ }, where

$$N = floor\left(\frac{n_2 - n_1}{\Delta n}\right) + 1$$

   Operation: enter n_1, enter n_2, enter Δn, press [CRLST]

   Example: [.][5][ENTER][3][.][5][ENTER][.][5][ENTER][CRLST] creates the list:
   {0.5 1 1.5 2 2.5 3 3.5}

(3) CLIST: creates a list with cumulative sums of the elements, i.e., if the original list is {x_1 x_2 x_3 ... x_N}, then CLIST creates the list:

$$\{x_1, x_1 + x_2, x_1 + x_2 + x_3, ..., \sum_{i=1}^{N} x_i\}$$

This type of procedure is useful when determining cumulative frequency or probability distributions.

   Operation: place the original list in level 1, press [CLIST]

   Example: {1 2 3 4 5}[ENTER][CLIST] produces {1 3 6 10 15}.

Applications of list-generating programs

Next, the programs CRLIST, and CLIST are used to generate complete probability distributions for the binomial variate, and cumulative distribution functions for the binomial, Poisson, and geometric distributions.

Generating tables of mass and cumulative distribution functions for discrete random variables

Discrete random variables can take only a specific number of values. Many discrete random variables take only integer numbers. For example, in a binomial distribution with parameters (n,p), the random variable X can only takes values of 0,1,2,...,n. A complete binomial distribution is obtained then by evaluating the probability mass function, pmf, for each value of the variable. Using lists and the programs defined above for discrete pmf's, we can generate complete probability distributions. Some examples are shown below. To facilitate calculations it is convenient that you have copies of the programs CRLIST and CLIST in the same sub-directory where you defined the discrete probability functions: bpd, BDF, Popd, PoDF, hpd, and gpd.

Binomial distribution

Suppose that you want to describe the complete probability distribution for a random variable X that follows the binomial distribution with parameters n =10 and p = 0.35. First, create the list for X = {0,1,2,...,10} by using:

[0][ENTER][1][0][ENTER][1][CRLST]	Result: {0 1 2 3 4 5 6 7 8 9 10}.
[ENTER]	Makes an extra copy of the list.
[1][0][ENTER][.][3][5][ENTER]	Enter values of n and p.
[bpd]	Calculates the list of values of the mdf for X.
[ENTER]	Makes an extra copy of the mdf's values.
[CLIST]	Calculates the cumulative distribution function CDF as a list.

The results obtained through this exercise can be listed as a table as follows:

x	mdf(x)	CDF(x)
0	0.0135	0.0135
1	0.0725	0.0860
2	0.1757	0.2616
3	0.2522	0.5138
4	0.2377	0.7515
5	0.1536	0.9051
6	0.0689	0.9740
7	0.0212	0.9951
8	0.0043	0.9995
9	0.0005	0.9999
10	0.0000	1.0000

Poisson distribution

The Poisson distribution is defined for any nonnegative integer, i.e., x = 0,1,2,...,∞. Therefore, it is not possible to completely describe this distribution as we did with the binomial distribution. We will still provide a table of values by limiting our output to a finite number of values of the discrete random variable X. Suppose that you want to describe the probability distribution for a random variable X that follows the Poisson distribution with parameter λ =1.2, for X = {0,1,2,...,10}. Use:

[0][ENTER][1][0][ENTER][1][CRLST]	Result: {0 1 2 3 4 5 6 7 8 9 10}.
[ENTER]	Makes an extra copy of the list.
[1][.][2][ENTER]	Enter value of λ.
[Popd]	Calculates the list of values of the mdf for X.
[ENTER]	Makes an extra copy of the mdf's values.
[CLIST]	Calculates the cumulative distribution function CDF as a list.

The results obtained through this exercise can be listed as a table as follows:

x	mdf(x)	CDF(x)
0	0.3012	0.3012
1	0.3614	0.6626
2	0.2169	0.8795
3	0.0867	0.9662
4	0.0260	0.9923
5	0.0062	0.9985
6	0.0012	0.9997
7	0.0001	0.9999
8	0.0000	0.9999
9	0.0000	0.9999
10	0.0000	0.9999

Geometric distribution

The geometric distribution is valid for $x = 1, 2, \ldots, \infty$. The following exercise produces a list of values for the geometric distribution with parameter $p = 0.3$ corresponding to $x = \{1, 2, 3, 4, 5\}$:

[1][SPC][5][SPC][1][CRLST]	([SPC] is an alternative to [ENTER] in this context).
[ENTER]	Makes an extra copy of the list.
[.][3][ENTER]	Enter value of $p = 0.3$.
[gpd]	Calculates the list of values of the mdf for X.
[ENTER]	Makes an extra copy of the mdf's values.
[CLIST]	Calculates the cumulative distribution function CDF as a list.

The results obtained through this exercise can be listed as a table as follows:

x	mdf(x)	CDF(x)
1	0.3000	0.3000
2	0.2100	0.5100
3	0.1470	0.6570
4	0.1029	0.7599
5	0.0720	0.8319

Functions and operations applied to lists with DOLIST, DOSUBS, STREAM, and SEQ

The keystroke sequence [←][PRG][LIST][PROC] lets you access a menu of functions for applying procedures to lists. Herein we describe the operation of the functions [DOLIS], [DOSUB], [STREA], and [SEQ].

DOLIST

This function applies a procedure term by term to a number of lists. For example, given the lists x = {1 2 3}, y = {3 4 5}, and z = {2 4 5}, calculate $z + 1/(y \cdot e^{-x})$ as a list.

If we were to apply this operation to numbers z, y, x, present in stack levels 3, 2, and 1, respectively, we will use the keystroke sequence [+/-][e^x][×][1/x][+]. Try this procedure using the values z = 2, y = 3, x = 1. Use:

[2][ENTER][3][ENTER][1][ENTER] [+/-] [↰][e^x] [×] [1/x] [+]

to get 2.90609.... Thus, we could build the procedure << NEG EXP * INV + >> to perform the calculation. Try it again with the values z = 2, y = 3, x = 1. Use:

[2][ENTER][3][ENTER][1][ENTER]   Enters 2, 3, 1 in stack levels 3, 2, 1, respectively.
[↱][<<>>] [+/-] [↰][e^x] [×] [1/x] [+] [ENTER] Enter program.
[↱][EVAL]   Evaluate program with available input.

The result is again 2.90609...

To apply this procedure to the three lists given above, we enter the lists as follows:

{2 4 5}[ENTER]{3 4 5}[ENTER]{1 2 3}[ENTER]   Enter the lists.
[↱][<<>>] [+/-] [↰][e^x] [×] [1/x] [+] [ENTER] Enter program.
[3][ENTER]   Enter number of lists involved in the operation.
[↰][PRG][LIST][PROC][DOLIS]   Apply procedure to the list.

Resulting in { 2.90609... 5.84726... 9.01710...}.

Defining a function instead of using DOLISTS

Alternatively, you may want to type in the function `'f(x,y,z) = z + 1/(y*EXP(-x))'`, and use the keystroke combination [↰][DEF] to generate the program `<< → x y z 'z + 1/(y*EXP(-x))' >>`. The program requires that the values of x, y, and z, be entered into levels 3, 2, and 1 of the stack before invoking the function soft key. To operate the program with the lists indicated above we will use, therefore:

{1 2 3}[ENTER]{3 4 5}[ENTER] {2 4 5} [ENTER]   Enter the lists
[f]   Calculate the function.

The result now is {2 4 5 0.9060... 1.8472... 4.0171...}, which is completely different from what we obtained earlier. The reason again, as in the early discussion about lists and programs, has to do with the use of the operator + with lists. The operator + represents concatenation of lists rather than addition. Instead we should use the function ADD. We need to re-write our program to read as follows:

`<< → x y z <<z '1/(y*EXP(-x))' →NUM ADD>> >>`.

This modification reproduces the result obtained earlier, namely, { 2.90609... 5.84726... 9.01710...}.

Note: be careful with the order in which you enter your input to a function or when using DOLISTS.

DOSUBS

This function is used to apply a given procedure to n consecutive elements of the list, generating a new list with the results of the procedure. The number of elements of the resulting list will depend on the type of procedure use as well as on the number of consecutive elements in the original list used in the procedure. For example, the procedure << + + 3 / >> applied to three consecutive elements in a lst will provide the average of those three elements. To apply the procedure shown above to the list {1 2 3 4 5 6 7} use:

{1 2 3 4 5 6 7} [ENTER]   Enter the list to be operated upon.
[3][ENTER]   Enter the number of consecutive elements in the list that
                                     will participate in the operation.
[↰][<<>>] [+] [+] [3] [÷] [ENTER]   Enter the procedure to be applied.
[↰][PRG][LIST][PROC][DOSUB]   Apply procedure to the list.

The result is the list {2 3 4 5 6}, which you can prove represents a list of the averages of three consecutive elements of the original list, namely, {(1+2+3)/3 (2+3+4)/3 (3+4+5)/3 (4+5+6)/3 (5+6+7)/3}. While there were 7 elements to the original list {1 2 3 4 5 6 7}, there are only five possible combinations of three consecutive elements, namely [1,2,3], [2,3,4], [3,4,5], [4,5,6], and [5,6,7]. Therefore, the resulting list contains only five elements {2 3 4 5 6}.

Moving averages

An operation as the one demonstrated above is referred to as a _moving average_. With moving averages data taken sequentially can be smoothed out by taking averages of n consecutive elements in the data list. The example shown above represents a three-element moving average.

STREAM

Depending on the procedure or operator used, the function STREAM can be used to apply a procedure to each element of a list, which results in decomposing the list into its elements, or to apply a procedure repeatedly to the list elements. In the latter case, the procedure is first applied to the first two elements of the list, then this result is combined with element number 3, the next result is again combined with element number 4, and so on, until the list is exhausted, and a single result is produced.

Examples of applying the STREAM function to lists are shown below:

{1 2 3 4} [ENTER]   Enter list
[↱][<<>>][√x][ENTER]   Enter procedure to apply to the list elements
[↰][PRG][LIST][PROC][STREAM]Use STREAM to apply procedure - decomposes list.

{1 2 3 4} [ENTER]   Enter list
[↱][<<>>][↰][x^2][ENTER]   Enter procedure to apply to the list elements
[↰][PRG][LIST][PROC][STREAM]Use STREAM to apply procedure -- decomposes list.

{1 2 3 4} [ENTER]   Enter list
[↱][<<>>][+]   Enter procedure to apply to the list elements
[↰][PRG][LIST][PROC][STREAM]Use STREAM to apply procedure.
Notice that this procedure, << + >>, adds all list elements. Same as using [↰][MTH][LIST][ΣLIST].

{1 2 3 4} [ENTER]   Enter list

[r→][<<>>][2][+]   Enter procedure to apply to the list elements
[←][PRG][LIST][PROC][STREAM]Use STREAM to apply procedure.

Notice that this procedure, << + >>, adds all list elements. Same as using [←][MTH][LIST][ΣLIST].

SEQ

The function SEQ (sequence) is used to generate a list of values by applying a procedure to an index. The user provides the procedure to be applied, the name, starting and ending values, and increment for the index.

As an example, suppose you want to generate the squares of the list of numbers {0.5 1.0 1.5 2.0 2.5}, you could use the following keystroke sequence to obtain the list of squares:

[r→][<<>>][ALPHA][←][J] [←][x^2][ENTER]   Enter procedure to apply to the list elements
[r→]['] [ALPHA][←][J] [ENTER]   Enter index name
[.][5][ENTER][2][.][5][ENTER][.][5][ENTER] Enter initial and final values and increment for index
[←][PRG][LIST][PROC][NXT][SEQ]   Generate sequence as a list

The result is {0.25 1.00 2.25 4.00 6.25}.

A second example aims at applying a linear function to the list {0.25 0.50 0.75 1.00}, as follows. The procedure to use now is entered as:

[r→][<<>>] [2] [SPC] [ALPHA][←][X] [×] [5] [+] [ENTER]

Next enter:

[r→]['] [ALPHA][←][X] [ENTER]   Enter index name
[.][2][5][ENTER][1][ENTER][.][2][5][ENTER] Enter initial and final values and increment for index
[←][PRG][LIST][PROC][NXT][SEQ]   Generate sequence as a list

The result is { 5.5   6.0   6.5   7.0 }.

Notes:

[1] The functions DOLIST, DOSUBS, STREAM, and SEQ are very useful when writing your own programs in User RPL language.

[2] Other functions available using the keystroke sequence [←][PRG][LIST][PROC] are [NSUB] and [ENDSUB], which are related to DOSUBS (see the HP 49 G Advanced User's Manual), as well as [REVLI] and [SORT], which are the same functions available under the menu [←][MTH][LIST].

7 HP 49 G programming basics

In this chapter we present the basics of programming the HP 49 G in User RPL language through a variety of examples. I recommend that you try the programming examples presented in chapter 6 before reading this chapter. It is not the intention of this chapter to teach everything about programming the calculator. It is instead aimed to providing programming tools and examples that can be used later in the development of simple programming for evaluating functions, data processing, and even graphs.

Examples of sequential programming

Some very simple programming examples in User RPL language were presented in Chapter 6. In general, a program is any sequence of calculator instructions enclosed between the program symbols << and >>. The examples presented in chapter 6 were can be classified basically into two types:

Programs generated by defining a function.

These are programs generated by entering a function definition of the form:

'function_name(x_1, x_2, …) = expression containing variables x_1, x_2, …)'

and using the keystroke sequence

[←][DEF].

The program is stored in a variable called function_name. When the program is recalled to the
   stack, by using

[→][function_name]

the program shows up as follows:

<< → x_1, x_2, … 'expression containing variables x_1, x_2, …'>>.

To evaluate the function for a set of input variables x_1, x_2, …, enter the variables into the stack in the appropriate order (i.e., x_1 first, followed by x_2, then x_2, etc.), and press the soft key labeled [function_name]. The calculator will return the value of the function function_name(x_1, x_2, …).

Example: Manning's equation for wide rectangular channel.

As an example, consider the following equation that calculates the unit discharge (discharge per unit width), q, in a wide rectangular open channel using Manning's equation:

$$q = \frac{C_u}{n} y_0^{5/3} \sqrt{S_0},$$

where C_u is a constant that depends on the system of units used [C_u = 1.0 for units of the International System (S.I.), and C_u = 1.486 for units of the English System (E.S.)], n is the Manning's resistance coefficient, which depends on the type of channel lining and other factors, y_0 is the flow depth, and S_0 is the channel bed slope given as a dimensionless fraction.

Note: Values of the Manning's coefficient, n, are available in tables as dimensionless numbers, typically between 0.001 to 0.5. The value of Cu is also used without dimensions. However, care should be taken to ensure that the value of y0 has the proper units, i.e., m in S.I. and ft in E.S. The result for q is returned in the proper units of the corresponding system in use, i.e., m^2/s in S.I. and ft^2/s in E.S. Manning's equation is, therefore, not dimensionally consistent.

Suppose that we want to create a function q(Cu, n, y0, S0) to calculate the unit discharge q for this case. We will proceed as follows:

[EQW] [ALPHA][↵][Q] [↵][( )] [ALPHA][C] [ALPHA][↵][U] [SPC] [ALPHA][↵][N] [SPC] [ALPHA][↵][Y][0] [SPC] [ALPHA][S][0] [▲][▲] [→][=] [ALPHA][C] [ALPHA][↵][U] [÷] [ALPHA][↵][N] [▶] [×][ALPHA][↵][Y][0] [y^x] [5] [÷] [3] [▶][▶] [×] [√x] [ALPHA][S][0] [ENTER]

The expression entered may look like this:

'q(Cu,n,y0,S0)=Cu/n*y0^(5/3)*√S0',

if you have not selected the textbook display option. Otherwise, it will look like the equation shown earlier.

To define the function use:

[↵][DEF]

Press [VAR], if needed, to recover the variable list. At this point there will be a variable called [q] in your soft key menu. To see the contents of q, use

[→][q].

The program generated by defining the function q(Cu,n,y0,S0) shows up as:

<< → Cu n y0 S0 'Cu/n*y0^(5/3)*√S0' >>.

This is to be interpreted as

"enter Cu, n, y0, S0, in that order, then calculate the expression between quotes."

For example, to calculate q for Cu = 1.0, n = 0.012, y0 = 2 m, and S0 = 0.0001, use:

[1][ENTER] [.][0][1][2][ENTER] [2][ENTER] [.][0][0][0][1][ENTER] [q]

The result is 2.6456684 (or, q = 2.6456684 m^2/s).

You can also separate the input data with spaces in a single stack line rather than using [ENTER]. For example, to calculate q given Cu = 1.486, n = 0.022, y0 = 3.2 m, and S0 = 1×10^{-3}, use:

[1][.][4][8][6] [SPC] [.][0][2][2] [SPC] [3][.][2] [SPC] [1][EEX][3][+/-] [q]

The result now is 14.8426948578 (or, q = 14.8426948578 ft^2/s).

Programs that mimic a sequence of calculator operations.

In this case, the terms to be involved in the sequence of operations are assumed to be present in the stack. The program is typed in by first opening the program symbols with [→][<<>>]. Next, the sequence of operations to be performed is entered. When all the operations have been typed in, press [ENTER] to complete the program. If this is to be a once-only program, you can at this point, press [→][EVAL] to execute the program using the input data available. If it is to be a permanent program, it needs to be stored in a variable name.

The best way to describe this type of programs is with an example:

Example: Velocity head for a rectangular channel.

Suppose that we want to calculate the velocity head, h_v, in a rectangular channel of width b, with a flow depth y, that carries a discharge Q. The specific energy is calculated as

$$h_v = \frac{Q^2}{2g(by)^2},$$

Where g is the acceleration of gravity (g = 9.806 m/s^2 in S.I. units or g = 32.2 ft/s^2 in E.S. units). If we were to calculate h_v for Q = 23 cfs (cubic feet per second = ft^3/s), b = 3 ft, and y = 2 ft, we would use:

$$h_v = \frac{23^2}{2 \cdot 32.2 \cdot (3 \cdot 2)^2}.$$

Using the RPN system in the HP 49 G, interactively, we can calculate this quantity as:

[2] [ENTER] [3] [×] [←][x^2] [3][2][.][2] [×] [2] [×] [2][3][←][x^2] [▶] [÷]

Resulting in 0.228174, or h_v = 0.228174... ft. (Since we didn't use units in the input, the result will have no units either).

Notice that we start by calculating the last term in the denominator, then entering the next term to the left. Next, we multiply those two terms. Next we enter 2 and multiply it with register x. Finally, we enter the numerator, swap the order, and divide the two terms. Although this is not the only way to obtain this result, this approach is the most efficient in order to translate the procedure into a program because it uses mainly register x in the calculations.

To put this calculation together as a program we need to have the input data (Q, g, b, y) in the stack in the order in which they will be used in the calculation. In terms of the variables Q, g, b, and y, the calculation just performed is written as (do not type the following):

$$y \text{ [ENTER] } b \text{ [×] } [\leftarrow][x^2] \quad g \text{ [×] } [2] \text{ [×] } Q \text{ [}\leftarrow][x^2] \text{ [▶] } [÷]$$

As you can see, y is used first, then we use b, g, and Q, in that order. Therefore, for the purpose of this calculation we need to enter the variables in the inverse order, i.e., (do not type the following):

Q [ENTER] g [ENTER] b [ENTER] y [ENTER]

For the specific values under consideration use:

[2][3][ENTER]   [3][2][.][2][ENTER]   [3][ENTER] [2][ENTER]

The program itself will contain only those keystrokes (or commands) that result from removing the input values from the interactive calculation shown earlier, i.e., removing Q, g, b, and y from (do not type the following):

$$y \text{ [ENTER] } b \text{ [×] } [\leftarrow][x^2] \quad g \text{ [×] } [2] \text{ [×] } Q \text{ [}\leftarrow][x^2] \text{ [▶] } [÷]$$

and keeping only the operations shown below (do not type the following):

$$[×] \text{ }[\leftarrow][x^2] \text{ }[×] \text{ }[2] \text{ }[×] \text{ }Q \text{ }[\leftarrow][x^2] \text{ }[▶] \text{ }[÷]$$

Note: When entering the program do not use the keystroke [▶], instead use the keystroke sequence: [←] [PRG][STACK][SWAP].

However, unlike the interactive use of the calculator performed earlier, we need to do some swapping of registers x and y within the program. To write the program, we use, therefore:

[→][<<>>]	Opens program symbols
[×]	Multiply y (register x) with b (register y)
[←][x²]	Square (b·y)
[×]	Multiply (b·y)² [register x] times g [register y]
[2] [×]	Enter a 2 [now in regx x] and multiply it with g· (b·y)² [now in reg. y]
[←][PRG][STACK][SWAP]	Swap Q [reg. y] with 2·g· (b·y)² [now in reg. y]
[←][x²]	Square Q
[←][PRG][STACK][SWAP]	Swap 2·g· (b·y)² [reg. y] with Q² [reg. x]
[÷]	Divide Q² [reg. y] by 2·g· (b·y)² [reg. x]
[ENTER]	Enter the program

The resulting program looks like this:

```
<< * SQ * 2 * SWAP SQ SWAP / >>
```

> Note: SQ is the function that results from the keystroke sequence [↰][x²].

Let's make an extra copy of the program and save it into a variable called hv:

[ENTER] [↱][`] [ALPHA][↰][H] [ALPHA][↰][V] [STO▶]

A new variable [hv] should be available in your soft key menu. (Press [VAR] to see your variable list.) Also, check that your stack looks like this:

```
5:                              23.
4:                              32.2
3:                              3.
2:                              2.
1:    << * SQ * 2 * SWAP SQ SWAP / >>
```

To evaluate the program use [↱][EVAL]. The result should be 0.228174..., as before.

The program is available for future use in variable [hv]. For example, for Q = 0.5 m³/s, g = 9.806 m/s², b = 1.5 m, and y = 0.5 m, use:

[.][5][SPC] [9][.][8][0][6][SPC] [1][.][5][SPC] [.][5] [hv]

> Note: [SPC] is used here as an alternative to [ENTER] for input data entry.

The result now is 2.26618623518E-2, i.e., hv = 2.26618623518×10⁻² m.

Since the equation programmed in [hv] is dimensionally consistent, we can use units in the input. Try the following:

```
[ . ][5] [↱][UNITS] [VOL][m^3] [1][↱][UNITS] [TIME][ s ] [÷]
[9][ . ][8][0][6] [↱][UNITS][LENG][ m ] [1][↱][UNITS] [TIME][ s ] [↰][x²] [÷]
[1][ . ][5] [↱][UNITS][LENG][ m ]
[ . ][5] [↱][UNITS][LENG][ m ]
[VAR][ hv ]
```

The result, now with units attached, is `2.26618623518E-2_m`.

The two types of programs presented in this section are sequential programs, in the sense that the program flow follows a single path, i.e., INPUT→ OPERATION →OUTPUT. Branching of the program flow is possible by using the commands in the menu [↰][PRG][BRCH], i.e., IF, CASE, START, FOR, DO, WHILE, and the functions IFT and IFTE (examples of the latter were also presented in chapter 6). More detail on program branching is presented below.

Interactive input in programs

In the sequential program examples shown in the previous section it is not always clear to the user the order in which the variables must be placed in the stack before program execution. For the case of the program q, written as

$$<< \rightarrow Cu\ n\ y0\ S0\ 'Cu/n*y0^{(5/3)}*\sqrt{S0}' >>,$$

it is always possible to recall the program definition into the stack ([┌→][q]) to see the order in which the variables must be entered, namely,

→ Cu n y0 S0.

However, for the case of the program [hv], its definition

<< * SQ * 2 * SWAP SQ SWAP / >>,

does not provide a clue of the order in which the data must be entered, unless, of course, you are extremely experienced with RPN and the User RPL language.

One way to check the result of the program as a formula is to enter symbolic variables, instead of numeric results, in the stack, and let the program operate on those variables. For this approach to be effective the calculator's CAS (Calculator Algebraic System) must be set to symbolic and exact modes. This is accomplished by typing

[MODE][CAS],

and ensuring that the check marks in the options

_Numeric and _Approx

are removed. Press [OK][OK] to return to normal calculator display. Press [VAR] to display your variables menu.

We will use this latter approach to check what formula results from using the program [hv] as follows: We know that there are four inputs to the program, thus, we use the symbolic variables S4, S3, S2, and S1 to indicate the stack levels at input:

[ALPHA][S][4][ENTER] [ALPHA][S][3][ENTER] [ALPHA][S][2][ENTER] [ALPHA][S][1][ENTER]

Next, press [hv]. The resulting formula may look like this

'SQ(S4)/(S3*SQ(S2*S1)*2)',

if your display is not set to textbook style, or like this,

$$\frac{SQ(S4)}{S3 \cdot SQ(S2 \cdot S1) \cdot 2},$$

if textbook style is selected. Since we know that the function SQ( ) stands for x^2, we interpret the latter result as

$$\frac{S4^2}{2 \cdot S3 \cdot (S2 \cdot S1)^2},$$

which indicates the position of the different stack input levels in the formula. By comparing this result with the original formula that we programmed, i.e.,

$$h_v = \frac{Q^2}{2g(by)^2},$$

we find that we must enter y in stack level 1 (S1), b in stack level 2 (S2), g in stack level 3 (S3), and Q in stack level 4 (S4).

Prompt with an input string

These two approaches for identifying the order of the input data are not very efficient. You can, however, help the user identify the variables to be used by prompting him or her with the name of the variables. From the various methods provided by the User RPL language, the simplest is to use an input string and the function INPUT ([↰][PRG][NXT][IN][INPUT]) to load your input data.

The following program prompts the user for the value of a variable a and places the input in stack level 1:

<< "Enter a:" {"↵:a:" {2 0} V } INPUT OBJ→ >>

To type this program use the following:

[↰][<<>>][↰][" "]	Open program symbol, open double quotes (string).
[ALPHA][↰][ALPHA]	Lock alpha keyboard in lower case
[ALPHA][ALPHA]	Lock keyboard in alphabetic mode
[↰][E][N][T][E][R][SPC][A][↰][::]	Type in Enter a:"
[ALPHA]	Unlock alphabetic keyboard
[▶][↰][{}][↰][" "]	Type in {"
[↰][↵][↰][::][ALPHA][A] [▶][▶]	Type in ↵:a:" (*)
[↰][{}][2][SPC][0] [▶]	Type in {2 0}
[ALPHA][↰][V] [▶]	Type in V }
[↰][PRG][NXT][IN][INPUT]	Enter INPUT
[NXT][PRG][TYPE][OBJ→]	Enter OBJ→
[ENTER]	Enter program in stack level 1

(*) The symbol ↵ indicates a line feed. This symbol is obtained by using [↰] with the [.] key.

Save the program in a variable called INPTa (for INPuT a) as follows:

[↰][`] [ALPHA][ALPHA] [I][N][P][T][↰] [A] [ENTER] [STO▶]

Press [VAR]. The variable [INPTa] should be available in your soft key menu.

Try running the program by pressing the soft key labeled [INTPa]. The result is a stack prompting the user for the value of a and placing the cursor right in front of the prompt :a: Enter a value for a, say 35, then press [ENTER]. The result is the input string

:a:35

in stack level 1.

A function with an input string

If you were to use this piece of code to calculate the function, f(a) = 2*a^2+3, you could modify the program to read as follows:

        << "Enter a:" {"↵:a:" {2 0} V } INPUT OBJ→ → a << '2*a^2+3'>> >>

To modify the program use:

[→][INTPa] [▼]	Copies contents of INTPa to register x and launches editor
[→][▼][◄]	Sends cursor to end of program
[→][→][ALPHA][←][A]	Type in → a
[→][<<>>][→]['] (')	Creates sub-program symbol and starts algebraic symbol
[2][×][ALPHA][←][A][y^x][2][+][3] [ENTER]	Type in algebraic expression 2*a^2+3

Save this new program under the name 'FUNCa' (FUNCtion of a):

        [→]['] [ALPHA][ALPHA] [F][U][N][C][←] [A] [ENTER] [STO▶]

Run the program by pressing [FUNCa]. When prompted to enter the value of a enter, for example, 2, and press [ENTER]. The result is simply the algebraic $2a^2+3$, which is an incorrect result. The HP 49 G provides functions for debugging programs to identify logical errors during program execution as shown below.

Debugging the program

To figure out why it did not work we use the DBUG program in the HP 49 G as follows:

[→]['] [FUNCa] [ENTER]	Copies program name to stack level 1
[←][PRG][NXT][NXT][RUN][DBUG]	Starts debugger
[SST↓]	Step-by-step debugging, result: "Enter a:"
[SST↓]	Result: {" ↵ a:" {2 0} V}
[SST↓]	Result: user is prompted to enter value of a
[2][ENTER]	Enter a value of 2 for a. Result: "↵:a:2"
[SST↓]	Result: a:2
[SST↓]	Result: empty stack, executing →a
[SST↓]	Result: empty stack, entering subprogram <<
[SST↓]	Result: '2*a^2+3'
[SST↓]	Result: '2*a^2+3', leaving subprogram >>
[SST↓]	Result: '2*a^2+3', leaving main program>>

Further pressing [SST↓] produces no more output since we have gone through the entire program, step by step. This run through the debugger did not provide any information on why the program is not calculating the value of $2a^2+3$ for a = 2. To see what is the value of a in the sub-program, we need to run the debugger again and evaluate a within the sub-program. Try the following:

[VAR]	Recovers variables menu
[r→]['] [FUNCa] [ENTER]	Copies program name to stack level 1
[←][PRG][NXT][NXT][RUN][DBUG]	Starts debugger
[SST↓]	Step-by-step debugging, result: "Enter a:"
[SST↓]	Result: {" ↵a:" {2 0} V}
[SST↓]	Result: user is prompted to enter value of a
[2][ENTER]	Enter a value of 2 for a. Result: "↵:a:2"
[SST↓]	Result: a:2
[SST↓]	Result: empty stack, executing →a
[SST↓]	Result: empty stack, entering subprogram <<

At this point we are within the subprogram << '2*a^2+3'>> which uses the local variable a. To see the value of a use:

[ALPHA][←][A] [r→][EVAL]   This indeed shows that the local variable a = 2

Let's kill the debugger at this point since we already know the result we will get. To kill the debugger press [KILL]. You receive an <!> Interrupted message acknowledging killing the debugger. Press [ON] to recover normal calculator display.

> Note: In debugging mode, every time we press [SST↓] the top left corner of the display shows the program step being executed. A soft key function called [SST] is also available under the [RUN] menu. This can be used to execute at once any sub-program called from within a main program. Examples of the application of [SST] will be shown later.

Fixing the program

Let's list the program in the stack once more:

[VAR]	Recovers variables menu
[r→][FUNCa]	Copies program to stack level 1

The only possible explanation for the failure of the program to produce a numerical result seems to be the lack of an EVAL function after the algebraic expression '2*a^2+3'. Let's edit the program by adding the missing EVAL function as follows:

[▼]	Launch editor
[r→][▼][◄][◄][◄]	Sends cursor to end of algebraic expression
[r→][EVAL]	Evaluate algebraic
[ENTER]	Enter edited program back into stack level 1
[←][FUNCa]	Shortcut to store level 1 into variable FUNCa

Run the program again by pressing [FUNCa]. When prompted for the value of a enter 2, and press [ENTER]. The result is now 11, the correct result.

The modified function program will look like this (use [r→][FUNCa] to recall its contents to the stack):

<< "Enter a:" {"↵:a:" {2 0} V } INPUT OBJ→ → a << '2*a^2+3' EVAL >> >>

Input string for programs requiring two or three input values

In this section we will create a sub-directory, within the directory HOME, to hold examples of input strings for one, two, and three input data values. These will be generic input strings that can be incorporated in any future program, taking care of changing the variable names according to the needs of each program.

Let's get started by creating a sub-directory called PTRICKS (Programming TRICKS) to hold programming tidbits that we can later borrow from to use in more complex programming exercises. To create the sub-directory, first make sure that you move to the HOME directory, by using [UPDIR] as many times as needed. Within the HOME directory, use the following keystrokes to create the sub-directory PTRICKS:

[→]['][ALPHA][ALPHA][P][T][R][I][C][K][S] [ENTER]   Enter directory name PTRICKS
[←][PRG][MEM][DIR][CRDIR]   Create directory
[VAR]   Recover variables listing in soft keys

Before we develop the programs for 2 and 3 variables, let's copy the program INPTa into PTRICKS as follows:

- Move to the directory where you defined INPTa (skip this step if INTPa is defined in the HOME directory)
- Copy contents of INPTa into stack level 1 by using [→][INPTa].
- Copy the name of the program to stack level 1 by using [→]['][INPTa][ENTER]
- Edit the text in level 1 to read INPT1 by using [▼][←][▶][◀][⇦][1][ENTER]
- Move within directory PTRICKS by pressing [PTRICKS] in the HOME directory (move back to the HOME directory before performing this step if needed)
- You should have the program in stack level 2 and the program name, INPT1, in stack level 1. Press [STO▶] to store the program into the appropriate variable name.
- Press [VAR] to recover the list of variables within PTRICKS. There should be only one variable listed, namely, INPT1.

> Note: A procedure similar to this can be used to easily copy a variable from one sub-directory to another. You can even copy the contents of a full directory to stack level 1 by using [→] followed by the soft key corresponding to the directory name. For example, moving back to the HOME directory you could copy the contents of PTRICKS onto stack level 1 by using [→][PTRICKS]. At this moment, you should get the following in stack level 1:
>
> ```
> DIR
> INPT1 << "Enter a:"
> {"↵:a:" {2 0} V
> } INPUT OBJ→ >>
> END
> ```
>
> This is the format in which a directory is listed in the stack. The listing starts with the word DIR to identify the contents of the stack as a directory, followed by the names and contents of all the variables available in the directory, and finishing with the word END.

A program may have more than 3 input data values, however, when using input strings we want to limit the number of input data values to 3 at a time for the simple reason that, in general, we have visible only 5 stack levels. If we use stack level 5 to give a title to the input string, and leave stack level 4 empty to facilitate reading the display, we have only stack levels 1 through 3 to define input variables. [Note: With the use of a smaller font we can get more than 5 levels in the stack, thus allowing form more than 3 variables to be input at a time, but with the standard HP 49 G display format the default is 5 stack levels.] If we have more than 3 variables to input, we can use a second, third, etc., input string to enter the remaining input values. Having available programs for entering 1, 2, and 3 input values per input string we can handle most inputs to our programs.

Input string program for two input values

The input string program for two input values, say a and b, looks as follows:

<< "Enter a and b: " {" ↵:a:↵:b: " {2 0} V } INPUT OBJ→ >>

This program can be easily created by modifying the contents of INPT1 as follows:

[→][INPT1]	Copy contents of INPT1 to stack level 1
[▼]	Start editor
[▶](10 times)	Move cursor to the right of the letter a
[ALPHA][←][ALPHA]	Lock alpha keyboard in lower case
[ALPHA][ALPHA]	Lock keyboard in alphabetic mode
[SPC][A][N][D][SPC][B]	Type in and b
[ALPHA]	Unlock alphabetic keyboard
[▶](10 times)	Move cursor to the right of the character :
[→][↵][←][::][ALPHA][B]	Type in ↵ :b:"
[ENTER]	Enter program in stack
[→][`][INPT1]	Enter variable name INPT1 in stack
[◀][⇦][2][ENTER]	Edit variable name to read INPT2
[STO▶]	Store new program in variable INPT2
[VAR]	To recover variables list

To check the operation of the new program, press [INTP2]. Enter a value for a, say 2, (careful here), then press [▼], not [ENTER], to move the cursor in front of the b prompt, and enter the value of b, say 3. Now, press [ENTER]. The result is the tagged values a:2 and b:3 in stack levels 2 and 1, respectively.

Application: evaluating a function of two variables

The ideal gas law was introduced in Chapter 3:

$$pV = nRT,$$

where:

- p = gas pressure (Pa),
- V = gas volume(m^3),
- n = number of moles (gmol),
- R = universal gas constant = 8.31451 J/(gmol*K), and
- T = absolute temperature (K).

We can define the pressure p as a function of two variables, V and T, as p(V,T) = nRT/V for a given mass of gas since n will also remain constant. Assume that n = 0.2 gmol, then the function to program is

$$p(V,T) = 8.31451 \cdot 0.2 \cdot \frac{T}{V} = (1.662902 \_ \frac{J}{K}) \cdot \frac{T}{V}.$$

We can define the function by typing the function definition as:

[EQW][ALPHA][↰][P][↰][( )][ALPHA][V][SPC][ALPHA][T] [▲][▲] [↱][=][↰][( )]

[1][.][6][6][2][9][0][2][↱][_][ALPHA][J][÷][ALPHA][K][▶][▶][▶]

[×] [ALPHA][T][÷][ALPHA][V][ENTER]

The function is defined by using [↰][DEF] which creates the variable [p]. The contents of this variable, which can be recalled by using [VAR][↱][P], are shown as

<< → V T '1.662902*(J/K)*(T/V)' >>

Note: the keystroke sequence [↱][_] did not have an effect in the equation writer, instead, a * (times) sign was included. This, however, does not produce the desire effect of defining J/K as units. Edit the program to read:

<< → V T '(1.662902_J/K)*(T/V)' >>

and store it back into variable [p]. The next step is to add the input string that will prompt the user for the values of V and T. To create this input stream we place the contents of variable INPT2 into the stack: [↱][INPT2], producing:

<< "Enter a and b: " {"↵:a:↵:b:" {2 0} V } INPUT OBJ→ >>

Modify it to read:

<< "Enter V and T: " {"↵:V:↵:T:" {2 0} V } INPUT OBJ→ >>

Press [ENTER] to create an additional copy of the program, and press [↱][EVAL] to run this modified version of INPT2. Enter values of V = 0.01_m^3 and T = 300_K. Before pressing [ENTER], the stack will look like this:

```
Enter V and T:
:V:0.01_m^3
:T:300_K
```

Press [ENTER] to get the result

```
3: <<Enter V and T: "...
2:           V:0.01_m^3
1:             T:300_K
```

This shows that the INPT2 program places the values of V and T in stack levels 2 and 1, respectively. This is the correct order for data input into the program [p], i.e., in →V T. To drop the contents of level 1 and 2, press the backspace key, [⇦], twice.

The next step is to copy the corresponding piece of code from the program listed in the stack to the function definition. This can be accomplished by using the following:

[▼]	Start the program editor
[▶][↵][BEGIN] (press the APPS key)	Position cursor at beginning of code to copy
[↵][▼][◀]	Position cursor at end of code to copy
[↵][END] (press the MODE key)	Highlight code to copy
[↵][COPY] (press the VAR key)	Copy highlighted code
[ENTER]	End editing of this program
[↵][p][▼][▶]	Copy contents of p, start the editor, position cursor
[↵][PASTE] (in the NXT key)	Paste code copied earlier
[ENTER]	Enter new program in stack

The combined program will look like this:

```
<< "Enter V and T: " {"↵  :V:↵  :T: " {2 0} V } INPUT OBJ→ →V T
'(1.662902_J/K)*(T/V)' >>
```

Store the new program into variable p by using [←][p].

To test the program, press [p]. Enter values of V = 0.01_m^3 and T = 300_K in the input string, then press [ENTER]. The result is 49887.06_J/m^3. The units of J/m^3 were proven earlier to be equivalent to Pascals (Pa), the preferred pressure unit in the S.I. system.

> Note: because we deliberately included units in the function definition, the input values must have units attach to them in input to produce the proper result.

Input string program for three input values

The input string program for three input values, say a, b, and c, looks as follows:

```
<< "Enter a, b and c: " {"↵  :a:↵  :b:↵  :c: " {2 0} V } INPUT OBJ→ >>
```

This program can be easily created by modifying the contents of INPT2 to make it look like shown immediately above. The resulting program can then be stored in a variable called INPT3. With this program we complete the collection of input string programs that will allow us to enter one, two, or three data values. Keep these programs as a reference and copy and modify them to fulfill the requirements of new programs you write.

Application: evaluating a function of three variables

Suppose that we want to program the ideal gas law including the number of moles, n, as an additional variable, i.e., we want to define the function

$$p(V,T,n) = (8.31451\_\frac{J}{K})\frac{n \cdot T}{V},$$

and modify it to include the three-variable input string. The procedure to put together this function is very similar to that used earlier in defining the function p(V,T). The resulting program will look like this:

`<< "Enter V, T, and n:" {" ↵ :V:↵ :T:↵ :n:" {2 0} V } INPUT OBJ→ →V T n '(8.34451_J/(K*mol))*(n*T/V)' >>`

Store this result back into the variable [p]. To run the program, press [p].

Enter values of V = 0.01_m^3, T = 300_K, and n = 0.8_mol. Before pressing [ENTER], the stack will look like this:

```
Enter V, T and n:
:V:0.01_m^3
:T:300_K
:n:0.8_mol
```

Press [ENTER] to get the result 200268.24_J/m^3, or 200268.24_Pa = 200.27 kPa.

Identifying output in programs

The simplest way to identify numerical program output is to "tag" the program results. A tag in the HP 49 G calculator is simply a string attached to a number. The string will be the name associated with the numeric result. For example, earlier on, when checking the operation of the input string programs INTPa (or INPT1) and INPT2, we obtained as results tagged numerical output such as :a:35.

Tagging a numerical result

To tag a numerical result you need to place the number in stack level 2 and the tagging string in stack level 2, then use the →TAG function ([←][PRG][TYPE][→TAG]). For example, to produce the tagged result B:5., use:

[5][ENTER]   Enter the numerical result
[→]["][ALPHA][B]   Enter the tagging string
[←][PRG][TYPE][→TAG]   Tag numerical result

Decomposing a tagged numerical result into a number and a tag

To decompose a tagged result into its numerical value and its tag, simply use the →OBJ function, available at [←][PRG][TYPE][→OBJ]. The result of decomposing a tagged number with →OBJ is to place the numerical value in stack level 2 and the tag in stack level 1. If you are interested in using the numerical value only, then you will drop the tag by using the backspace key [⇐].

"De-tagging" a tagged quantity

"De-tagging" means to extract the numerical quantity out of a tagged quantity. This function is accessed through the keystroke combination: [←][PRG][TYPE][NXT][DTAG]. For example, given the tagged quantity a:2, DTAG returns the numerical value 2.

Examples: tagging output from function

Example 1 - tagging output from function FUNCa

Let's modify the function FUNCa, defined earlier, to produce a tagged output. Use [→][FUNCa] to recall the contents of FUNCa to the stack. The original function program reads

<< "Enter a: " {"↵:a: " {2 0} V } INPUT OBJ→ → a << '2*a^2+3' EVAL >> >>

Modify it to read:

<< "Enter a: " {"↵:a: " {2 0} V } INPUT OBJ→ → a << '2*a^2+3' EVAL "F" →TAG >> >>

Store the program back into FUNCa by using [←][FUNCa]. Next, run the program by pressing [FUNCa]. Enter a value of 2 when prompted, and press [ENTER]. The result is now the tagged result F:11.

Example 2 - tagging input and output from function FUNCa

In this example we modify the program FUNCa so that the output includes not only the evaluated function, but also a copy of the input with a tag.

Use [→][FUNCa] to recall the contents of FUNCa to the stack:

<< "Enter a: " {"↵:a: " {2 0} V } INPUT OBJ→ → a << '2*a^2+3' EVAL >> >>

Modify it to read:

<< "Enter a: " {"↵:a: " {2 0} V } INPUT OBJ→ → a << '2*a^2+3' EVAL "F" →TAG a SWAP>>
>>

(Recall that the function SWAP is available by using [←][PRG][STACK][SWAP].) Store the program back into FUNCa by using [←][FUNCa]. Next, run the program by pressing [FUNCa]. Enter a value of 2 when prompted, and press [ENTER]. The result is now two tagged numbers a:2. in stack level 2, and F:11. in stack level 1.

> Note: Because we use an input string to get the input data value, the local variable a actually stores a tagged value (:a:2, in the example above). Therefore, we do not need to tag it in the output. All what we need to do is place an a before the SWAP function in the subprogram above, and the tagged input is placed in the stack. It should be pointed out that, in performing the calculation of the function, the tag of the tagged input a is dropped automatically, and only its numerical value is used in the calculation.

To see the operation of the function FUNCa, step by step, you could use the DBUG function as follows:

[→]['] [FUNCa] [ENTER]	Copies program name to stack level 1
[←][PRG][NXT][NXT][RUN][DBUG]	Starts debugger
[SST↓]	Step-by-step debugging, result: "Enter a:"
[SST↓]	Result: {" ↵a:" {2 0} V}
[SST↓]	Result: user is prompted to enter value of a
[2][ENTER]	Enter a value of 2 for a. Result: "↵:a:2"
[SST↓]	Result: a:2
[SST↓]	Result: empty stack, executing →a
[SST↓]	Result: empty stack, entering subprogram <<
[SST↓]	Result: '2*a^2+3'
[SST↓]	Result: 11.,
[SST↓]	Result: "F"
[SST↓]	Result: F: 11.
[SST↓]	Result: a:2.
[SST↓]	Result: swap levels 1 and 2
[SST↓]	leaving subprogram >>
[SST↓]	leaving main program>>

Example 3 - tagging input and output from function p(V,T)

In this example we modify the program [p] so that the output tagged input values and tagged result.

Use [→][p] to recall the contents of the program to the stack:

<< "Enter V, T, and n:" {" ↵ :V:↵ :T:↵ :n:" {2 0} V } INPUT OBJ→ →V T n
            '(8.34451_J/(K*mol))*(n*T/V)' >>

Modify it to read:

<< "Enter V, T and n:" {"↵ :V:↵   :T:" {2 0} V } INPUT OBJ→ →V T n <<V T n
            '(8.34451_J/(K*mol))*(n*T/V)' EVAL "p" →TAG >> >>

> Note: Notice that we have placed the calculation and tagging of the function p(V,T,n), preceded by a recall of the input variables V T n, into a sub-program [the sequence of instructions contained within the inner set of program symbols << >>].   This is necessary because without the program symbol separating the two listings of input variables, namely V T N << V T n, the program will assume that the input command
>
> →V T N V T n
>
> requires six input values while only three are available. The result would have been the generation of an error message and the interruption of the program execution.
>
> To include the subprogram mentioned above in the modified definition of program [p], will require you to use [→][<< >>] at the beginning and end of the sub-program. Because the program symbols occur in pairs whenever [→][<< >>] is invoked, you will need to erase the closing program symbol (>>) at the beginning, and the opening program symbol (<<) at the end, of the sub-program.
>
> To erase any character while editing the program, place the cursor to the right of the character to be erased and use the backspace key [⇐].

Store the program back into variable p by using [↵][p]. Next, run the program by pressing [p]. Enter values of V = 0.01_m^3, T = 300_K, and n = 0.8_mol, when prompted. Before pressing [ENTER] for input, the stack will look like this:

```
Enter V, T and n:

:V:0.01_m^3
:T:300_K
:n:0.8_mol
```

After execution of the program, the stack will look like this:

```
5:
4:                    V:0.01_m^3
3:                       T:300_K
2:                      n:0.8_mol
1:       p: 200268.24_J/m^3
```

In summary: The common thread in the three examples shown here is the use of tags to identify input and output variables. If we use an input string to get our input values, those values are already pre-tagged and can be easily recall into the stack for output. Use of the →TAG command allows us to identify the output from a program.

Using a message box for output

A message box is a fancier way to present output from a program. The message box command in the HP 49 G is obtained by using [↵][PRG][NXT][OUT][MSGBO]. The message box command requires that the output string to be placed in the box be available in stack level 1. To see the operation of the MSGBOX command try the following exercise:

[→]["][ALPHA][→][T][ALPHA][↵][::][1][.][2][→][_]   Starts string "θ:1.2_
[ALPHA][↵][ALPHA] [ALPHA][ALPHA][R][A][D][ENTER]   Completes string "θ:1.2_rad"
[↵][PRG][NXT][OUT][MSGBO]   Places string in a message box
[OK]   To cancel message box output

You could use a message box for output from a program by using a tagged output, converted to a string, as the output string for MSGBOX. To convert any tagged result, or any algebraic or non-tagged value, to a string, use the function →STR available at [↵][PRG][TYPE][→STR].

Example - using a message box for program output

The function [p], from the last example, can be modified to read:

<< "Enter V, T and n: " {"↵ :V:↵ :T:↵ :n: " {2 0} V } INPUT OBJ→ →V T n <<V T n '(8.34451_J/(K*mol))*(n*T/V)' EVAL "p" →TAG →STR MSGBOX>> >>

You know the drill:

- Store the program back into variable p by using [↵][p].
- Run the program by pressing [p].
- Enter values of V = 0.01_m^3, T = 300_K, and n = 0.8_mol, when prompted.

As in the earlier version of [p], before pressing [ENTER] for input, the stack will look like this:

```
Enter V, T and n:
:V:0.01_m^3
:T:300_K
:n:0.8_mol
```

The first program output is a message box containing the string:

```
:p:
'200268.24_J/m^3'
```

Press [OK] to cancel message box output. The stack will now look like this:

```
5:
4:
3:            V:0.01_m^3
2:              T:300_K
1:             n:0.8_mol
```

Including input and output in a message box - string concatenation

We could modify the program so that not only the output, but also the input, is included in a message box. For the case of program [p], the modified program will look like:

<< "Enter V, T and n: " {"↵" :V:↵ :T:↵ :n: " {2 0} V } INPUT OBJ→ →V T n <<V →STR "↵ " + T →STR "↵ " + n →STR "↵ " +'(8.34451_J/(K*mol))*(n*T/V)' EVAL "p" →TAG →STR + + + MSGBOX>> >>

Notice that you need to add the following piece of code after each of the variable names V, T, and n, within the sub-program:

$$\rightarrow STR \text{ "↵ " } +$$

To get this piece of code typed in the first time use:

[↵][PRG][TYPE] [→STR] [→][" "] [→][↵] [▶] [+]

Because the functions for the TYPE menu remain available in the soft menu keys, for the second and third occurrences of the piece of code (→STR "↵ " +) within the sub-program (i.e., after variables T and n, respectively), all you need to use is:

[→STR] [→][" "] [→][↵] [▶] [+]

You will notice that after typing the keystroke sequence [→][↵] the result is that a new line is generated in the stack.

The last modification that needs to be included is to type in the plus sign three times after the call to the function at the very end of the sub-program.

> Note: The plus sign (+) in this program is used to concatenate strings. Concatenation is simply the operation of joining individual character strings.

To see the program operating:

- Store the program back into variable p by using [↰][p].
- Run the program by pressing [p].
- Enter values of V = 0.01_m^3, T = 300_K, and n = 0.8_mol, when prompted.

As in the earlier version of [p], before pressing [ENTER] for input, the stack will look like this:

```
Enter V, T and n:
:V:0.01_m^3
:T:300_K
:n:0.8_mol
```

The first program output is a message box containing the string:

```
:V:     '.01_m^3'
:T:     '300_K'
:n:     '.8_mol'
:p:
'200268.24_J/m^3'
```

Press [OK] to cancel message box output.

Incorporating units within a program

As you have been able to observe from all the examples for the different versions of program [p] presented in this chapter, attaching units to input values may be a tedious process. You could have the program itself attach those units to the input and output values. We will illustrate these options by modifying yet once more the program [p], as follows.

Recall the contents of program [p] to the stack by using [↱][p], and modify them to look like this:

> Note: I've separated the program arbitrarily into several lines for easy reading. This is not necessarily the way that the program shows up in the calculator's stack. The sequence of commands is correct, however. Also, recall that the character ↵ does not show in the stack, instead it produces a new line.

```
<< "Enter V,T,n [S.I.]:" {"↵ :V:↵ :T:↵ :n: " {2 0} V } INPUT OBJ→ →V T n
<<V '1_m^3' * { }+ T '1_K' * + n '1_mol' * + EVAL →V T n
<<V "V" →TAG →STR "↵ "+ T "T" →TAG →STR "↵ "+ n "n" →TAG →STR "↵ "+
'(8.34451_J/(K*mol))*(n*T/V)' EVAL "p" →TAG →STR + + + MSGBOX>> >> >>
```

This new version of the program includes an additional level of sub-programming (i.e., a third level of program symbols << >>, and some steps using lists, i.e.,

$$V \ \text{'1\_m^3'} \ * \ \{ \} + T \ \text{'1\_K'} \ * \ + \ n \ \text{'1\_mol'} \ * \ + \ \text{EVAL} \ \rightarrow V \ T \ n$$

The interpretation of this piece of code is as follows. (We use input string values of :V:0.01, :T:300, and :n:0.8):

1. V   : The value of V, as a tagged input (e.g., V:0.01) is placed in the stack.

2. '1_m^3'   : The S.I. units corresponding to V are then placed in stack level 1, the tagged input for V is moved to stack level 2.

3. *   : By multiplying the contents of stack levels 1 and 2, we generate a number with units (e.g., 0.01_m^3), but the tag is lost.

4. { }   : An empty list is place on stack level 1, and the previous result is moved to stack level 2.

5. +   : The contents of stack level 2 are 'added' to the empty list resulting in the list {0.01_m^3}.

6. T '1_K' * + :   : These operations result in the value of T, with its proper units, being 'added' to the list under consideration, resulting in the list being expanded to {0.01_m^3 300_K }

7. n '1_mol' * +   : These operations result in the value of n, with its proper units, being 'added' the list under consideration, resulting in the list being expanded to {0.01_m^3 300_K 0.8_mol }

8. EVAL   : This command has the effect of decomposing the list and placing its elements in order in the stack.

9. → V T n   : The values of V, T, and n, located respectively in stack levels 3, 2, and 1, are passed on to the next level of sub-programming.

To see this version of the program in action do the following:

- Store the program back into variable p by using [←][p].
- Run the program by pressing [p].
- Enter values of V = 0.01, T = 300, and n = 0.8, when prompted (no units required now).

Before pressing [ENTER] for input, the stack will look like this:

```
Enter V,T,n [S.I.]:
:V:0.01
:T:300
:n:0.8
```

Press [ENTER] to run the program. The output is a message box containing the string:

```
:V:    '.01_m^3'
:T:    '300_K'
:n:    '.8_mol'
:p:    '200268.24_J/m^3'
```

Press [OK] to cancel message box output.

Note: The actual way that the program shows up in the stack when editing it is the following:

```
<<
"Enter V,T,n [S.I.]: "
{ "
:V:
:T:
:n: " {2.
0.} V } INPUT OBJ→ →
V T n
   <<V '1_m^3' * { }
+ T '1_K' * + n '1_
mol' * + EVAL →V T n
      <<V "V" →TAG →STR
"
" + T "T" →TAG →STR
"
" + n "n" →TAG →STR
"
" + '(8.34451_J/(K*
mol))*(n*T/V)' EVAL
"p" →TAG →STR ı + + +
MSGBOX
         >>
     >>
>>
```

Styling note: You may have noted in all these exercises that unit objects, such as the four objects listed in the message box shown above, keep the quotation marks when transformed into strings. Thus, when shown in a message box, the output may not be aesthetically pleasant. Pure numerical results can easily be translated into strings and thus, results without units will produce better-looking results than those with units. Thus, my recommendation is to use simple tagged output when attaching units to the results, and use message boxes if the output is entirely numerically.

Example - Message box output without units

Let's modify the program [p] once more to eliminate the use of units throughout it. The unitless program will look like this:

<< "Enter V,T,n [S.I.]: " {"↵ :V:↵ :T:↵ :n: " {2 0} V } INPUT OBJ→ →V T n
<<V DTAG { } + T DTAG + n DTAG + EVAL →V T n
<<"V=" V →STR + "↵ " + "T=" T →STR + "↵ " + "n=" n →STR + "↵ " +
'8.34451*n*T/V' EVAL →STR "p=" SWAP + + + MSGBOX>> >> >>

And when run with the input data V = 0.01, T = 300, and n = 0.8, produces the message box output:

```
V=.01
T=300.
n=.8
p=200268.24
```

Press [OK] to cancel the message box output.

Relational and logical operators

So far we have worked with sequential programs only. The User RPL language provides statements that allow branching and looping of the program flow. Many of these make decisions based on whether a logical statement is true or not. In this section we present some of the elements used to construct such logical statements, namely, relational and logical operators.

Relational operators

Relational operators are those operators used to compare the relative position of two objects. For example, dealing with real numbers only, relational operators are used to make a statement regarding the relative position of two or more real numbers. Depending on the actual numbers used, such a statement can be true (represented by the numerical value of 1. in the calculator), or false (represented by the numerical value of 0. in the calculator).

The relational operators available for the HP 49 G calculator are:

Operator	Meaning	Example
==	"is equal to"	'x==2'
≠	"is not equal to"	'3 ≠ 2'
<	"is less than"	'm<n'
>	"is greater than"	'10>a'
≥	"is greater than or equal to"	'p ≥ q'
≤	"is less than or equal to"	'7≤12'

All of the operators, except == (which can be created by typing [→][=][→][=]), are available in the keyboard as secondary functions in the keys [+/-], [X], and [1/X]. They are also available in the first menu obtained by using:

[←][PRG][TEST].

The examples shown above represent that use relational operators are elementary logical statements that could be true (1.), false (0.), or could simply not be evaluated. To determine whether a logical statement is true or not, place the statement in stack level 1, and press [→][EVAL]. Examples:

'2<10' [→][EVAL], result: 1. (true)

'2>10' [→][EVAL], result: 0. (false)

In the next example it is assumed that the variable m is not initialized (it has not been given a numerical value):

'2==m' [→][EVAL], result: '2==m'

The fact that the result from evaluating the statement is the same original statement indicates that the statement can not be evaluated uniquely.

Logical operators

Logical operators are logical particles that are used to join or modify simple logical statements. The logical operators available in the HP 49 G can be easily accessed through the keystroke sequence:

[←][PRG][NXT][TEST][NXT].

The available logical operators are: AND, OR, XOR (exclusive or), NOT, and SAME. The operators will produce results that are true or false, depending on the truth-value of the logical statements affected. The operator NOT (negation) applies to a single logical statements. All of the others apply to two logical statements.

Tabulating all possible combinations of one or two statements together with the resulting value of applying a certain logical operator produces what is called the <u>truth table of the operator</u>. The following are truth tables of each of the operators shown above. The symbol between parentheses is the symbol for the particular logical operator used traditionally in mathematical logic. The (logical) variables p and q represent logical statements that are either true (1.) or false (0.):

NOT(~ , ¬): produces the opposite possibility to that of the statement being negated:

Mathematical logic		HP 49 G implementation	
p	~p	P	NOT P
T	F	1	0
F	T	0	1

AND (^): produces a true result only if both logical statements being linked are true:

Mathematical logic			HP 49 G implementation		
p	q	p^q	P	Q	P AND Q
T	T	T	1	1	1
T	F	F	1	0	0
F	T	F	0	1	0
F	F	F	0	0	0

OR (∨): produces a true result as long as at least one of the statements being linked is true:

Mathematical logic			HP 49 G implementation		
p	q	p ∨ q	P	Q	P OR Q
T	T	T	1	1	1
T	F	T	1	0	1
F	T	T	0	1	1
F	F	F	0	0	0

XOR (◊): produces a true result only if the two statements being linked have different truth values:

Mathematical logic		
p	q	p◊q
T	T	F
T	F	T
F	T	T
F	F	F

HP 49 G implementation		
P	Q	P XOR Q
1	1	0
1	0	1
0	1	1
0	0	0

SAME : this is a non-standard logical operator used to determine if objects in stack levels 1 and 2 are identical. If they are identical, a value of 1. (true) is returned, if not, a value of 0. (false) is returned. For example, try the following exercise:

[→]['][↵][x²][2][ENTER]   Enter 'SQ(2)'
[4][ENTER]   Enter 4
[↵][PRG][TEST][NXT][SAME]   Result: 0 (false).

Please notice that the use of SAME implies a very strict interpretation of the word "identical." For that reason, SQ(2) is not identical to 4, although they both evaluate to 4.

Program branching

Branching of a program flow implies that the program makes a decision among two or more possible flow paths. The User RPL language provides a number of commands that can be used for program branching. The menus containing these commands are accessed through the keystroke sequence:

[↵][PRG][BRCH]

This menu shows sub-menus for the program constructs

[IF][CASE][START][FOR][DO][WHILE]

The program constructs IF...THEN..ELSE...END, and CASE...THEN...END will be referred to as program branching construct. The remaining constructs (START, FOR, DO, WHILE) are appropriate for controlling repetitive processing within a program and will be referred to as Program Looping. The latter types of program constructs are presented in more detail in a later section.

What IF...?

In this section we presents examples using the constructs IF...THEN...END and IF...THEN...ELSE...END.

IF...THEN...END

The IF...THEN...END is the simplest of the IF program constructs. The general format of this construct is:

IF *logical_statement* THEN *program_statements* END.

The operation of this construct is as follows:

1. Evaluate logical_statement.
2. If logical_statement is true, perform program _statements and continue program flow after the END statement.
3. If logical_statement is false, skip program_statements and continue program flow after the END statement.

To type in the particles IF, THEN, ELSE, and END, use:

$$[\leftarrow][PRG][BRCH][\ IF\ ].$$

The functions [IF][THEN][ELSE][END] are available in that menu to be typed selectively by the user. Alternatively, to produce an IF...THEN...END construct directly on the stack, use:

$$[\leftarrow][PRG][BRCH][\leftarrow][\ IF\ ].$$

This will create the following input in the stack:

```
2:
1:
IF ←
THEN
END
```

With the cursor ← in front of the IF statement prompting the user for the logical statement that will activate the IF construct when the program is executed.

Example: Type in the following program:

<< → x << IF 'x<3' THEN 'x^2' EVAL END "Done" MSGBOX >> >>

and save it under the name 'f1'.

To type the program in you could use the following:

[→][<<>>][→][→] [ALPHA][←][X]	Start first program level, type in →x
[→][<<>>][←][PRG][BRCH][←][IF]	Start second program level, generate IF.THEN.ELSE
[→]['] [ALPHA][←][X] [→][<] [3] [▼]	Type in logical statement for the IF
[ALPHA][←][X] [y^x] [2] [▼][→][EVAL]	Type in program statements to follow if logical statement true
[→]["] [ALPHA][D] [ALPHA][←][O]	Type in "Do
[ALPHA][←][N] [ALPHA][←][E] [▶]	Type in ne"
[←][PRG][NXT][OUT][MSGBO][←]	Type in MSGBOX
[ENTER]	Enter program in stack
[→]['] [ALPHA][←][F][1]	Type in name of program in stack
[STO▶]	Store program

Press [VAR] and verify that variable [f1] is indeed available in your variable menu. Try the operation of the program by using the following input:

[0] [f1]   Results: Message box: Done   [OK]   Stack level 1: 0. (i.e., x^2) [⇐]
(clear stack)

[1][.][2][f1]   Results: Message box: [Done   ] [OK]   Stack level 1: 1.44 (i.e., x^2) [⇦]
[3][.][5][f1] Results: Message box: [Done   ] [OK]   Stack level 1: empty - no action taken
[1][0][f1]   Results: Message box: [Done   ] [OK]   Stack level 1: empty - no action taken

These results confirm the correct operation of the IF...THEN...END construct. The program, as written, calculates the function $f_1(x) = x^2$, if x < 3 (and not output otherwise).

IF...THEN...ELSE...END

The IF...THEN...ELSE...END construct permits two alternative program flow paths based on the truth value of the logical_statement. The general format of this construct is:

IF *logical_statement* THEN *program_statements_if_true* ELSE *program_statements_if_false* END.

The operation of this construct is as follows:

1. Evaluate logical_statement.
2. If logical_statement is true, perform program statements_if_true and continue program flow after the END statement.
3. If logical_statement is false, perform program statements_if_false and continue program flow after the END statement.

To produce an IF...THEN...ELSE...END construct directly on the stack, use:

[⇦][PRG][BRCH][⇨][IF].

This will create the following input in the stack:

```
1:
IF ⬅
THEN
ELSE
END
```

Example: Type in the following program:

`<< → x << IF 'x<3' THEN 'x^2' ELSE '1-x' END EVAL "Done" MSGBOX >> >>`

and save it under the name 'f2'.

Try the operation of the program by using the following input:

[0] [f2]   Results: Message box: [Done   ] [OK]   Stack level 1: 0. (i.e., x^2) [⇦]
(clear stack)
[1][.][2][f2]   Results: Message box: [Done   ] [OK]   Stack level 1: 1.44 (i.e., x^2) [⇦]
[3][.][5][f2] Results: Message box: [Done   ] [OK]   Stack level 1: -2.5 (i.e., 1-x) [⇦]
[1][0][f2]   Results: Message box: [Done   ] [OK]   Stack level 1: -9 (i.e., 1-x) [⇦]

These results confirm the correct operation of the IF...THEN...ELSE...END construct. The program, as written, calculates the function

$$f_2(x) = \begin{cases} x^2, \text{ if } x < 3 \\ 1-x, \text{ otherwise} \end{cases}.$$

> Note: For this particular case, a valid alternative would have been to use an IFTE function of the form: 'f2(x) = IFTE(x<3,x^2,1-x)'

Nested IF...THEN...ELSE...END constructs

In most computer programming language where the IF...THEN...ELSE...END construct is available, the general format used for program presentation is the following:

```
IF logical_statement THEN
        program_statements_if_true
ELSE
        program_statements_if_false
END.
```

In designing a calculator program that includes IF constructs, you could start by writing by hand the pseudo-code for the IF constructs as shown above. For example, for program [f2], you could write

```
IF x<3 THEN
        x²
ELSE
        1-x
END.
```

While this simple construct works fine when your function has only two branches, you may need to nest IF...THEN...ELSE...END constructs to deal with function with three or more branches. For example, consider the function

$$f_3(x) = \begin{cases} x^2, \text{ if } x < 3 \\ 1-x, \text{ if } 3 \leq x < 5 \\ \sin(x), \text{ if } 5 \leq x < 3\pi \\ \exp(x), \text{ if } 3\pi \leq x < 15 \\ -2, \text{ elsewhere} \end{cases}.$$

Here is a possible way to evaluate this function:

```
IF x<3 THEN
        x²
ELSE
        IF x<5 THEN
                1-x
        ELSE
                IF x<3π THEN
                        sin(x)
                ELSE
                        IF x<15 THEN
                                exp(x)
                        ELSE
                                -2
                        END
                END
        END
END
```

A complex IF construct like this is called a set of <u>nested</u> IF...THEN...ELSE...END constructs. In most computer programming language such set of nested IF constructs is simplified to read:

```
IF x<3 THEN
        x²
ELSE IF x<5 THEN
        1-x
ELSE IF x<3π    THEN
        sin(x)
ELSE IF x<15 THEN
        exp(x)
ELSE
        -2
END
```

This construct is known as an IF...THEN...ELSEIF construct. Such a construct is, however, not possible in User RPL language. Therefore, you will have to type in the earlier construct with the four END statements as shown.

A possible way to evaluate f3(x), based on the nested IF construct shown above, is to write the program:

```
<< →x << IF 'x<3' THEN 'x^2' ELSE IF 'x<5' THEN '1-x' ELSE IF 'x<3*π' THEN 'SIN(x)' ELSE IF 'x<15' THEN 'EXP(x)' ELSE -2 END END END END EVAL >> >>
```

Store the program in variable [f3] and try the following evaluations:

1.5   [f3] Result: 2.25 (i.e., x^2)
2.5   [f3] Result: 6.25 (i.e., x^2)
4.2   [f3] Result: -3.2 (i.e., 1-x)

5.6 [f3] Result: -0.631266... (i.e., sin(x), with x in radians)
12   [f3] Result: 162754.791419 (i.e., exp(x))
23   [f3] Result: -2. (i.e., -2)

Just in CASE ...

The CASE construct can be used to code several possible program flux paths, as in the case of the nested IF constructs presented earlier. The general format of this construct is as follows:

```
CASE
Logical_statement₁ THEN program_statements₁ END
Logical_statement₂ THEN program_statements₂ END
.
.
.
Logical_statement THEN program_statements END
Default_program_statements (optional)
END
```

When evaluating this construct, the program tests each of the logical_statements until it finds one that is true. The program executes the corresponding program_statements, and passes program flow to the statement following the END statement.

The CASE, THEN, and END statements are available for selective typing by using

[←][PRG][BRCH][CASE].

If you are in the BRCH menu, i.e., ([←][PRG][BRCH]) you can use the following shortcuts to type in your CASE construct (The location of the cursor is indicated by the symbol ♦):

- [←][CASE] : Starts the case construct providing the prompts: CASE ♦ THEN END END
- [→][CASE] : Completes a CASE line by adding the particles THEN ♦ END

Example - program f₃(x) using the CASE statement

The function is defined by the following 5 expressions:

$$f_3(x) = \begin{cases} x^2, & \text{if } x < 3 \\ 1-x, & \text{if } 3 \leq x < 5 \\ \sin(x), & \text{if } 5 \leq x < 3\pi \\ \exp(x), & \text{if } 3\pi \leq x < 15 \\ -2, & \text{elsewhere} \end{cases}$$

Using the CASE statement in User RPL language we can code this function as:

```
<< →x << CASE 'x<3' THEN 'x^2' END 'x<5' THEN '1-x' END 'x<3*π' THEN 'SIN(x)' END 'x<15' THEN 'EXP(x)' END -2 END EVAL >> >>
```

Store the program into a variable called [f3c]. Then, try the following exercises:

1.5	[f3c] Result: 2.25 (i.e., x^2)
2.5	[f3c] Result: 6.25 (i.e., x^2)
4.2	[f3c] Result: -3.2 (i.e., 1-x)
5.6	[f3c] Result: -0.631266... (i.e., sin(x), with x in radians)
12	[f3c] Result: 162754.791419 (i.e., exp(x))
23	[f3c] Result: -2. (i.e., -2)

As you can see, f3c produces exactly the same results as f3. The only difference in the programs is the branching constructs used. For the case of function f₃(x), which requires five expressions for its definition, the CASE construct may be easier to code than a number of nested IF...THEN...ELSE...END constructs.

Program loops

Program loops are constructs that permit the program the execution of a number of statements repeatedly. For example, suppose that you want to calculate the summation of the square of the integer numbers from 0 to n, i.e.,

$$S = \sum_{k=0}^{n} k^2.$$

Well, with the HP 49 G calculator, all that you have to do is use the [←][Σ] key within the equation editor and load the limits and expression for the summation (examples of summations were presented in Chapter 6). However, in order to illustrate the use of programming loops, we will calculate this summation with our own User RPL codes. There are four different commands that can be used to code a program loop in User RPL, these are START, FOR, DO, and WHILE. The commands START and FOR use an index or counter to determine how many times the loop is executed. The commands DO and WHILE rely on a logical statement to decide when to terminate a loop execution. Operation of the loop commands is described in detail in the following sections.

START

The START construct uses a couple of values of an index to repeat a number of statements. There are two versions of the START construct: START...NEXT and START ... STEP. The START...NEXT version is used when the index increment is equal to 1, and the START...STEP version is used when the index increment is determined by the user.

Commands involved in the START construct are available through: [←][PRG][BRCH][START].

Within the BRCH menu ([←][PRG][BRCH]) the following keystrokes are available to generate START constructs (the symbol indicates cursor position):

⁃ [←][START] : Starts the START...NEXT construct: START ◆ NEXT

⁃ [→][START] : Starts the START...STEP construct: START ◆ STEP

START...NEXT

The general form of this statement is:

`start_value end_value START program_statements NEXT`

Because for this case the increment is 1, in order for the loop to end you should ensure that `start_value < end_value`. Otherwise you will produce what is called an infinite (never-ending) loop.

Example - calculating of the summation S defined above

The START...NEXT construct contains an index whose value is inaccessible to the user. Since for the calculation of the sum the index itself (k, in this case) is needed, we must create our own index, k, that we will increment within the loop each time the loop is executed. A possible implementation for the calculation of S is the program:

`<< 0. DUP →n S k << 0. n START k SQ S + 1. k + 'k' STO 'S' STO NEXT S "S" TAG >> >>`

Type the program in, and save it in a variable called [S1].

Here is a brief explanation of how the program works:

1. This program needs an integer number as input. Thus, before execution, that number (n) is in stack level 1. The program is then executed.
2. A zero is entered, moving n to stack level 2.
3. The command DUP, which can be typed in as [ALPHA][ALPHA][D][U][P][ALPHA], copies the contents of stack level 1, moves all the stack levels upwards, and places the copy just made in stack level 1. Thus, after DUP is executed, n is in stack level 3, and zeroes fill stack levels 1 and 2.
4. The piece of code `→n S k` stores the values of n, 0, and 0, respectively into local variables n, S, k. We say that the variables n, S, and k have been initialized (S and k to zero, n to whatever value the user chooses).
5. The piece of code `0. n START` identifies a START loop whose index will take values of 0, 1, 2, ..., n
6. The sum S is incremented by k^2 in the piece of code that reads: `k SQ S +`
7. The index k is incremented by 1 in the piece of code that reads: `1. k +`
8. At this point, the updated values of S and k are available in stack levels 2 and 1, respectively. The piece of code `'k' STO` stores the value from stack level 1 into local variable k. The updated value of S now occupies stack level 1.
9. The piece of code `'S' STO` stores the value from stack level 1 into local variable k. The stack is now empty.
10. The particle NEXT increases the index by one and sends the control to the beginning of the loop (step 6).
11. The loop is repeated until the loop index reaches the maximum value, n.
12. The last part of the program recalls the last value of S (the summation), tags it, and places it in stack level 1 to be viewed by the user as the program output.

To see the program in action, step by step, you can use the debugger as follows (use n = 2). Let SL1 mean stack level 1:

[VAR][2][→]['][S1][ENTER]	Place a 2 in level 2, and the program name, 'S1', in level 1
[←][PRG][NXT][NXT][RUN][DBUG]	Start the debugger. SL1 = 2.
[SST↓]	SL1 = 0., SL2 = 2.
[SST↓]	SL1 = 0., SL2 = 0., SL3 = 2. (DUP)
[SST↓]	Empty stack (-> n S k)
[SST↓]	Empty stack (<< - start subprogram)
[SST↓]	SL1 = 0., (start value of loop index)
[SST↓]	SL1 = 2.(n), SL2 = 0. (end value of loop index)
[SST↓]	Empty stack (START - beginning of loop)

--- loop execution number 1 for k = 0

[SST↓]	SL1 = 0. (k)
[SST↓]	SL1 = 0. (SQ(k) = k^2)
[SST↓]	SL1 = 0.(S), SL2 = 0. (k^2)
[SST↓]	SL1 = 0. (S + k^2)
[SST↓]	SL1 = 1., SL2 = 0. (S + k^2)
[SST↓]	SL1 = 0.(k), SL2 = 1., SL3 = 0. (S + k^2)
[SST↓]	SL1 = 1.(k+1), SL2 = 0. (S + k^2)
[SST↓]	SL1 = 'k', SL2 = 1., SL3 = 0. (S + k^2)
[SST↓]	SL1 = 0. (S + k^2) [Stores value of SL2 = 1, into SL1 = 'k']
[SST↓]	SL1 = 'S', SL2 = 0. (S + k^2)
[SST↓]	Empty stack [Stores value of SL2 = 0, into SL1 = 'S']
[SST↓]	Empty stack (NEXT - end of loop)

--- loop execution number 2 for k = 1

[SST↓]	SL1 = 1. (k)
[SST↓]	SL1 = 1. (SQ(k) = k^2)
[SST↓]	SL1 = 0.(S), SL2 = 1. (k^2)
[SST↓]	SL1 = 1. (S + k^2)
[SST↓]	SL1 = 1., SL2 = 1. (S + k^2)
[SST↓]	SL1 = 1.(k), SL2 = 1., SL3 = 1. (S + k^2)
[SST↓]	SL1 = 2.(k+1), SL2 = 1. (S + k^2)
[SST↓]	SL1 = 'k', SL2 = 2., SL3 = 1. (S + k^2)
[SST↓]	SL1 = 1. (S + k^2) [Stores value of SL2 = 2, into SL1 = 'k']
[SST↓]	SL1 = 'S', SL2 = 1. (S + k^2)
[SST↓]	Empty stack [Stores value of SL2 = 1, into SL1 = 'S']
[SST↓]	Empty stack (NEXT - end of loop)

--- loop execution number 3 for k = 2

[SST↓]	SL1 = 2. (k)
[SST↓]	SL1 = 4. (SQ(k) = k^2)
[SST↓]	SL1 = 1.(S), SL2 = 4. (k^2)
[SST↓]	SL1 = 5. (S + k^2)
[SST↓]	SL1 = 1., SL2 = 5. (S + k^2)
[SST↓]	SL1 = 2.(k), SL2 = 1., SL3 = 5. (S + k^2)
[SST↓]	SL1 = 3.(k+1), SL2 = 5. (S + k^2)
[SST↓]	SL1 = 'k', SL2 = 3., SL3 = 5. (S + k^2)
[SST↓]	SL1 = 5. (S + k^2) [Stores value of SL2 = 3, into SL1 = 'k']
[SST↓]	SL1 = 'S', SL2 = 5. (S + k^2)
[SST↓]	Empty stack [Stores value of SL2 = 0, into SL1 = 'S']
[SST↓]	Empty stack (NEXT - end of loop)

--- for n = 2, the loop index is exhausted and control is passed to the statement following NEXT

[SST↓]	SL1 = 5 (S is recalled to the stack)

[SST↓]	SL1 = "S", SL2 = 5 ("S" is placed in the stack)
[SST↓]	SL1 = S:5 (tagging output value)
[SST↓]	SL1 = S:5 (leaving sub-program >>)
[SST↓]	SL1 = S:5 (leaving main program >>)

The step-by-step listing is finished. The result of running program [S1] with n = 2, is S:5.

Check also the following results:

[VAR]
[3][S1]   Result: S:14   [4][S1]   Result: S:30
[5][S1]   Result: S:55   [8][S1]   Result: S:204
[1][0][S1]   Result: S:385   [2][0][S1]   Result: S:2870
[3][0][S1]   Result: S:9455   [1][0][0][S1]   Result: S:338350

START...STEP

The general form of this statement is:

```
start_value end_value START program_statements increment NEXT
```

The start_value, end_value, and `increment` of the loop index can be positive or negative quantities. For `increment > 0`, execution occurs as long as the index is less than or equal to end_value. For `increment < 0`, execution occurs as long as the index is greater than or equal to end_value.

Example - generating a list of values

Suppose that you want to generate a list of values of x from x = 0.5 to x = 6.5 in increments of 0.5. You can write the following program:

```
<< → xs xe dx << xs DUP xe START DUP dx + dx STEP DROP xe xs - dx / ABS 1. +
→LIST >> >>
```

and store it in variable [GLIST].

In this program , xs = starting value of the loop, xe = ending value of the loop, dx = increment value for loop. The program places values of xs, xs+dx, xs+2·dx, xs+3·dx, ... in the stack. Then, it calculates the number of elements generated using the piece of code:

```
xe xs - dx / ABS 1. +
```

Finally, the program puts together a list with the elements placed in the stack.

- Check out that the program call 0.5 2.5 0.5 [GLIST] produces the list {0.5 1. 1.5 2. 2.5}.
- To see step-by-step operation use the program DBUG for a short list, for example:

[VAR][1][SPC][1][.][5][SPC][.][5]	Enter parameters 1 1.5 0.5
[↱]['][GLIST][ENTER]	Enter the program name, 'S1', in level 1
[↰][PRG][NXT][NXT][RUN][DBUG]	Start the debugger.

Use [SST↓] to step into the program and see the detailed operation of each command.

FOR

As in the case of the START command, the FOR command has two variations: the FOR...NEXT construct, for loop index increments of 1, and the FOR...STEP construct, for loop index increments selected by the user. Unlike the START command, however, the FOR command does require that we provide a name for the bop index (e.g., j, k, n). We need not to worry about incrementing the index ourselves, as done in the examples using START. The value corresponding to the index is available for calculations.

Commands involved in the FOR construct are available through: [←][PRG][BRCH][FOR].

Within the BRCH menu ([←][PRG][BRCH]) the following keystrokes are available to generate FOR constructs (the symbol indicates cursor position):

⬇ [←][FOR]   : Starts the FOR...NEXT construct: FOR ← NEXT

⬇ [→][FOR]   : Starts the FOR...STEP construct: FOR ← STEP

FOR...NEXT

The general form of this statement is:

`start_value end_value FOR loop_index program_statements NEXT`

To avoid an infinite loop, make sure that `start_value < end_value`.

Example - calculate the summation S using a FOR...NEXT construct

The following program calculates the summation

$$S = \sum_{k=0}^{n} k^2.$$

Using a FOR...NEXT loop:

`<< 0. →n S << 0. n FOR k k SQ S + 'S' STO NEXT S "S" TAG >> >>`

Store this program in a variable [S2]. Verify the following exercises:

[VAR]
[3][S2]	Result: S:14	[4][S2]	Result: S:30
[5][S2]	Result: S:55	[8][S2]	Result: S:204
[1][0][S2]	Result: S:385	[2][0][S2]	Result: S:2870
[3][0][S2]	Result: S:9455	[1][0][0][S2]	Result: S:338350

You may have noticed that the program is much simpler than the one stored in [S1]. There is no need to initialize k, or to increment k within the program. The program itself takes care of producing such increments.

FOR...STEP

The general form of this statement is:

```
start_value end_value FOR loop_index program_statements increment STEP
```

The start_value, end_value, and `increment` of the loop index can be positive or negative quantities. For `increment > 0`, execution occurs as long as the index is less than or equal to end_value. For `increment < 0`, execution occurs as long as the index is greater than or equal to end_value.

Example - generate a list of numbers using a FOR...STEP construct

Type in the program:

```
<< → xs xe dx << xe xs - dx / ABS 1. + → n << xs xe FOR x x dx STEP n →LIST >>
>> >>
```

and store it in variable [GLIS2].

- Check out that the program call 0.5 2.5 0.5 [GLIS2] produces the list {0.5 1. 1.5 2. 2.5}.
- To see step-by-step operation use the program DBUG for a short list, for example:

[VAR][1][SPC][1][.][5][SPC][.][5]	Enter parameters 1 1.5 0.5
[↱]['][GLIS2][ENTER]	Enter the program name, 'S1', in level 1
[↰][PRG][NXT][NXT][RUN][DBUG]	Start the debugger.

Use [SST↓] to step into the program and see the detailed operation of each command.

DO

The general structure of this command is:

```
DO program_statements UNTIL logical_statement END
```

The DO particle starts an indefinite loop repeating the `program_statements` and checking for the value of `logical_statement` at the end of each repetition. The program sends the control to the statement following END when `logical_statement` becomes 0 (false). The `logical_statement` must contain the value of an index whose value is changed in the program_statements.

Example 1 - calculate the summation S using a DO...UNTIL...END construct

The following program calculates the summation

$$S = \sum_{k=0}^{n} k^2.$$

Using a DO...UNTIL...END loop:

```
<< 0. →n S << DO n SQ S + 'S' STO n 1 - 'n' STO UNTIL 'n<0' END S "S" TAG >> >>
```

Store this program in a variable [S3]. Verify the following exercises:

[VAR]
[3][S3]   Result: S:14   [4][S3]   Result: S:30
[5][S3]   Result: S:55   [8][S3]   Result: S:204
[1][0][S3]   Result: S:385   [2][0][S3]   Result: S:2870
[3][0][S3]   Result: S:9455   [1][0][0][S3]   Result: S:338350

Example 2 - generate a list using a DO...UNTIL...END construct

Type in the following program

```
<< → xs xe dx << xe xs - dx / ABS 1. + xs → n x << xs DO 'x+dx' EVAL DUP 'x' STO
UNTIL 'x xe' END n →LIST >> >> >>
```

and store it in variable [GLIS3].

- Check out that the program call 0.5 2.5 0.5 [GLIS3] produces the list {0.5 1. 1.5 2. 2.5}.
- To see step-by-step operation use the program DBUG for a short list, for example:

[VAR][1][SPC][1][.][5][SPC][.][5]   Enter parameters 1 1.5 0.5
[↱]['][GLIS2][ENTER]   Enter the program name, 'S1', in level 1
[↰][PRG][NXT][NXT][RUN][DBUG]   Start the debugger.

Use [SST↓] to step into the program and see the detailed operation of each command.

WHILE

The general structure of this command is:

```
WHILE logical_statement REPEAT program_statements END
```

The WHILE particle checks that whether the `logical_statement` is true and, if so, repeats the `program_statements`. If not, program control is passed to the statement right after END. The `program_statements` must include a loop index that gets modified before the `logical_statement` is checked at the beginning of the next repetition.

Example - calculate the summation S using a WHILE...REPEAT...END construct

The following program calculates the summation

$$S = \sum_{k=0}^{n} k^2.$$

Using a WHILE...REPEAT...END loop:

```
<< 0. →n S << WHILE 'n≥0' REPEAT n SQ S + 'S' STO n 1 - 'n' STO END S "S" TAG >> >>
```

Store this program in a variable [S4]. Verify the following exercises:

[VAR]
[3][S4]   Result: S:14   [4][S4]   Result: S:30
[5][S4]   Result: S:55   [8][S4]   Result: S:204
[1][0][S4]   Result: S:385   [2][0][S4]   Result: S:2870
[3][0][S4]   Result: S:9455   [1][0][0][S4]   Result: S:338350

Example 2 - generate a list using a WHILE...REPEAT...END construct

Type in the following program

```
<< → xs xe dx << xe xs - dx / ABS 1. + xs → n x << xs WHILE 'x<xe' REPEAT 'x+dx'
EVAL DUP 'x' STO END n →LIST >> >> >>
```

and store it in variable [GLIS3].

- Check out that the program call 0.5 2.5 0.5 [GLIS3] produces the list {0.5 1. 1.5 2. 2.5}.
- To see step-by-step operation use the program DBUG for a short list, for example:

[VAR][1][SPC][1][.][5][SPC][.][5]   Enter parameters 1 1.5 0.5
[↱]['][GLIS2][ENTER]   Enter the program name, 'S1', in level 1
[↰][PRG][NXT][NXT][RUN][DBUG]   Start the debugger.

Use [SST↓] to step into the program and see the detailed operation of each command.

Procedural programming and object-oriented programming

In this chapter we have covered a number of relatively simple programming examples that include the basic ideas of sequential, branching, and looping programs. This type of programming in which the programmer is controlling the program flow at an elementary level is called procedural programming.

In the last decade or so, there has been an emphasis in the development and use of object-oriented programming. In this approach the user takes advantage of pre-programmed functions (objects) for which he or she provides the appropriate input, and from which he or she receives a pre-specified type of output.

Most of the functions provided by the HP 49 G calculator emphasize object-oriented programming (OOP). For example, to solve a quadratic equation in the HP 49 G calculator, you need to enter a quadratic expression, the variable to solve for, and then use the function QUAD. The user is not concerned with the details of how the roots of the equation were calculated. All what he or she cares is that the proper type of input is provided to the function or object, and that the calculator will return the desired output.

Concluding remarks on programming

We emphasize once again that the examples and techniques shown in this chapter are very elementary, and that it would be almost impossible to include every aspect of the art of programming the HP 49 G calculator in this chapter, let alone in the rest of this book. In the upcoming chapters we will discuss the utilization of the calculator functions in specific mathematical problems, such as algebra, differential and integral calculus, linear algebra, vector calculus, etc. Additional programming examples will be provided as we discuss those subjects. In the meantime, the user is encouraged to explore routine calculations that can be easily addressed through calculator programs.

8 Algebra and related subjects

The Calculator Algebraic System or CAS allows the user to manipulate algebraic expressions including expansion, factorization, operations with fractions, substitutions, polynomial manipulation, and solution of single and multiple algebraic equations. This chapter introduces you to the basic of algebraic manipulation with the HP 49 G calculator. Before getting into the details of algebraic operations, however, we will formally define the logarithmic, exponential, hyperbolic and trigonometric functions, so as to take advantage of the HP 49 G algebraic menu commands that use such functions.

The CAS and RPN mode

The CAS allows the symbolic manipulation of algebraic objects (simply referred to as algebraics), but it does not mean that you need to operate your calculator in algebraic mode. This chapter will show you how to use algebraic manipulations with your calculator in RPN mode. To get started, make sure that your calculator CAS is set to exact mode (i.e., uncheck the _Approx mode indicator in the CAS MODES display. This display is accessed by using [MODES][CAS]).

Some transcendental functions

In Chapter 4 we presented some simple calculations utilizing exponential, logarithmic, trigonometric, and hyperbolic functions. Sometimes, these are referred to as transcendental functions, for they transcend the realm of algebraic functions. In this section we will introduce these formally and explain their relationships in order to take advantage of the many functions available in the HP 49 G for expanding, factoring, and replacing one type of function with others.

Trigonometric functions

Consider a unit radius circle (referred to as the unit circle) centered at the origin in a two-dimensional Cartesian coordinate system, as shown in the left-hand side figure below.

Let θ be the length of an arc of the circle measured from point A(1,0) and reaching to point P(x,y) on the unit circle. We define the following <u>trigonometric functions</u>: sine (sin), cosine (cos), tangent (tan), cotangent (cot), secant (sec), cosecant (csc):

$$\sin \theta = y, \cos \theta = x, \tan \theta = y/x, \cot \theta = x/y, \sec \theta = 1/x, \csc \theta = 1/y.$$

Recall, that the coordinates x and y used in the definition above <u>must</u> be on the unit circle, and that θ, as defined here, is an angle measured in radians. The angle θ is defined as positive if it is measured counterclockwise from the point the positive x-axis, and as negative if measured clockwise from that axis. The figure also shows typical values of the angle θ corresponding to the main axes directions.

Extending the definitions to a circle of radius r≠1, as shown in the right-hand side figure above, we define the angle in radians corresponding to an arc of length s as,

$$\theta = s/r.$$

The trigonometric functions, based on the coordinates of point P at the end of the arc, are defined as:

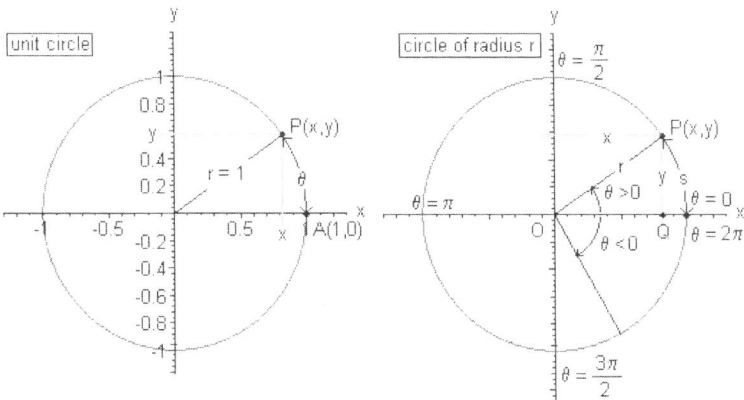

$$\sin\theta = y/r,\ \cos\theta = x/r,\ \tan\theta = y/x,\ \cot\theta = x/y,\ \sec\theta = r/x,\ \csc\theta = r/y.$$

The trigonometric functions sine, cosine, and tangent have their own main-function key in the HP 49 keyboard (fifth row, third, fourth, and fifth columns), i.e., [SIN], [COS], and [TAN]. The left-hand option ([↰]) for those keys represent the <u>inverse trigonometric</u> functions,

$$\text{asin}(x) = \sin^{-1}(x),\ \text{acos}(x) = \cos^{-1}(x),\ \text{and atan}(x) = \tan^{-1}(x).$$

The trigonometric functions cotangent, secant, and cosecant, can be calculated using the trigonometric identities:

$$\cot x = 1/\sin x,\ \sec x = 1/\cos x,\ \csc x = 1/\sin x.$$

Trigonometric identities

Trigonometric identities are relationships between the trigonometric functions of an angle that can be used to simplify algebraic expressions involving such functions. Some identities follow from the definition of the functions themselves, for example,

$$\tan\theta = \sin\theta/\cos\theta,\ \cot\theta = \cos\theta/\sin\theta.$$

Other identities follow from applying the Pythagorean Theorem to the right triangle formed by points OPQ, i.e., $x^2+y^2=r^2$. Dividing the expression by r^2 and using the definitions of $\sin\theta$ and $\cos\theta$ given above, we get

$$\sin^2\theta + \cos^2\theta = 1.$$

If we divide this identity by $\cos^2\theta$ and use the identities $1/\cos\theta = \sec\theta$, and $\sin\theta/\cos\theta = \tan\theta$, we get

$$\tan^2\theta + 1 = \sec^2\theta.$$

If we divide the identity $\sin^2\theta + \cos^2\theta = 1$ by $\sin^2\theta$, and use the identities $1/\sin\theta = \csc\theta$, and $\cos\theta/\sin\theta = \cot\theta$, we get

$$1 + \cot^2\theta = \csc^2\theta.$$

Some identities are related to trigonometric functions of the sum and difference of angles, for example,

$$\sin(\alpha+\beta) = \sin\alpha\cos\beta + \cos\alpha\sin\beta$$
$$\sin(\alpha-\beta) = \sin\alpha\cos\beta - \cos\alpha\sin\beta$$
$$\cos(\alpha+\beta) = \cos\alpha\cos\beta - \sin\alpha\sin\beta$$
$$\cos(\alpha-\beta) = \cos\alpha\cos\beta + \sin\alpha\sin\beta$$
$$\tan(\alpha+\beta) = (\tan\alpha + \tan\beta)/(1 - \tan\alpha\tan\beta)$$
$$\tan(\alpha-\beta) = (\tan\alpha - \tan\beta)/(1 + \tan\alpha\tan\beta)$$

From the latter set of identities we can obtain some identities related to the double angle, i.e.,

$$\sin 2\theta = 2\sin\theta\cos\theta$$
$$\cos 2\theta = \cos^2\theta - \sin^2\theta = 1 - 2\sin^2\theta = 2\cos^2\theta - 1$$
$$\tan 2\theta = 2\tan\theta/(1 - \tan^2\theta)$$

And, from these, in turn, we can get some identities related to the half-angle, i.e.,

$$\sin^2(\theta/2) = (1 - \cos\theta)/2$$
$$\cos^2(\theta/2) = (1 + \cos\theta)/2$$
$$\tan^2(\theta/2) = (1 - \cos\theta)/(1 + \cos\theta).$$

What is a logarithm?

Let a, b, and r be three real numbers. If $a = b^r$, then r is said to be the logarithm of base b of a, and written

$$r = \log_b(a).$$

For example, since $32 = 2^5$, then $\log_2(32) = 5$.

Logarithms of base 10

Because our numerical system is based on powers of 10, logarithms of base 10 were the most commonly used logarithms before calculators and computers became readily available. Anyone who was serious about performing complex calculations before, say, the early 1970's, would keep in his library a reliable table of logarithms. This meant, of course, logarithms of base 10. When referring to the logarithm of base 10 of a number a, people would simply write 'log a' or 'log(a)'. For that reason, the HP 49 G provides the function LOG, as the [→] option for the [EEX] key. The inverse of the LOG function is, of course, the function 10^x. This function is available as the [→] option for the [EEX] key.

Natural logarithms and the exponential function

Natural logarithms have for base the irrational number e = 2.718281828.... [An irrational number, as opposite to a rational number, is a number that cannot be expressed as the ration of two integers. Examples of other famous irrational numbers are $\sqrt{2}$, $\sqrt{3}$, and π.] They are called natural logarithms because they follow naturally from some properties of the function 1/x. As a matter of fact, you could define the natural logarithm of a real number r>1, as the area under the curve f(x) = 1/x, between x = 1 and x = r, as illustrated in the figure below. The natural logarithm of a number r is written as 'ln r' or 'ln(r)'.

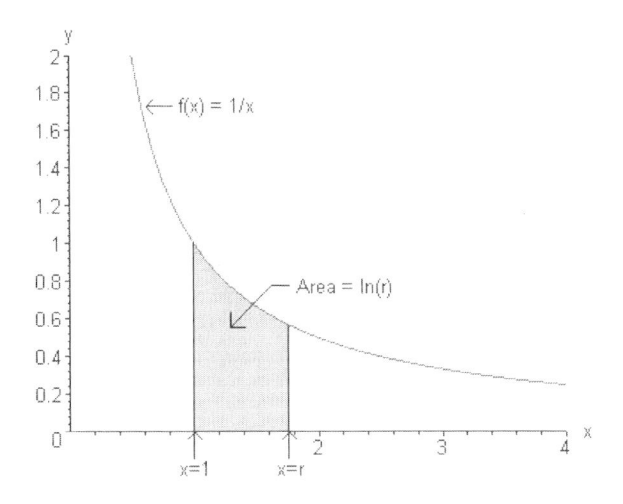

The inverse of the ln(x) function is the exponential function, $\exp(x) = e^x$. The exponential and natural logarithm functions are defined as the left-shift [←] and right-shift [→] options, respectively, corresponding to the [y^x] key in your calculator's keyboard.

Properties of the exponential function

The exponential function, $\exp(x) = e^x$, has the following properties:

$\exp(0) = e^0 = 1$	$\exp(-x) = e^{-x} = 1/e^x = 1/\exp(x)$
$\exp(x)+\exp(y) = e^{x+y} = \exp(x)\cdot\exp(y) = e^x\cdot e^y$	$\exp(x)-\exp(y) = e^{x-y} = \exp(x)/\exp(y) = e^x/e^y$
$[\exp(x)]^n = (e^x)^n = e^{xn} = \exp(nx)$	$[\exp(x)]^{1/n} = (e^x)^{1/n} = e^{xn} = \exp(nx)$

Properties of logarithms

Let x, y, b, be real numbers, then

$\log_b b = 1$,	$\log_b 1 = 0$,
$\log_b(x\cdot y) = \log_b x + \log_b y$	$\log_b(x/y) = \log_b x - \log_b y$
$\log_b(y^x) = x\log_b y$	$\log_b(y^{1/x}) = \log_b y / x$

In terms of natural logarithms, with b = e, these properties are written as:

$\ln e = 1$,	$\ln 1 = 0$,
$\ln(x\cdot y) = \log_b x + \log_b y$	$\ln(x/y) = \log_b x - \log_b y$
$\ln(y^x) = x\log_b y$	$\ln(y^{1/x}) = \log_b y / x$

Converting logarithms of different bases

Let x, y, a, b, and r, be real numbers. Let $x = \log_a r$ and $y = \log_b r$. Therefore, $r = a^x$, and $r = b^y$, and $a^x = b^y$. Taking, logarithms of base a on both sides of this equation, we get $\log_a(a^x) = \log_a(b^y)$, or $x\cdot\log_a a = y\cdot\log_a b$, and, $\log_a r = \log_b r \cdot \log_a b$. From this result it follows that,

$$\log_b r = \frac{\log_a r}{\log_a b}.$$

For example, let b = e, a = 10, then

$$\ln r = \frac{\log r}{\log e} = \frac{\log r}{0.4343} = 2.303\cdot \log r.$$

and,

$$\log r = 0.4343 \cdot \ln r.$$

Hyperbolic functions

Hyperbolic functions are defined in terms of exponential functions as follows:

- Hyperbolic sine (sinh):

$$\sinh x = \frac{1}{2}(e^x - e^{-x}).$$

- Hyperbolic cosine (cosh):

$$\cosh x = \frac{1}{2}(e^x + e^{-x}).$$

- Hyperbolic tangent (tanh):

$$\tanh x = \frac{\sinh x}{\cosh x} = \frac{e^x - e^{-x}}{e^x + e^{-x}}.$$

The inverse hyperbolic functions are defined as:

- Hyperbolic arcsin (asinh)

$$a\sinh(x) = \ln(x + \sqrt{x^2 + 1})$$

- Hyperbolic arccos (acosh)

$$a\cosh(x) = \ln(x + \sqrt{x^2 - 1})$$

- Hyperbolic arctan (atanh)

$$\operatorname{atanh}(x) = \ln\left(\frac{1+x}{1-x}\right)$$

These hyperbolic functions are available in the HP 49 G through the keystroke sequence [↰][MTH][HYP].

Other hyperbolic functions not available directly in the calculator are the hyperbolic cotangent (coth), hyperbolic secant (sech), and hyperbolic cosecant (csch), defined as:

$$\coth x = 1/\tanh x, \quad \operatorname{sech} x = 1/\cosh x, \quad \operatorname{csch} x = 1/\sinh x.$$

Euler formula and complex arguments for logarithmic, exponential, and hyperbolic functions

So far we have defined logarithmic, exponential, and hyperbolic functions for real numbers only. We can extend the definition of these functions to complex arguments by using Euler's formula:

$$e^{i\theta} = \cos\theta + i\cdot\sin\theta.$$

So that, if we use as argument for the functions previously defined the complex variable $z = x + i\cdot y = r\cdot e^{i\theta}$, we can extend their definitions as follows:

- natural logarithm: $\ln(z) = \ln(r\cdot e^{i\theta}) = r + i\theta = (x^2+y^2)^{1/2} + i\cdot\operatorname{atan}(y/x)$.
- exponential: $\exp(z) = e^z = e^{x+iy} = e^x \cdot e^{iy} = e^x(\cos y + i\cdot\sin y)$.
- hyperbolic sine: $\sinh(z) = (e^z - e^{-z})/2 = \sinh x \cdot \cos y + i\cdot\cosh x \cdot \sin y$
- hyperbolic cosine: $\cosh(z) = (e^z + e^{-z})/2 = \cosh x \cdot \cos y + i\cdot\sinh x \cdot \sin y$
- hyperbolic tangent: $\tanh(z) = (e^z - e^{-z})/(e^z + e^{-z}) =$
  $(\sinh x \cdot \cosh x + i\cdot\sin y\cdot\cos y)/(\cosh^2 x + \cos^2 y - 1)$

Euler's formula also permits us to find expressions for the trigonometric functions in terms of complex arguments,

- sine: $\sin(z) = -i\cdot(e^{iz} - e^{-iz})/2 = \sin x \cdot \cosh y + i\cdot\cos x \cdot \sinh y$

- cosine: $\cos(z) = (e^{iz} + e^{-iz})/2 = \cos x \cdot \cosh y - i\cdot\sin x \cdot \sinh y$

- tangent: $\tan(z) = -i\cdot(e^{iz} - e^{-iz})/(e^{iz} + e^{-iz}) =$
  $(\sin x \cdot \cos x + i\cdot\sinh y \cdot \cosh y)/(\cos^2 x + \sinh^2 y)$

Inverse trigonometric and hyperbolic functions with complex arguments can be calculated by the expressions, where $z = x + i\cdot y$:

- arcsine: $\text{asin}(z) = -i \cdot \ln(\sqrt{1-z^2} + i \cdot z)$

- arccosine: $\text{acos}(z) = -i \cdot \ln(z + i \cdot \sqrt{1-z^2})$

- arctangent: $\text{atan}(z) = i/2 \cdot \ln((1-i \cdot z)/(1+i \cdot z))$

- hyperbolic arcsine: $\text{asinh}(z) = \ln(z + \sqrt{z^2+1})$

- hyperbolic arccosine: $\text{acosh}(z) = \ln(z + \sqrt{z^2-1})$

- hyperbolic arctangent: $\text{atanh}(z) = \ln((1+z)/(1-z))$.

The advantage of having in your hands a calculator that can calculate standard functions with complex arguments, is that you don't have to evaluate the definitions shown above to obtain values such as trigonometric and hyperbolic functions of complex variables. The following exercises will let you check the calculator's ability to compute standard functions using complex arguments. Before attempting thee exercises, check that your calculator CAS is set to √Complex. Use your HP 49 G calculator to obtain the values of:

ln(-5):	[5][+/-][↰][LN] [↰][→NUM]	Result: (1,6094,3.1416)
ln(3-2i)	[↰][()][3][↰][,][2][+/-] [↰][LN]	Result: (1.2825, -0.5880)
exp(iπ/2):	[↰][i][SPC][↰][π][×][2][÷][↰][e^x]	Result: 'i'
sin(2-3i):	[↰][()][2][↰][,][3][+/-][SIN]	Result: (9.1544, 4.1689)
cos(-5+i):	[↰][()][5] [+/-] [↰][,][1][COS]	Result: (0.4377,-1.269)
tan((1+i)/2)):	[1] [SPC][↰][i][+][2][÷][TAN][↰][→NUM]	Result: (0.4039,0.5641)
cot(-i/2):	[↰][i][+/-][2][÷][TAN][↰][→NUM][1/x]	Result: (0.,2.1640)
sec(2*i):	[2][↰][i][×][COS][↰][→NUM][1/x]	Result: (0.2658,0)
csc(-2+3i):	[↰][()][2][+/-][↰][,][3][SIN][1/x]	Result: (-9.0473E-02, 4.1200E-02)
arcsin(5):	[5][↰][ASIN] [↰][→NUM]	Result: (1.5707,-2.2924)
arcsin(-5-2i):	[↰][()][5] [+/-] [↰][,][2][+/-][↰][ASIN]	Result: (-1,1842,-2.3705)
arccos(-10):	[1][0][+/-][↰][ACOS] [↰][→NUM]	Result: (3.1416,-2.9932)
arccos(-2+4:i)	[↰][()][2] [+/-] [↰][,][4][↰][ACOS]	Result: (2.0247,-2.1986)
arctan(-7+i):	[↰][()][7] [+/-] [↰][,][1][↰][ATAN]	Result: (-1.4316,1.9617E-2)

For the next exercises, activate the menu [↰][MTH][HYP]:

sinh(-5+3i):	[↰][()][5] [+/-] [↰][,][3][SINH]	Result: (73.46,10.47)
cosh(7-2i):	[↰][()][7] [↰][,][3] [+/-][COSH]	Result: (73.46,10.47)
tanh(-2+7i):	[↰][()][2] [+/-] [↰][,][7][TANH]	Result: (-0.9944,3.6094E-2)
arcsinh(-2+0.5i):	[↰][()][2] [+/-] [↰][,][.][5][ASINH]	Result: (-1.4657,0.2211)
arccosh(-7+6i):	[↰][()][7] [+/-] [↰][,][6][ACOSH]	Result: (2.9140,2.4300)
arctanh(-3-i):	[↰][()][3] [↰][,][1][ENTER][+/-][ATANH]	Result: (2.9140,2.4300)

> Note: The menu [←][MTH][HYP][NXT] contains two additional functions related to the exponential and natural logarithm functions: [EXPM] and [LNP1] defined as
>
> $$EXPM(x) = EXP(x)-1,$$
>
> and
>
> $$LNP1(x) = LN(x+1).$$

These functions are useful when writing programs to save programming steps.

Operations with algebraic objects

An algebraic object, or simply, algebraic, is any number, variable name or algebraic expression that can be operated upon, manipulated, and combined according to the rules of algebra. An algebraic object is an object of type 9 in the calculator. Examples of algebraic objects are the following:

- A number: 12.3, 15.2_m, 'π', 'e', 'i'
- A variable name: 'a', 'ux', 'width', etc.
- An expression: 'π*D^2/4', 'f*(L/D)*(V^2/(2*g))', 'y+Q^2/(2*g*A(y)^2)'
- An equation: 'p*V=n*R*T', 'Q=(Cu/n)*A(y)*R(y)^(2/3)*So^0.5'

Entering algebraic objects

Algebraic objects can be created by typing the object between single quotes directly into stack level 1 or by using the equation writer [EQW]. For example, to enter the algebraic object ''π*D^2/4' directly into stack level 1 use:

[→]['][←][π][×][ALPHA][D][y^x][2][÷][4][ENTER]

Or, if using the equation writer:

[EQW] [←][π][×][ALPHA][D][y^x][2] [▲][▲][▲] [÷][4][ENTER]

Basic operations

Algebraic objects can be added, subtracted, multiplied, divided (except by zero), raised to a power, used as arguments for a variety of standard functions (exponential, logarithmic, trigonometry, hyperbolic, etc.), as you would any real or complex number. To demonstrate basic operations with algebraic objects, let's create a couple of objects, say 'π*R^2' and 'g*t^2/4', and place them in a list so that they are available for algebraic manipulation at any time. Let's get started:

[→]['][←][π][×][ALPHA][R][y^x][2][ENTER]   Enter 'π*R^2'

[→]['][ALPHA][←][G][×][ALPHA][←][T][y^x][2][÷][4][ENTER]   Enter 'g*t^2/4'
[2][←][PRG][LIST][→LIST]   Create a list with two algebraics

The list will look like this in your calculator's display: { 'π*R^2' 'g*t^2/4' }

Press [ENTER] several times, say 10 times, to make copies of the list of algebraic objects, so that we can use them for demonstrating operations. To decompose the list into its components, simply use [→][EVAL].

Now, try the following exercises:

[→][EVAL][+]   To add the algebraics, result: 'π*R^2 + g*t^2/4'
[⇐]   Drop result from display

[→][EVAL][-]   To subtract the algebraics, result: 'π*R^2 - g*t^2/4'
[⇐]   Drop result from display

[→][EVAL][×]   To multiply the algebraics, result: 'π*R^2 *(g*t^2/4)'
[⇐]   Drop result from display

[→][EVAL][÷]   To divide the algebraics, result: 'π*R^2 /(g*t^2/4)'
[⇐]   Drop result from display

[→][EVAL][2][y^x]   To square an algebraic, result: '(g*t^2/4)^2'
[→][ALG][EXPAN]   To 'expand' the square, result: '.0625*t^4.*g^2'
[⇐]   Drop result from display

[√x]   Square root of second algebraic, result: '√(π*R^2)'
[→][ALG][EXPAN]   To 'expand' the square root, result: '1.77245385091*R'.
                           Note: the constant π was replaced by its numerical value
[⇐]   Drop result from display

[→][EVAL][SIN]   Sine function, result: 'SIN(g*t^2/4)'
[▶]   Swap levels 1 and 2
[←][MTH][HYP][TANH] Hyperbolic tangent function, result: 'TANH(π*R^2)'
[+]   Add the two previous results, result: 'SIN(g*t^2/4)+ TANH(π*R^2)'
[⇐]   Drop result from display

[→][EVAL][→][LN]   Natural log function, 'LN(g*t^2/4)'
[▶]   Swap levels 1 and 2
[←][e^x]   Exponential function, result: 'EXP(π*R^2)'
[÷]   Divide the two previous results, result: 'LN(g*t^2/4)/ EXP(π*R^2)'
[⇐]   Drop result from display

As you can see operations with algebraic objects is not different than operations with numbers.

NOTE: Make sure your calculator is in the EXACT mode before attempting the following exercises.

Expanding expressions

Who among you does not still remember with great joy those long hours spent expanding algebraic expressions such as $(x+y)^3$ or $(a+b+c)^2$ in High School? Well, now you can let the calculator have all the fun by using the keystroke combination: [↱][ALG](the key for number 4)[EXPAN]. Try the following exercises:

[↱]['][↰][( )] [ALPHA][X] [+][ALPHA][Y] [▶][y^x][2] [ENTER]   Enter '(X+Y)^2'
[↱][ALG][EXPAN]   Result: 'X^2+2*X*Y+Y^2'

Nice, isn't it? No need to remember anymore that "... the square of X + Y is equal to the square of X plus twice the product of X times Y plus the square of Y."

Now, let's use the calculator to verify the following expansions. At this point I will assume that you know how to enter algebraic expressions in the stack, therefore, I will skip the detail keystroke sequences and provide the resulting algebraic only:

'(X+Y)^3'[EXPAN]   Result: 'X^3+3*Y*X^2+3*Y^2*X+Y^3'

'a*(X+Y)'[EXPAN]   Result: 'a*X+a*Y'

'(2+A)*A^2*(1+A)'[EXPAN]   Result: 'A^4+3*A^3+2*A^2'

'(a+b+c)^2'[EXPAN]   Result: 'a^2+(2*b+2*c)*a+(b^2+2*c*b+c^2)'

'2*(a+b)*(2+a)'[EXPAN]   Result: '2*a^2+(2*b+4)*a+4*b'

Notice that, in the last two exercises, the EXPAND function does not give you the full expansion than one expects. This may be a shortfall of the current version of the calculator's CAS, or a deliberate attempt by the author(s) of the CAS to let you finish the calculation by hand. (The exercises in this chapter were developed using ROM version 1.16). The moral of the story is that , in some instances, the calculator will not carry operations to the ultimate expected expressions. (Even if you press [EXPAN] repeatedly). Keep a paper and pencil handy, and use it, when necessary, to finish some calculations by hand. For example,

$$(a+b+c)^2 = a^2+b^2+c^2+2ab+2ac+2cb,$$

and

$$2(a+b)(2+a) = 2a^2+4a+2ab+4b.$$

Factoring expressions

The function FACTOR, available through [↱][ALG][FACTO], can be used to obtain factorization of algebraic expressions. As in the case of EXPAN, however, there will be instances when the calculator will not be able to factor an expression down to its simplest factors, in which case, it will need some help from you.

Try the following <u>exercises</u>:

'T^2+5*T+6' [FACTO]   Result: '(T+3)*(T+2)'

'H^2-4' [FACTO]   Result: '(H+2)*(H-2)'

'Z^4-8*Z^2+16' [FACTO]   Result: '(Z+2)^2*(Z-2)^2'

Note: If you work in Approx mode, your result would have been:'(Z+2)*(Z-2)*(Z+2)*(Z-2)'.

`M^3+M^2-6*M` [FACTO]   Result: `(M+0)*(M-2)*(M+3)` = M(M-2)(M+3)

`X^2+Y^2+Z^2+2*X*Y+2*X*Z+2*Y*Z` [FACTO]   Result: `(X+(Y+Z))*(X+(Y+Z))`=$(X+Y+Z)^2$

Complex mode factorization

If your CAS is set to ✓Complex mode, FACTOR will try to factor an algebraic expression down to their simplest factors. For example, working with real numbers only, the expression `pH2+4` cannot be factored at all:

`pH^2+4` [FACTO]   Result: `pH^2+4`

However, in CAS Complex mode you will get:

`pH^2+4` [FACTO]   Result: `(pH,2)*(pH-2*i)`

The first factor in this result, `(pH,2)`, is the complex number `pH+2*i` represented as an ordered pair.

Try a few more examples of Complex mode factorization:

`xK^3+2*xK^2+4*xK+8` [FACTO]   Result: `(xK+2)*(xK,2)*(xK-2*i)`

`Y^4+(m+2)*Y^3+(2*m+4)*Y^2+(4*m+8)*Y+8*m` [FACTO]
                                    Result: `(Y+2)*(Y+m)*(Y,2)*(Y-2*i)`

`p^3-β*p^2+2*α*p-2*α*β` [FACTO]   Result: `(p-b)*(p+i*√(2*a))*(p-i*√2*√a)`

`SIN(q)^2+2` [FACTO]   Result: `(SIN(q)+i*√2)*(SIN(q)-i*√2)`

Substituting expressions in algebraic objects

Substitution of algebraic expressions or numerical values in algebraic objects can be accomplished in two different ways. The first way uses the command [SUBS], available through the keystroke sequence [→][ALG][SUBS]. Application of this command requires placing the algebraic object, where the substitution will take place, in stack level 2, and the substitution, in the form `variable = value` in stack level 1. For example, enter the algebraic object: `x^2+x+1` [ENTER], and then enter `x = y+1`[ENTER], and press [SUBS]. The result is: `(y+1)^2+(y+1)+1`. Press [EXPAN] to expand and simplify the expression. The result is now: `y^2+3*y+3`. This mode of substitution allows substituting only one variable at a time. Other examples follow:

`h = 30 + 2.5*t - 16.6*t^2` [ENTER] `t = 4` [ENTER][→][ALG][SUBS][→][EVAL] Results in `h=-225.6`
`h = h0 + v0*t-g*t^2/2` [ENTER] `t = tf` [ENTER] [SUBS] Results in `h = h0 + v0*tf-g*tf^2/2`
Next, enter `tf = 2` [SUBS] to get `h = h0 + v0*2-g*2^2/2`.
Next, enter `h0 = 10` to get `h = 10 + v0*2-g*2^2/2`
Next, enter `h0 = 10` to get `h = 10 + v0*tf-g*2^2/2`
Next, enter `v0 = 5` to get `h = 10 + 5*2-g*2^2/2`
Next, enter `g = 9.806` to get `h = 10 + 5*2-9.806*2^2/2`
Finally, to evaluate h, use [→][EVAL]. The result is `h=.388`
`V = Q/A` [ENTER] `A = (b+m*y)*y` [SUBS] Results in `V=Q/((b+m*y)*y)`.

A second form of substitution can be accomplished by using the [→][|] (TOOL) key. This requires that the algebraic object, where the substitution will take place, be placed in stack level 2, while stack level 1 contains a list, of the form { variable1 value1 variable2 value2 ...}. For example, enter the algebraic object 'a+b^2+c'[ENTER], and then enter the list {a -1 b 'x+1' c 'π' }[ENTER]. Press [→][|] [ENTER] to obtain '-1+((x+1)^2+π)'. Using [EXPAN] produces 'π+(x^2+2*x)'. This mode of substitution allows substituting more than one variable at a time. The function [|] follows from the substitution notation:

$$(a+b^2+c)\Big|_{\{a=-1, b=x+1, c=\pi\}}$$

Other examples:

'h = h0 + v0*t-g*t^2/2' [ENTER] { t 2 v0 5 h0 10 g 9.806 } [→][|] [ENTER] Produces:
'h = (2*10+(2*2*5-2^2*9.806))/2'. Use [→][EVAL] to obtain 'h =.388'

'F = CONST(G)*m1*m2/r^2' [ENTER] { m1 10_kg m2 20_kg r 12_m } [→][|] [ENTER] Produces:
'F=((6.67259E-11_m^3/(s^2*kg))*(20_kg)*10_kg)/(12_m)^2' Use [→][EVAL] to obtain
'F =.9.26748611111E-11_m*kg/s^2'. The units are m*kg/s^2' = N (newtons).

Substitution by using HP variables

The two substitution approaches presented above require that the original expression (where the substitution is to take place) and the substitution expressions be available in the stack. A different approach to substitution consists in defining the substitution expressions in variables and placing the name of the variables in the original expression. For example, store the following variables:

'(b+m*y)*y' [ENTER] 'A' [ENTER] [STO▶]
'b+2*y*√(1+m^2)' [ENTER] 'P' [ENTER] [STO▶]
'A/P' [ENTER] 'R' [STO▶]

Then enter the expression:

'Q = Cu*R^(2/3)*A*√S/n' [ENTER] [ENTER]

(to keep an additional copy of the original expression). Enter [→][EVAL] to evaluate the expression. The result is:

'Q = (y*b+y^2*m)*√S*Cu*((y*b+y^2*m)/(b+2*y*√(m^2+1)))^.666666666667/n'

As you can see, the resulting expression is given in terms of the most primitive variables, i.e., b, m, y, which come from the definitions of A, P, and R.

Now, drop the last result from the stack by using the backspace key [], and purge the variables defined:

{ A P Q } [TOOL][PURGE]

With the expression 'Q = Cu*R^(2/3)*A*√S/n' in stack level 1, try [→][EVAL]. The result is 'Q = √S*A*Cu*R^.666666666667/n'

In the latter case, because there is no definition available for A or R, no substitution occurs. The only change, after evaluating the expression, is in the order of the terms in the expression, and the replacement of (2/3) by its approximation .666666666667, if the approximate mode is selected. In exact mode, the value (2/3) is kept.

Expansion and factorization using transcendental functions

The HP 49 G calculator offers a number of functions that can be used to replace expressions containing logarithmic, exponential, trigonometric, and hyperbolic functions in terms of trigonometric identities or in terms of exponential functions. The menus containing functions to replace trigonometric functions can be obtained directly from the keyboard by pressing the left-shift key followed by the [8] key, i.e., [↱][TRIG]. The combination of this key with the right-shift key, i.e., [↰][EXP&LN], produces a menu that lets you replace expressions in terms of exponential or natural logarithm functions. In the next sections we cover those menus in more detail.

Expansion and factorization in terms of exponential and logarithmic functions

Here we use the EXP&LN menu obtained from the keystroke combination [↰][EXP&LN]. This produces the following menu:

[EXPLN][EXPM][LIN][LNCOL][LNP1][TEXPA]

And, after pressing [NXT], it shows as the second menu:

[TSIMP][   ][   ][   ][   ][   ]

We have described the operation of the functions [EXPM] (=exp(x)+1) and [LNP1] (= ln(x+1)) in an earlier note. Here, we describe the operation of the remaining functions in this menu.

⌙ The function [EXPLN] transforms trigonometric functions into expressions involving exponential and natural logarithms, without linearizing the expression. Consider the following examples:

'SIN(X)^2+1' [EXPLN]   Result: '((EXP(i*X)-1/EXP(i*X))/(2*i))^2+1'

'SIN(X)+TAN(X)'   Result:
        '(EXP(i*X)-1/EXP(i*X))/(2*i)+(EXP(i*(2*x))1)/(i*(exp(i*(2*x))+1)'

'SIN(X+Y)*COS(X)'[EXPLN] Result:
        '(EXP(I*(X+Y))-1/EXP(I*(X+Y))/(2*I)+(EXP(I*x)+1/EXP(I*X))/2'

'ASIN(X)' [EXPLN] Result: 'i*LN(EXP(LN(X^2-1)/2)+X)+π/2'

> Note: The expression obtained above include the function pair EXP and LN, which are inverse of each other, therefore, it can be simplified, by hand, to read $i \cdot \ln((x^2-1)^{1/2}+x)+\pi/2$. Earlier on, we defined the arcsin(z) as $-i \cdot \ln(\sqrt{(1-z^2)}+i \cdot z)$. Although, the two expressions are different, you can check that they produce the same result by calculating the two expressions with the same complex argument. For example, with the expression obtained earlier in stack level 1, enter the following:
>
> {X (2.,3.)} [ENTER][→][|][ENTER] [→][→NUM], to get (0.57065, 1.98338).
>
> Next, type in:
>
> '-i*LN(√(1-X^2)+i*X)' [ENTER] {X (2.,3.)} [ENTER][→][|][ENTER] [→][→NUM],
>
> to get (0.57065, 1.98338).

⬇ The function [LIN] linearizes expressions containing exponential or logarithmic terms. For example:

'SIN(X)' [LIN] results in '-(i/2*EXP(i*x))+i/2*EXP(-(i*X))'

'π*EXP(2*X)*EXP(Y)' [LIN] results in 'π*EXP(2*X+Y)'

'SIN(LN(X^2-1))' [LIN] results in '-(i/2*EXP(i*LN(X^2-1)))+i/2*EXP(-(i*LN(X^2-1)))'

'SIN(X)-COSH(X)' [LIN] results in '-1/2*EXP(X)+ −1/2*EXP(-X)-i/2*EXP(i*X)+i/2*EXP(-(i*X))'

⬇ The function [LNCOL] (LNCOLLECT) collects logarithmic terms in an expression. Examples:

'LN(X) + LN(X-1)-LN(X+1)' [LNCOL] results in 'LN(X*(X-1)/(X+1))'

'(1/2)*LN(X^2-1)+LN(X)' using [LNCOL] on this expression has no effect. However, if you press [LIN], the expression gets transformed into '(LN(X^2-1)-2*LN(X))/2'. Using [LNCOL] on this new version of the expression produces 'LN(X^2-1)/X^2)/2'. Since ln(a)/n = ln(a^(1/n)), we can write the latter result as

$$\frac{1}{2} \cdot \ln\left(\frac{X^2-1}{X^2}\right) = \ln\left(\frac{X^2-1}{X^2}\right)^{1/2} = \ln\left(\sqrt{\frac{X^2-1}{X^2}}\right) = \ln\left(\frac{\sqrt{X^2-1}}{X}\right)$$

'2+(1/3)*LN(Y)-LN(k)' [LIN] results in '(LN(Y)-(3*LN(k)-6)/3'. Using [LNCOL] on this result, produces '(6+LN(Y/k^3))/3'. This result can be simplified even further if we replace 6 with a LN expression. Since 6 = LN(EXP(6))=LN(403.43), we can write the latter expression as '(LN(403.43)+LN(Y/k^3))/3'. Press [LNCOL] to get ''(LN(403.43*(Y/k^3))/3'.

'(1/2)*LN(Y)-(1/3)*LN(X+1)' [LIN] results in '-((2*LN(X+1)-3*LN(Y))/6'. Using [LNCOL] on this result produces '-LN((X+1)^2/Y^3)/6'.

'LOG(X)-LOG(3)-(1/2)*LOG(Y)' [LIN] results in '(2*LN(X)-(LN(Y)+2*LN(3)))/(2*LN(10))'. Using [LNCOL] on this result produces '-(LN(Y*9/X^2)/LN(100))'.

✦ The function [TEXPA] (TEXPAND) expands expressions containing transcendental functions (i.e., exponential, logarithmic, hyperbolic, and trigonometric functions). Some examples follow:

'SIN(EXP(X*Y))' [TEXPA] results in 'COS(LN(Y))*SIN(LN(X))+SIN(LN(Y))*COS(LN(X))'.

'COS(A-B)' [TEXPA] results in 'COS(B)*COS(A)+SIN(B)*SIN(A)'

'LN((2*X-1)/(2*X+1)^(1/3))' [TEXPA] results in 'LN(2*X-1)-LN(XROOT(3,2*X+1))'. Now, apply the function [LNCOL] to get '-LN(XROOT(3,2*X+1)/(2*X-1))'.

> Note: XROOT(x,y) represents $^x\sqrt{y} = y^{1/x}$.

'EXP(X+Y^2-A)' [TEXPA] results in 'EXP(X)*EXP(Y^2)/EXP(A)'.

✦ The function [TSIMP] (visible if you press [NXT]) simplifies expressions involving natural logarithms and exponential functions. For example, expressions involving trigonometric or hyperbolic functions get replaced by their equivalent in exponential functions. On the other hand, expressions involving inverse trigonometric and hyperbolic functions get replaced by their equivalent in logarithmic functions. Some examples of the use of [TSIMP] are shown below:

'SIN(X)' [TSIMP] results in 'EXP(i*X)-1/EXP(i*X))/(2*i)'. Press [NXT][LIN] to transform the latter expression into '-(i/2*EXP(I*X)+i/2*EXP(-(I*X))'.

'ATAN(X)' [TSIMP] produces
'i/2*((LN(X^2+1)/2,ATAN(1/X)+ - 2*(π/2))+-1*(LN(X^2+1)/2,ATAN(X)+-1*(π/2)))'
If we use the function [LIN] on this expression it gets simplified to
'(-(i*LN((i*X+1)/(i*X-1)))+ (i *LN(-((X+1)/(X-i)))+π)/4'.

'SIN(X) + COS(X)' [TSIMP] results in '(EXP(i*X)-1/EXP(i*X))/(2*I)+(EXP(i*X)+1/EXP(i*X))/2' . Using the function [LIN] of this result produces '(1-i)/2*EXP(i*X)+(1+i)/2*EXP(-(i*X))'.

'ASIN(X)+LN(X)' [TSIMP] results in 'i*LN(X+ √ (X^2-1))+π/2+LN(X)'. Next, use [LIN] to produce '(2*i*LN(X+EXP(LN(X^2-1)/2))+(2*LN(X)+π))/2'. Using [LNCOL] produces
'(2*i*LN(X+EXP(LN(X^2-1)/2))+ π + LN(X^2))/2'.

Expansion and factorization in terms of trigonometric functions

In this section we present functions that let you convert trigonometric functions into other trigonometric functions. These menus are accessed through the keystroke combination [┌→][TRIG], which produces the following:

Menu 1:   [HYP][ACOS2][ASIN2][ASIN2][ATAN2][HALFT].

Use [NXT] to get

Menu 2:   [SINCO][TAN2S][TAN2S][TCOLL][TEXPA][TLIN]

Use [NXT] once more to get

Menu 3:   [TRIG][TRIGC][TRIGS][TRIGT][TSIMP][   ]

⬇ The [HYP] menu provides access to the hyperbolic function menu as you would using [↤][MTH][HYP].

⬇ The functions [TEXPA] and [TSIMP] were presented in the previous section.

⬇ The function [ACOS2] (ACOS2S) replaces occurrences of ACOS(X) with '$\pi/2$ - ASIN(X)'. Examples:

'ACOS(X^2-1)' [ACOS2]   Results in '$\pi/2$ - ASIN(X^2-1)'.
'ACOS(SIN(3*π/2))' [ACOS2]   Results in '$\pi/2$ + $\pi/2$'
'ACOS(1/X)' [ACOS2]   Results in '$\pi/2$ - ASIN(1/X)'.

⬇ You will notice that in the first menu there are two keys labeled [ASIN2]. In fact, the two [ASIN2] labels represent the functions [ASIN2C] and [ASIN2T], in that order. The function ASIN2C replaces occurrences of ASIN(X) with '$\pi/2$ - ACOS(X)', while the function ASIN2T replaces occurrences of ASIN(X) with 'ATAN(X/√ (1-X^2))'. Examples:

'ASIN(Y/2)' [ASIN2C]   Results in '$\pi/2$ - ACOS(Y/2)'.
'ASIN(Y/2)' [ASIN2T]   Results in 'ATAN(Y/2/(√ -(Y^2-4)/2))'
'ASIN(X)+ASIN(Y)' [ASIN2C]   Results in '$\pi/2$-ACOS(X)+($\pi/2$-ACOS(Y))'
'ASIN(TAN(X))' [ASIN2T]   Results in 'ATAN(TAN(X) /(√ -(TAN(X)^2-1))

⬇ The function [ATAN2](ATAN2S) replaces occurrences of ATAN(X) with ASIN(X/√ (X^2+1)). Examples:

'ATAN(SIN(π/2))' [ATAN2S]   Results in 'ASIN(1/√2)'. With [→][EVAL] we get $\pi/4$.
'ATAN(ABS(X))' [ATAN2S]   If in complex mode, results in
                                    'ASIN(√ (RE(X)^2+IM(X)^2)/ √ (RE(X)^2+IM(X)^2+1))'
                               If in real mode, results in 'ASIN(ABS(X)/ √ (X^2+1))'
'ATAN(X+Y)' [ATAN2S]   Results in 'ASIN((X+Y)/ √ (X^2+2*Y*X+(Y^2-1)))'

⬇ The function [HALFT] (HALFTAN) replaces occurrences of SIN(X), COS(X), and TAN(X), with expressions involving TAN(X/2).

'SIN(X)' [HALFT]   Results in '2*TAN(X/2)/(SQ(TAN(X/2))+1)'
'COS(X)' [HALFT]   Results in '1-SQ(TAN(X/2)))/(SQ(TAN(X/2)+1)'
'TAN(X)' [HALFT]   Results in '2*TAN(X/2)/(1-SQ(TAN(X/2)))'
'SIN(X)+COS(X)' [HALFT]   Results in

          '2*TAN(X/2)/(SQ(TAN(X/2))+1) + (1-SQ(TAN(X/2))/(SQ(TAN(X/2))+1)'

⚜ The function [SINCO] (SINCOS) converts natural logarithm and exponential expressions with complex arguments into expressions involving the functions sine and cosine. Set your calculator to complex mode before trying these exercises:

'EXP((2,3))' [SINCO]   Results in 'EXP(2)*(COS(3)+i*SIN(3))'
'EXP(Z)' [SINCO]   Results in 'EXP(RE(Z))*(COS(IM(Z))+ i*SIN(IM(Z)))'
'EXP(i*π)' [SINCO]   Results in '-1'
'LN(X+I*Y)' [SINCO]   Results in '(LN(X^2+Y^2)/2, ATAN(Y/X)+0*(π/2))'

⚜ There are two [TAN2S] functions listed in Menu 2. They represent the functions TAN2SC and TAN2SC2, in that order. The function TAN2SC replaces occurrences of TAN(X) with 'SIN(X)/COS(X)'. Examples:

'TAN(Y/2)' [TAN2S] (TAN2SC)   Results in 'SIN(Y/2)/COS(Y/2)'
'TAN(X)^2'[TAN2S] (TAN2SC)   Results in '(SIN(X)/COS(X))^2'

⚜ The operation of TAN2SC2 depends on whether flag -116 is set [Prefer Sin( )], in which case TAN(X) gets replaced by '(1-COS(2*X))/SIN(2*X))'. If flag -116 is clear [Prefer Cos( )], TAN(X) gets replaced by 'SIN(2*X)/(1+COS(2*X))'. Examples:

Set flag -116, as follows: [1][1][6][+/-][ENTER][ALPHA][S][ALPHA][F][ENTER].

'TAN(X)'   [TAN2S](TAN2SC2)   Results in '(1-COS(2*X))/SIN(2*X)'
'TAN(π/6)' [TAN2S](TAN2SC2)   Results in '(1-1/2)/ (√3/2)'
'TAN(X/2)' [TAN2S](TAN2SC2)   Results in '(1-COS(X))/SIN(X)'

Clear flag -116, as follows: [1][1][6][+/-][ENTER][ALPHA][C][ALPHA][F][ENTER].

'TAN(X)'   [TAN2S](TAN2SC2)   Results in 'SIN(2*X)/(1+COS(2*X))'.
'TAN(π/6)' [TAN2S](TAN2SC2)   Results in '√3/2/(1+1/2)'
'TAN(X/2)' [TAN2S](TAN2SC2)   Results in 'SIN(X)/(1+COS(X))'

⚜ The function [TCOLL] (TCOLLECT) collects terms involving sine and cosine of the same arguments. Examples:

'SIN(X)*COS(Y)' [TCOLL] Results in '1/2*SIN(X-Y)+1/2*SIN(X+Y)'
'SIN(X)^2+2*SIN(X)*COS(Y)+COS(Y)^2' [TCOLLECT] Results in
                                    'SIN(2*X)+ - 1/2*COS(2*X)+1+1/2*COS(2*Y)'
'SIN(X)^2' [TCOLL] Results in '-1/2*COS(2*X)+1/2'

⚜ The function [TLIN] linearizes trigonometric function expressions without collecting those with the same argument. Examples:

'SIN(X)^2+2*SIN(X)*COS(Y)+COS(Y)^2' [TLIN] Results in
                 'SIN(X-Y)+SIN(X+Y) - 1/2*COS(2*X)+1+1/2*COS(2*Y)'
'SIN(X)^2' [TLIN]   Results in '-1/2*COS(2*X)+1/2' (Same result as with [TCOLL]
'SIN(X)*COS(X)^2' [TLIN] Results in '1/4*SIN(3*X)+1/4*SIN(X)'

⚜ The function [TRIG] replaces exponential and logarithmic functions with complex arguments with equivalent trigonometric functions. Examples:

'LN(X+i*Y)' [TRIG]   Results in '(LN(X^2+Y^2)+2*I*ATAN(Y/X))/2'
'EXP(X+i*Y)' [TRIG]   Results in

`EXP(RE(X)-IM(Y))*COS(RE(Y)+IM(X))+I*EXP(RE(X)-IM(Y))*SIN(RE(Y)+IM(X))'

`EXP(Z)' [TRIG]          Results in `EXP(RE(Z))*COS(IM(Z))+I*EXP(RE(Z))*SIN(IM(Z))'

⬇ The function [TRIGC] (TRIGCOS) uses the identity `SIN(X)^2+COS(X)^2=1' to simplify expressions while returning cosine terms only if possible. Examples:

`SIN(X)^2-COS(X)^2' [TRIGC] Results in `-(2*COS(X)^2-1)'
`(SIN(X)-TAN(X))^2' [TRIGC] Results in `-(COS(X)^2-(TAN(X)^2+1)+2*TAN(X)*SIN(X))'
`(SIN(X)+COS(X))^2' [TRIGC] Results in `1+2*COS(X)*SIN(X)'

⬇ The function [TRIGS] (TRIGSIN) uses the identity `SIN(X)^2+COS(X)^2=1' to simplify expressions while returning sine terms only if possible.

`SIN(X)^2-COS(X)^2' [TRIGS] Results in `2*SIN(X)^2-1'
`(SIN(X)-TAN(X))^2' [TRIGC] Results in `SIN(X)^2-2*TAN(X)*SIN(X)+TAN(X)^2'
`(SIN(X)-COS(X))^2' [TRIGC] Results in `1-2*COS(X)*SIN(X)'

⬇ The function [TRIGT] (TRIGTAN) replaces terms involving SIN(X) and COS(X) with expressions involving TAN(X).

`SIN(X)^2-COS(X)^2' [TRIGC] Results in `(TAN(X)^2-1)/(TAN(X)^2+1)'
`(SIN(X)-TAN(X))^2' [TRIGC] Results in
          `(TAN(X)^4+2*TAN(X)^2-2*TAN(X)^2*(COS(X)*(SQ(TAN(X))+1)))/(TAN(X)^2+1)'
`(SIN(X)+COS(X))^2' [TRIGC] Results in `(TAN(X)^2+2*TAN(X)+1)/(TAN(X)^2+1)'

Fraction Manipulation

Fractions can be expanded and factored by using [↪][ALG][EXPAND] and [FACTOR]. For example:

`(1+X)^3/((X-1)(X+3))' [EXPAN] Results in `(X^3+3*X^2+3*X+1)/(X^2+2*X-3)'
`X^2*(X+Y)/(2*X-X^2)^2' [EXPAN] Results in `(X+Y)/(X^2-4*X+4)'
`X+(X+Y)/(X^2-1)' [EXPAN] Results in `(X^3+Y)/(X^2-1)'
`4+2*(X-1)+3/((X-2)*(X+3))-5/X^2' [EXPAN] Results in
                              `(2*X^5+4*X^4-10*X^3-14*X^2-5*X)/(X^4+X^3-6*X^2)'
`(3*X^3-2*X^2)/(X^2-5*X+6)' [FACTO] Results in `X^2*(3*X-2)/((X-2)*(X-3))'
`(X^3-9*X)/(X^2-5*X+6)' [FACTO] Results in `X*(X+3)/(X-2)'
`(X^2-1)/(X^3*Y-Y)' [FACTO] Results in `(X+1)/((X^2+X+1)*Y)'

SIMP2

There are a couple of functions in the menu [↩][ARITH] (the key for [1]) [NXT] that apply to fractions: [SIMP2] and [PROPF]. The function [SIMP2] takes as arguments two numbers or polynomials, in stack levels 2 and 1, representing the numerator and denominator, respectively, of a rational fraction, and returns the simplified numerator and denominator in stack levels 2 and 1, respectively. For example:

`X^3-1'[ENTER]`X^2-4*X+3'[ENTER][↩][ARITH][NXT][SIMP2] Results in:

```
2:            'X^2+X+1'
1:                'X-3'
```

The operation of [SIMP2] is not different from [FACTO] if you write the full fraction, i.e.,

'(X^3-1)/(X^2-4*X+3)' [FACTO], to get '(X^2+X+1)/(X-3)'

PROPFAC

The function PROPFAC ([↰][ARITH][NXT][PROPF]) converts a rational fraction into a "proper" fraction, i.e., an integer part added to a fractional part, if such decomposition is possible. For example:

'5/4' [↰][ARITH][NXT][PROPF] Results in '1+1/4'
'(x^2+1)/x^2' [PROPF] Results in '1+1/x^2'

PARTFRAC

The function PARTFRAC ([↰][ARITH][POLY][NXT][NXT][PARTF]) decomposes a rational fraction into the partial fractions that produce the original fraction. For example:

'(2*X^6-14*X^5+29*X^4-37*X^3+41*X^2-16*X+5)/(X^5-7*X^4+11*X^3-7*X^2+10*X)'
[↰][ARITH][POLY][NXT][NXT][PARTF] Results in '2*X+(1/2/(X-2)+5/(X-5)+1/2/X+X/(X^2+1))'
This technique is useful in calculating integrals (see chapter on calculus) of rational fractions.
If you have the Complex mode active, the result will be:
'2*X+(1/2/(X+i)+1/2/(X-2)+5/(X-5)+1/2/X+1/2/(X-i))'

FCOEF

The function FCOEF is used to obtain rational fraction, given the roots and poles of the fraction. This function is accessible through: [↰][ARITH][POLY][FCOEF].

Note: If a rational fraction is given as F(X) = N(X)/D(X), the roots of the fraction result from solving the equation N(X) = 0, while the poles result from solving the equation D(X) = 0.

The input for the function is a vector listing the roots followed by their multiplicity (i.e., how many times a given root is repeated), and the poles followed by their multiplicity represented as a negative number. For example, if we want to create a fraction having roots 2 with multiplicity 1, 0 with multiplicity 3, and -5 with multiplicity 2, and poles 1 with multiplicity 2 and -3 with multiplicity 5. The input is:

[2 1 0 3 -5 2 1 -2 -3 -5][FCOEF]. The result is: '(X—5)^2*X^3*(X-2)/9X—3)^5*(X-1)^2'

If you press [↱][EVAL], you will get:

'(X^6+8*X^5+5*X^4-50*X^3)/(X^7+13*X^6+61*X^5+105*X^4-45*X^3-297*X62-81*X+243)'

FROOTS

The function FROOTS ([FROOT]) obtains the roots and poles of a fraction. This function is accessible through [┐][ARITH][POLY][NXT][FROOT]. As an example, if you kept the latest result in your stack, press [FROOT] to obtain: [1 -2 -3 -5 0 3 2 1 -5 2]. The result shows poles followed by their multiplicity as a negative number, and roots followed by their multiplicity as a positive number. In this case, the poles are (1, -3) with multiplicities (2,5) respectively, and the roots are (0, 2, -5) with multiplicities (3, 1, 2), respectively. Another example is: `(X^2-5*X+6)/(X^5-X^2)' [FROOT]   Results in [0 -2 1 -1 3 1 2 1], i.e., poles = 0 (2), 1(1), and roots = 3(1), 2(1). If you have had Complex mode selected, then the results would be: [0 -2 1 -1 '-((1+i*√3)/2' -1].

Modular arithmetic

Modular arithmetic is used in some operations with polynomials. Therefore, we deemed necessary to present the subject in this chapter.

Consider a counting system of integer numbers that periodically cycles back on itself and starts again, such as the hours in a clock. Such counting system is called a ring. Because the number of integers used in a ring is finite, we can refer to arithmetic operations in such system as finite arithmetic operations or, simply, finite arithmetic. Let our system of finite integer numbers consists of the numbers 0, 1, 2, 3, ..., n-1, n. We can also refer to the arithmetic of this counting system as modular arithmetic of modulus n. In the case of the hours of a clock, the modulus is 12. (If working with modular arithmetic using the hours in a clock, however, we would have to use the integer numbers 0, 1, 2, 3, ..., 10, 11, rather than 1, 2, 3,...,11, 12).

Operations in modular arithmetic

Addition in modular arithmetic of modulus n, which is a positive integer, follow the rules that if j and k are any two nonnegative integer numbers, both smaller than n, if j+k≥ n, then j+k is defined as j+k-n. For example, in the case of the clock, i.e., for n = 12, 6+9 "=" 3. To distinguish this 'equality' from infinite arithmetic equalities, the symbol ≡ is used in place of the equal sign, and the relationship between the numbers is referred to as a congruence rather than an equality. Thus, for the previous example we would write 6+9 ≡ 3 (mod 12), and read this expression as "six plus nine is congruent to three, modulus twelve." If the numbers represent the hours since midnight, for example, the congruence 6+9 ≡ 3 (mod 12), can be interpreted as saying that "six hours past the ninth hour after midnight will be three hours past noon."   Other sums that can be defined in modulus 12 arithmetic are: 2+5 ≡ 7 (mod 12); 2+10 ≡ 0 (mod 12); 7+5 ≡ 0 (mod 12); etcetera.

The rule for subtraction will be such that if j - k < 0, then j-k is defined as j-k+n. Therefore, 8-10 ≡ 2 (mod 12), is read "eight minus ten is congruent to two, modulus twelve."   Other examples of subtraction in modulus 12 arithmetic would be 10-5 ≡ 5 (mod 12); 6-9 ≡ 9 (mod 12); 5 - 8 ≡ 9 (mod 12); 5 -10 ≡ 7 (mod 12); etcetera.

Multiplication follows the rule that if j·k > n, so that j·k = mn + r, where m and r are nonnegative integers, both less than n, then j·k ≡ r (mod n). The result of multiplying j times k in modulus n arithmetic is, in essence, the integer remainder of j·k/n in infinite arithmetic, if j·k>n.   For example, in modulus 12 arithmetic we have 7·3 = 21 = 12 + 9, (or, 7·3/12 = 21/12 = 1

+ 9/12, i.e., the integer reminder of 21/12 is 9). We can now write 7·3 ≡ 9 (mod 12), and read the latter result as "seven times three is congruent to nine, modulus twelve."

The operation of division can be defined in terms of multiplication as follows, r/k ≡ j (mod n), if, j·k ≡ r (mod n). This means that r must be the remainder of j·k/n. For example, 9/7 ≡ 3 (mod 12), because 7·3 ≡ 9 (mod 12). Some divisions are not permitted in modular arithmetic. For example, in modulus 12 arithmetic you cannot define 5/6 (mod 12) because the multiplication table of 6 does not show the result 5 in modulus 12 arithmetic. This multiplication table is shown below:

6*0 (mod 12)	0	6*6 (mod 12)	0
6*1 (mod 12)	6	6*7 (mod 12)	6
6*2 (mod 12)	0	6*8 (mod 12)	0
6*3 (mod 12)	6	6*9 (mod 12)	6
6*4 (mod 12)	0	6*10 (mod 12)	0
6*5 (mod 12)	6	6*11 (mod 12)	6

Formal definition of a finite arithmetic ring

The expression
$$a \equiv b \pmod{n}$$

is interpreted as "a is congruent to b, modulo n," and holds if (b-a) is a multiple of n. With this definition the rules of arithmetic simplify to the following:

If
$$a \equiv b \pmod{n} \text{ and } c \equiv d \pmod{n},$$
then
$$a+c \equiv b+d \pmod{n},$$
$$a-c \equiv b - d \pmod{n},$$
$$a \times c \equiv b \times d \pmod{n}.$$

For division, follow the rules presented earlier.

For example, 17 ≡ 5 (mod 6), and 21 ≡ 3 (mod 6). Using these rules, we can write:

17 + 21 ≡ 5 + 3 (mod 6)   =>   38 ≡ 8 (mod 6)   => 38 ≡ 2 (mod 6)
17 - 21 ≡ 5 - 3 (mod 6)   =>   -4 ≡ 2 (mod 6)
17 × 21 ≡ 5 × 3 (mod 6)   =>   357 ≡ 15 (mod 6)   => 357 ≡ 3 (mod 6)

Notice that, whenever a result in the right-hand side of the "congruence" symbol produces a result that is larger than the modulo (in this case, n = 6), you can always subtract a multiple of the modulo from that result and simplify it to a number smaller than the modulo. Thus, the results in the first case 8 (mod 6) simplifies to 2 (mod 6), and the result of the third case, 15 (mod 6) simplifies to 3 (mod 6). Confusing? Well, not if you let the calculator handle those operations. Thus, read the following section to understand how finite arithmetic rings are operated upon in your calculator.

Finite arithmetic rings in the HP 49 G

All along we have defined our finite arithmetic operation so that the results are always positive. The modular arithmetic system in the HP 49 G calculator is set so that the ring of modulus n includes the numbers -n/2+1, ...,-1, 0, 1,...,n/2-1, n/2, if n is even, and -(n-1)/2, -(n-3)/2,...,-1,0,1,...,(n-3)/2, (n-1)/2, if n is odd. For example, for n = 8 (even), the finite arithmetic ring in the HP 49 G includes the numbers:

$$(-3,-2,-1,0,1,3,4),$$

while for n = 7 (odd), the corresponding HP 49 G finite arithmetic ring is given by

$$(-3,-2,-1,0,1,2,3).$$

Modular arithmetic in the HP 49 G

To launch the modular arithmetic menu in the HP 49 G use [↰][ARITH][MODUL] (ARITH is the left-shift key corresponding to the key for number [1]). The available menu includes:

[ADDTM][DIVMO][DIV2M][EXPAN][FACTO][GCDMO]

Press [NXT] to get the second modular arithmetic menu, which shows the functions:

[INVMO][MOD][MODST][MULTM][POWMO][SUBTM]

The use of this functions is presented next.

Setting the modulus (or MODULO)

The calculator creates a variable called MODULO that is placed in the HOME directory and will store the magnitude of the modulus to be used in modular arithmetic. To get the calculator to create this variable, if it does not already exists in the HOME directory, launch the modular arithmetic menu, by using [↰][ARITH][MODUL]. Then, enter a number, say [8], and press the soft menu key corresponding to [EXPAN]. If the MODULO variable you should get the result -1. To check out your MODULO variable, press [VAR] and enter [↱][MODUL]. The default value of 3 should be placed in the stack indicating that such value currently occupies the variable MODULO.

To change the value of MODULO, you can either store the new value directly in the variable MODULO by placing the value in the stack and then using [↰][MODUL]. If you have already activated the modular arithmetic menu, move to the second menu by using [NXT], place the new value in the stack, and press the soft key labeled [MODST] (MODSTO) to store the new modulus value.

For example, to perform arithmetic operations of modulus 12, use:

[1][2][ENTER] [↰][ARITH][MODUL][NXT][MODST]

Modular arithmetic operations with numbers

To add, subtract, multiply, divide, and raise to a power using modular arithmetic you will use the functions ADDTM, SUBTM, MULTM, DIV3M and DIVMO (these two for division), and POWMO. In

RPN mode you need to enter the two numbers to operate upon, separated by an [ENTER] or an [SPC] entry, and then press the corresponding modular arithmetic function. For example, having stored 12 as our modulus, try the following operations:

[6][SPC][5][ADDTM]   Result: -1, i.e., 6+5 ≡ -1 (mod 12)
[6][SPC][6][ADDTM]   Result: 0, i.e., 6+6 ≡ 0 (mod 12)
[6][SPC][7][ADDTM]   Result: 1, i.e., 6+7 ≡ 1 (mod 12)
[1][1][SPC][5][ADDTM]   Result: 4, i.e., 11+5 ≡ 4 (mod 12)
[8][SPC][1][0][ADDTM]   Result: -6, i.e., 8+10 ≡ -6 (mod 12)

[5][SPC][7][SUBTM]   Result: -2, i.e., 5 - 7 ≡ -2 (mod 12)
[8][SPC][4][SUBTM]   Result: 4, i.e., 8 - 4 ≡ 4 (mod 12)
[5][SPC][1][0]][SUBTM]   Result: -5, i.e., 5 -10 ≡ -5 (mod 12)
[1][1][SPC][8][SUBTM]   Result: 3, i.e., 11 - 8 ≡ 3 (mod 12)
[8][SPC][1][2][SUBTM]   Result: -4, i.e., ≡ 8 - 12 (mod 12)

[6][SPC][8][MULTM]   Result: 0, i.e., 6·8 ≡ 0 (mod 12)
[9][SPC][8][MULTM]   Result: 0, i.e., 9·8 ≡ 0 (mod 12)
[3][SPC][2][MULTM]   Result: 6, i.e., 3·2 ≡ 6 (mod 12)
[5][SPC][6][MULTM]   Result: 6, i.e., 5·6 ≡ 6 (mod 12)
[1][1][SPC][3][MULTM]   Result: -3, i.e., 11·3 ≡ -3 (mod 12)

[1][2][SPC][3][DIVMO]   Result: 4, i.e., 12/3 ≡ 4 (mod 12)
[1][2][SPC][8][DIVMO]   Result: No solution in ring, i.e., 12/8 (mod 12) does not exist
[2][5][SPC][5][DIVMO]   Result: 5, i.e., 25/5 ≡ 5 (mod 12)
[6][4][SPC][1][3][DIVMO]   Result: 4, i.e., 64/13 ≡ 4 (mod 12)
[6][6][SPC][6][DIVMO]   Result: -1, i.e., 66/6 ≡ -1 (mod 12)

[1][2][SPC][5][DIV2M]   Result: , 0 and 0 , i.e., 2/3(mod 12) ≡ 0 (mod 12) with remainder = 0
[2][6][SPC][1][2][DIV2M]   Result: No solution in ring, i.e., 26/12 (mod 12) does not exist
[1][2][5][SPC][1][7][DIV2M] Result: 1 and 0, i.e., 125/17 (mod 12) ≡ 1 with remainder = 0
[6][8][SPC][7][DIV2M]   Result: -4 and 0, i.e., 68/7 ≡ 0 (mod 12) with remainder = 0
[7][SPC][5][DIV2M]   Result: -1 and 0, i.e., 7/5 ≡ -1 (mod 12) with remainder = 0

Note: DIVMO provides the quotient of the modular division j/k (mod n), while DIMV2M provides no only the quotient but also the remainder of the modular division j/k (mod n).

[2][SPC][3][POWMO]   Result: -4, i.e., 2^3 ≡ -4 (mod 12)
[3][SPC][5][POWMO]   Result: 3, i.e., 3^5 ≡ 3 (mod 12)
[5][SPC][1][0][POWMO]   Result: 1, i.e., 5^{10} ≡ 1 (mod 12)
[1][1][SPC][8][POWMO]   Result: 1, i.e., 11^8 ≡ 1 (mod 12)
[6][SPC][2][POWMO]   Result: 0, i.e., 6^2 ≡ 0 (mod 12)
[9][SPC][9][POWMO]   Result: -3, i.e., 9^9 ≡ -3 (mod 12)

In the examples of modular arithmetic operations shown above, we have used numbers that not necessarily belong to the ring, i.e., numbers such as 66, 125, 17, etc. The calculator will convert those numbers to ring numbers before operating on them. You can also convert any number into a ring number by using the function [EXPAND] (full name is EXPANDMOD). For example,

[1][2][5][EXPAND]   Result: 5, or 125 ≡ 5 (mod 12)
[1][7][EXPAND]   Result: 5, or 17 ≡ 5 (mod 12)
[6][6][EXPAND]   Result: 6, or 66 ≡ 6 (mod 12)

The modular inverse of a number

Let a number k belong to a finite arithmetic ring of modulus n, then the modular inverse of k, i.e., 1/k (mod n), is a number j, such that $j \cdot k \equiv 1$ (mod n). The modular inverse of a number can be obtained by using the function [INVMOD] in the second menu resulting from [←][ARITH][MODUL].

For example, in modulus 12 arithmetic:

[6][INVMOD]   Result: No solution in ring, i.e., 1/6 (mod 12) does not exist.
[5][INVMOD]   Result: 5, i.e, $1/5 \equiv 5$ (mod 12)
[7][INVMOD]   Result: -5, i.e., $1/7 \equiv -5$ (mod 12)
[3][INVMOD]   Result: No solution in ring, i.e, 1/3 (mod 12) does not exist.
[1][1][INVMOD]   Result: -1, i.e., $1/11 \equiv -1$ (mod 12)

The MOD function

The MOD function is used to obtain the ring number of a given modulus corresponding to a given integer number. To operate this function in RPN mode, enter the integer number to be converted to a ring number followed by the modulus, and separated by an [ENTER] or a [SPC]. Then press [MOD]. On paper this operation is written as
$$m \bmod n = p$$

and is read as "m modulo n is equal to p".

For example, to calculate 15 mod 8, enter

[1][5][SPC][8][MOD].

The result is 7, i.e., 15 mod 8 = 7.

Try the following exercises:

[1][8][SPC][1][1][MOD]   Result: 7, i.e., 18 mod 11 = 7
[2][3][SPC][2][MOD]   Result: 1, i.e., 23 mod 2 =   1
[4][0][SPC][1][3][MOD]   Result: 1, i.e., 40 mod 13 = 1
[2][3][SPC][1][7][MOD]   Result: 6, i.e., 23 mod 17 = 6
[3][4][SPC][6][MOD]   Result: 4, i.e., 34 mod 6 = 4

One practical application of the MOD function for programming purposes is to determine when an integer number is odd or even, since n mod 2 = 0, if n is even, and n mode 2 = 1, if n is odd. It can also be used to determine when an integer m is a multiple of another integer n, for if that is the case m mod n = 0.

Other modular arithmetic functions

Other functions included in the modular arithmetic menu are [GCDMO] (GCDMOD), or greatest common divisor under modular arithmetic, and [FACTO] (FACTORMOD), or factorization under modular arithmetic. These functions are used on polynomial expressions to produce the GCM of two polynomials, or to factor a polynomial. Examples of such applications are presented in the next section.

Polynomials

Polynomials are algebraic expressions consisting of one or more terms containing decreasing powers of a given variable. For example, 'X^3+2*X^2-3*X+2' is a third-order polynomial in X, while 'SIN(X)^2-2' is a second-order polynomial in SIN(X). The HP 49 G calculator includes a large number of functions for manipulating polynomials, besides the EXPAND and FACTOR functions, already introduced.

Polynomial manipulation functions are available in the calculator menu: [↰][ARITH][POLY]:

Menu 1:   [ABCUV][CHINR][DIV2][EGCD][FACTO][FCOEF]

Press [NXT]

Menu 2:   [PROOT][GCD][HERMI][HORNE][LAGRA][LCM]

Press [NXT]

Menu 3:   [LEGEN][PARTF][PCOEF][PROOT][PTAYL][QUOT]

Press [NXT]

Menu 4:   [REMAI][   ][   ][   ][   ][ARITH]

Modular arithmetic with polynomials

> The same way that we defined a finite-arithmetic ring for numbers in a previous section, we can define a finite-arithmetic ring for polynomials with a given polynomial as modulo. For example, we can write a certain polynomial P(X) as P(X) = X (mod X^2), or another polynomial Q(X) = X + 1 (mod X-2). Some polynomial functions apply to such algebraic rings, as presented below.
>
> A polynomial, P(X) belongs to a finite arithmetic ring of polynomial modulus M(X), if there exists a third polynomial Q(X), such that (P(X) - Q(X)) is a multiple of M(X). We then would write:
>
> $$P(X) \equiv Q(X) \ (mod \ M(X)).$$
>
> The later expression is interpreted as "P(X) is congruent to Q(X), modulo M(X)".

ABCUV

Given polynomials A(X), B(X), and C(X), in stack levels 3, 2, and 1, respectively, [ABCUV] returns two polynomials, U(X) and V(X), in stack levels 2 and 1, respectively, so that:

C(X) = U(X)*A(X) + V(X)*B(X).

For example, with A(X) = X^2-1, B(X) = X^2+1, C(X) = X^3+2*X^2-7, use:

'X^2-1' [ENTER] 'X^2+1' [ENTER] 'X^3+2*X^2-7' [ENTER][ABCUV] gives

```
2:       '(X^3+2*X-7)/-2'
1:       '-(X^3+2*X-7)/-2'
```

CHINREM

CHINREM stands for CHINese Remainder. The operation coded in this command solves a system of two congruences using the Chinese Remainder Theorem. This command can be used with polynomials, as well as with numbers. The input consists of two vectors [expression_1, modulo_1] and [expression_2, modulo_2], in stack levels 2 and 1, respectively. The output is a vector containing [expression_3, modulo_3], where modulo_3 is related to the product (modulo_1)·(modulo_2). Example:

['X+1', 'X^2-1'][ENTER]['X+1','X^2'][ENTER] [CHINREM] Results in [`'X+1',-(X^4-X^2)]

DIV2

Given polynomials P(X) in stack level 2, and Q(x) in stack level 1, the function [DIV2] returns the quotient and residual of P(X)/Q(X) in stack levels 2 and 1, respectively. For example:

'X^3-1'[ENTER]'X-5'[ENTER][DIV2] Results in:

```
2:       'X^2+5*X+25'
1:            124
```

EGCD

EGCD stands for Extended Greatest Common Divisor. Given two polynomials, A(X) and B(X), in stack levels 2 and 1, respectively, it produces the polynomials C(X), U(X), and V(X), in stack levels 3, 2, and 1, respectively, so that

$$C(X) = U(X)*A(X) + V(X)*B(X).$$

For example, for A(X) = X^2+1, B(X) = X^2-1, use:

'X^2+1'[ENTER] 'X^2-1' [ENTER] [EGCD] Results in

```
3:        2
2:        1
1:       -1
```

i.e., 2 = 1*(X^2+1')-1*(X^2-1).

Try another example:

'X' [ENTER] 'X^3-2*X+5' [ENTER] [EGCD] Results in

```
3:        5
2:     '-(X^2-2)'
1:        1
```

i.e., 5 = - (X^2-2)*X + 1*(X^3-2*X+5)

FACTOR

The function FACTOR [FACTO] has been presented earlier in the chapter.

FCOEF

The function FCOEF was presented in an earlier section.

FROOTS

The function FROOTS was presented in an earlier section.

GCD

The function GCD (Greatest Common Denominator) can be used to obtain the greatest common denominator of two polynomials or of two lists of polynomials of the same length. The two polynomials or lists of polynomials will be placed in stack levels 2 and 1 before using GCD. The results will be a polynomial or a list representing the greatest common denominator of the two polynomials or of each list of polynomials. Examples follow:

'X^3-1'[ENTER]'X^2-1'[ENTER][GCD] Results in: 'X-1'
{'X^2+2*X+1','X^3+X^2'}[ENTER] {'X^3+1','X^2+1'}[ENTER][GCD] Results in {'X+1' 1}

HERMITE

The function HERMITE [HERMI] uses as argument an integer number, k, and returns the Hermite polynomial of k-th degree. A Hermite polynomial, $He_k(x)$ is defined as

$$He_0 = 1, \quad He_n(x) = (-1)^n e^{x^2/2} \frac{d^n}{dx^n}(e^{-x^2/2}), \quad n = 1, 2, ...$$

An alternate definition of the Hermite polynomials is

$$H_0^* = 1, \quad H_n^*(x) = (-1)^n e^{x^2} \frac{d^n}{dx^n}(e^{-x^2}), \quad n = 1, 2, ...$$

Where d^n/dx^n = n-th derivative with respect to x.

Examples: The Hermite polynomials of orders 3 and 5 are given by:

[3][HERMI] which produces '8*X^3-12*X' ,
and

[5][HERMI] which gives '32*x^5-160*X^3+120*X'

> Note: To check which of the two definitions of Hermite polynomials given above is the one that the calculator uses, enter the following:
>
> [EQW][↰][( )][1][+/-][▲][▲][▲][y^x][3][▲][▲][×] [↰][e^x][ALPHA][X][y^x][2][▶][▶][▶][×]
> [↱][∂][ALPHA][X] [▶][↱][∂][ALPHA][X] [▶][↱][∂][ALPHA][X] [▶]
> [↰][e^x][ALPHA][X][+/-][y^x][2] [ENTER]
>
> Resulting in '(-1)^3*EXP(x^2)*∂x(∂x(∂x(EXP(-x^2))))'. Press [ENTER] to keep an extra copy of the expression in the stack. Then, enter [↱][EVAL]. The result is '(8*X^3-12*X)*EXP(x^2)*EXP(-x^2)'. Pressing [↱][EVAL], once more, we get the result '8*X^3-12*X', which is the same produced with [5][HERMI]. Therefore, we verified, at least for n = 3, that the second definition, i.e., that of H*n(x), is the one used for the calculator to generate Hermite's polynomials.
>
> From this exercise we learn that the keystroke combination [↱][∂] generates derivative symbols in the equation writer. More details on the use of derivatives is presented in Chapter ....

HORNER

The function HORNER [HORNE] produces the Horner division, or synthetic division, of a polynomial P(X) by the factor (X-a). The input to the function are the polynomial P(X), in stack level 2, and the number a, in stack level 1. The function returns the quotient polynomial Q(X) that results from dividing P(X) by (X-a), the value of a, and the value of P(a), in stack levels 3, 2, and 1, respectively. In other words, P(X) = Q(X)(X-a)+P(a). For example,

'X^3+2*X^2-3*X+1' [ENTER] 2 [ENTER] [HORNE] produces the result:

```
3:          'X^2+4*X+5'
2:                    2
1:                   11
```

We could, therefore, write $X^3+2X^2-3X+1 = (X^2+4X+5)(X-2)+11$.

A second example, would be:

'X^6-1'[ENTER] [5][+/-] [ENTER] [HORNE], which results in :

```
3:'X^5-5*X^4+25*X^3-
125*X^2+625*X-3125'
2:                   -5
1:                15624
```

i.e., $X^6-1 = (X^5-5*X^4+25X^3-125X^2+625X-3125)(X+5)+15624$.

The variable VX, or "Why do you use only X in your examples?"

The first time you activate any of the CAS functions, i.e., algebraic, calculus, etc., a variable called VX is created in your HOME directory that takes, by default, the value of 'X'. This is the name of the preferred independent variable for algebraic and calculus applications. For that reason, I have used X as the unknown variable in all the examples in this section. If you use other independent variable names, for example, with HORNER, the CAS will not work properly.

The variable VX is a permanent inhabitant of the HOME directory. There are other CAS variables in the HOME directory that you may or may not have currently in your calculator. Some of these variables are REALASSUME ([REALA]), MODULO ([MODUL]), CASINFO ([CASIN]), etc. You can purge all of them, except VX. Try it.

You can change the value of VX by storing a new algebraic name in it, e.g., 'x', 'y', 'm', etc. In the rest of this section, and in the rest of the book, I will assume that 'X' will be the standard value of VX and I will continue using X as the preferred independent variable for CAS applications.

Also, avoid using the variable VX in your programs or equations, so as to not get it confused with the CAS' VX. If you need to refer to the x-component of velocity, for example, you can use vx or Vx.

LAGRANGE

The function LAGRANGE [LAGRA] requires as input a matrix having two rows and n columns. The matrix stores data points of the form [[x$_1$ x$_2$ ... x$_n$] [y$_1$ y$_2$ ... y$_n$]]. Application of the function LAGRANGE produces the polynomial expanded from

$$p_{n-1}(x) = \sum_{j=1}^{n} \frac{\prod_{k=1, k \neq j}^{n}(x - x_k)}{\prod_{k=1, k \neq j}^{n}(x_j - x_k)} \cdot y_j.$$

For example, for n = 2, we will write:

$$p_1(x) = \frac{x - x_2}{x_1 - x_2} \cdot y_1 + \frac{x - x_1}{x_2 - x_1} \cdot y_2 = \frac{(y_1 - y_2) \cdot x + (y_2 \cdot x_1 - y_1 \cdot x_2)}{x_1 - x_2}.$$

Check this result with your calculator:

[['x1' 'x2']['y1' 'y2']][ENTER][LAGRA]. The result is: '((y1-y2)*X+(y2*x1-y1*x2))/(x1-x2)'.

Other examples:

[[1 2 3][2 8 15]][ENTER][LAGRA] produces '(X^2+9*X-6)/2'

[[0.5 1.5 2.5 3.5 4.5][12.2 13.5 19.2 27.3 32.5]] [ENTER] [LAGRA] produces
'-(.1375*X^4+ -.7666666666667*X^3+ -.74375*X^2 = 1.991666666667*X-12.92265625)'.

Entering matrices directly in the stack

Matrices are introduced in a letter chapter. They are basically arrays of numbers, or symbols, in rows and columns. The matrix corresponding to the latter problem's input would be written on paper as

$$\begin{bmatrix} 0.5 & 1.5 & 2.5 & 3.5 & 4.5 \\ 12.2 & 13.5 & 19.2 & 27.3 & 32.5 \end{bmatrix}$$

In the calculator stack, you can enter the matrix by opening a pair of square brackets and using sets of square brackets to represent the rows of the matrix. For the example under consideration, this is:

[[0.5 1.5 2.5 3.5 4.5][12.2 13.5 19.2 27.3 32.5]].

LCM

The function LCM (Least Common Multiple) obtains the least common multiple of two polynomials or of lists of polynomials of the same length. Examples:

'2*X^2+4*X+2' [ENTER] 'X^2-1' [ENTER] [LCM] produces '(2*X^2+4*X+2)*(X-1)'.

'X^3-1'[ENTER] 'X^2+2*X' [ENTER] [LCM] produces '(X^3-1)*(X^2+2*X)'

LEGENDRE

A Legendre polynomial of order n is a polynomial function that solves the differential equation

$$(1-x^2) \cdot \frac{d^2 y}{dx^2} - 2 \cdot x \cdot \frac{dy}{dx} + n \cdot (n+1) \cdot y = 0.$$

To obtain the n-th order Legendre polynomial, enter the order of the polynomial, then press [LEGEN] (LEGENDRE). Examples:

[3][LEGEN] produces: '(5*X^3-3*X)/2'

[5][LEGEN] produces: '(63*X^5-70*X^3+15*X)/8'

Checking the solution to Legendre's equation

To verify that these polynomials indeed satisfy Legendre's equation, enter the general expression of the equation in a variable to be called LEGEQ, as follows:

[EQW][↵][( )] [1][-][ALPHA][X][y^x][2][▶][▶][▶][▶] [×][↵][∂][ALPHA][X] [▶] [↵][∂][ALPHA][X][▶] [ALPHA][Y][↵][( )][ALPHA][X][▶][▶][▶][▶][-][2][×][ALPHA][X][×][↵][∂][ALPHA][X][▶] [ALPHA][Y][↵][( )][ALPHA][X][▶] [▶][▶][▶][▶][+][ALPHA][↵][N] [×] [↵][( )] [ALPHA][↵][N] [+] [1] [▶][▶][▶][×] [ALPHA][Y] [↵][( )][ALPHA][X][▶] [▶][▶][▶][▶][▶][↵][=][0] [ENTER]

[`][ALPHA][ALPHA][L][E][G][E][Q][ENTER][STO▶]

Next, bring the equation back to the stack, by pressing [VAR][LEGEQ]. Replace the value of n with 3, i.e.,
[↵][{}][ALPHA][↵][N][SPC][3][ENTER][↵][|][ENTER]. The result is:
'-((X^2-1)*d1d1Y(X)+(2*X*d1Y(X)-(3^2+3)*Y(X)))=0'

Note: d1Y(X) now represents dY/dX, and d1d1Y(X) represents d^2Y/dX^2.

Use [↵][ALG][EXPAN] to get: '-((X^2-1)*d1d1Y(X)+(2*X*d1Y(X)-12*Y)=0'

Next, we generate the third-order Legendre polynomial:

[3][↵][ARITH][POLY][NXT][NXT][LEGEN] to get '(5*X^3-3*X)/2'

Next, enter 'Y(X)' by using: [↵][`][ALPHA][Y] [↵][( )][ALPHA][X][ENTER]

Press [▶] to exchange stack levels 1 and 2, and then [↵][=] to get: 'Y(X) =`(5*X^3-3*X)/2'

Press [↵][ALG][SUBST] to replace Y(X) with the Lagrange polynomial, and then press [↵][EVAL]. The result is: ' 0 = 0'.

PARTFRAC

The function PARTFRAC was presented in an earlier section.

PCOEF

Given an array containing the roots of a polynomial, the function PCOEF generates an array containing the coefficients of the corresponding polynomial. The coefficients correspond to decreasing order of the independent variable. For example:

[-2 -1 0 1 1 2] [PCOEF] produces: [1. -1. -5. 5. 4. -4. 0.], which represents the polynomial $X^6-X^5-5X^4+5X^3+4X^2-4X$.

PROOT

Given and array containing the coefficients of a polynomial, in decreasing order, the function PROOT provides the roots of the polynomial. As PCOEF, this is a numerical solution originally developed for the HP 48 G/G+/GX series calculator. Example:

From $X^2+5X-6 =0$, we enter [1 -5 6][ENTER][PROOT]. The function provides the result: [2. 3.].

Direct access to polynomials numerical solution using NUM.SLV

⬥ The function PCOEF corresponds to the numerical solution for polynomial coefficients given its roots (first developed for the HP 48 G/G+/GX series calculator). This solution is also accessible using an input form by using the keystroke sequence: [↪][NUM.SLV] (NUMerical SoLVer). This will generate a dropdown menu. Use the down-arrow key, [▼], twice to highlight 3. Solve poly.... Press [OK]. The input form will show the field corresponding to the polynomial coefficients highlighted. Press [▼] to move to the Roots: field. In this field, enter the array or vector of roots, i.e.,

[↩][[ ]][2][+/-][SPC][1][+/-][SPC][0][SPC][1][SPC][1][SPC][2][ENTER] (or [OK])

This will send the cursor back to the Coefficients field. Now, press [SOLVE]. The result is the vector [1. -1. -5. 5. 4. ... in the Coefficients field. To see the entire array, press [EDIT]. This triggers the matrix writer (a sort of spreadsheet used to enter array data) showing a matrix with one column. To see the elements of the array use the right-arrow or left-arrow keys to move about the array. Verify that the array is [1. -1. -5. 5. 4. -4. 0.]. Press [ENTER] when done. Press [ENTER] to return to normal calculator display. You will notice that the polynomial coefficient array has been copied to the stack.

⬥ The NUM.SLV menu can be used to produce the polynomial and place it into the stack by using a procedure similar to that used above, but ending by pressing the key [SYMB] after using [SOLVE]. Repeat the exercise above an try this option. Press [ENTER], when done, to return to normal calculator display. You should get the expression: 'X^6+-1*X^5+-5*X^4+5*X^3+4*X^2+-4*X'.

⬥ Start the polynomial solution from the NUM.SLV once more. This time we will provide the polynomial coefficients in the proper field, say [1 -5 6], corresponding to $X^2-5X+6 = 0$, and find the roots by pressing [SOLVE] when the Roots: field is highlighted. The solution should be [2. 3.].

PTAYL

Given a polynomial P(X) and a number a (stack levels 2 and 1, respectively), the function PTAYL is used to obtain an expression Q(X-a) = P(X), i.e., to develop a polynomial in powers of (X- a). This is also known as a Taylor polynomial, from which the name of the function, Polynomial & TAYLor, follow.

For example, 'X^3-2*X+2'[ENTER] [2] [ENTER] [PTAYL] produces 'X^3+6*X^2+10*X+6'. In actuality, you should interpret this result to mean '(X-2) ^3+6*(X-2) ^2+10*(X-2) +6'. Let's check by using the substitution: 'X = x - 2' [↪][ALG][SUBST] [EXPAN]. We recover the original polynomial, but in terms of lower-case x rather than upper-case x.

QUOTIENT and REMAINDER

The functions QUOTIENT [QUOT] and REMAINDER [REMAI] provide, respectively, the quotient Q(X) and the remainder R(X), resulting from dividing two polynomials, $P_1(X)$ and $P_2(X)$. In other words,, they provide the values of Q(X) and R(X) from $P_1(X)/P_2(X) = Q(X) + R(X)/P_2(X)$. For example,

'X^3-2*X+2'[ENTER] 'X-1' [ENTER] [QUOT] results in 'X^2+X-1'
'X^3-2*X+2'[ENTER] 'X-1' [ENTER] [REMAI] results in 1.

Thus, we can write: $(X^3-2X+2)/(X-1) = X^2+X-1 + 1/(X-1)$.

Note: you could get the latter result by using PARTFRAC:
'(X^3-2*X+2)/(X-1)' [ENTER] [PARTF] results in 'X^2+X-1 + 1/(X-1)'.

The following functions are not accessible through the POLY menu:

EPSX0 and the CAS variable EPS

The variable ε (epsilon) is typically used in mathematical textbooks to represent a very small number. The HP 49 G CAS will create a variable EPS, with default value $0.0000000001 = 10^{-10}$, when you use the EPSX0 function. You can change this value, once created, if you prefer a different value for EPS. The function EPSX0, when applied to a polynomial in stack level 1, will replace all coefficients whose absolute value is less than EPS with a zero. Examples:

'X^3-1.2E-12*X^2+1.2E-6*X+6.2E-11'[CAT][ALPHA][E]. Use the up and down arrow keys to find EPSX0, then press [OK]. The result is 'X^3-0*X^2+.0000012*X+0'. Use [→][ALG][EXPAN] to obtain 'X^3+.0000012*X+0'.

PEVAL

The functions PEVAL (Polynomial EVALuation) can be used to evaluate a polynomial

$$p(x) = a_n \cdot x^n + a_{n-1} \cdot x^{n-1} + \ldots + a_2 \cdot x^2 + a_1 \cdot x + a_0,$$

given an array of coefficients $[a_n\ a_{n-1}\ \ldots\ a_2\ a_1\ a_0]$ and a value of x_0, placed in stack levels 2 and 1, respectively. The result is the evaluation $p(x_0)$. Examples:

[1. 5. 6. 1][ENTER][5][ENTER][CAT][ALPHA][P]. Use up and down arrows to find PEVAL, press [OK]. The result is 281.

TCHEBYCHEFF

Given an integer number, n>0, in stack level 1, the function TCHEBYCHEFF generates the Tchebycheff (or Chebyshev) polynomial of the first kind, order n, defined as

$$T_n(X) = \cos(n \cdot \arccos(X)).$$

If the integer n is negative (n < 0), the function TCHEBYCHEFF generates the Tchebycheff polynomial of the second kind, order n, defined as

$$T_n(X) = \sin(n \cdot \arccos(X))/\sin(\arccos(X)).$$

This function is not available in the POLY menu. To invoke it you have the use the catalog, i.e., [CAT][ALPHA][T], then use the down-arrow key [▼] nine times until the command TCHEBYCHEFF is highlighted. Press [OK] to activate the command. Examples:

[5][CAT][ALPHA][T] [▼][▼][▼][▼][▼][▼][▼][▼][▼] [OK] produces '16*X^5-20*x^3+5*X'

At this point the TCHEBYCHEFF command is readily available in the catalog. So, for the next example, use:

[5][+/-][CAT][OK] to obtain '16*X^4-12*X^2+1'.

9 Vectors

In the HP 49 G calculator a vector is simply a sequence of numerical or algebraic objects, of any size, enclosed by square brackets. If using the RPN mode, the objects are separated by spaces. In algebraic mode, objects are separated by commas. Examples of vectors are:

[1 5 -3], ['a' 'a^2' 'a^3'], ['x' 5 7 3 2 1], [0 1 1 2 3 5 8 13 21].

HP 49 G vectors of 2 and 3 dimensions can be used to represent Cartesian vectors in 2 and 3 dimensions if the RECT coordinate system is selected.

Entering vectors

Vectors can be entered by typing it directly into the stack, by using the matrix writer, or by creating a vector with the function [-> ARRY] if the elements are listed in the stack. The details of the different approaches to entering vectors are presented below.

Typing a vector directly into the stack

Vectors can be entered directly into the stack by activating the square bracket symbol in the keyboard, [↰][[ ]], and entering terms separated with spaces [SPC]. If the terms you enter are algebraic expressions, even a single non-numeric character, these expressions must be typed between single quotes. In other words, before typing an algebraic expression as element of a vector, enter [↱][']. To separate algebraics, you may want to use [▶] to move the cursor outside of the algebraic before pressing [SPC]. Try the following examples.

[↰][[ ]] [1][SPC][5][SPC][3][+/-][ENTER] to obtain [1 5 -3];

[↰][[ ]] [↱]['] [ALPHA][↰][A] [▶] [SPC] [↱]['] [ALPHA][↰][A] [y^x][2] [▶] [SPC] [↱]['][ALPHA][↰][A] [y^x][3] [ENTER] to obtain ['a' 'a^2' 'a^3'];

Using the matrix writer (MTRW)

Vectors can also be entered by using the matrix writer [↰][MTRW]. This command generates a species of spreadsheet corresponding to rows and columns of a matrix (see chapter 10). For a vector we are interested in filling only elements in the top row. By default, the cell in the top row and first column is selected. At the bottom of the spreadsheet you will find the following soft menu keys:

[EDIT][VEC■][←WID][WID→][GO→■][GO↓]

⬇ The [EDIT] key is used to edit the contents of a selected cell in the matrix writer.

⬇ The [VEC] key, when selected, will produce a vector, as opposite to a matrix of one row and many columns.

Vectors vs. matrices

> To see the [VEC] key in action, try the following exercises:
>
> (1) With [VEC■] selected, enter [3][SPC][5][SPC][2][ENTER][ENTER]. This produces [3. 5. 2.].
>
> (2) With [VEC] deselected, enter [3][SPC][5][SPC][2][ENTER][ENTER]. This produces [[3. 5. 2.]].
>
> Although these two results differ only in the number of brackets used, for the calculator they represent mathematical animals of different species. The first one is a vector with three elements, and the second one a matrix with one row and three columns. There are differences in the way that mathematical operations take place on a vector as opposite to a matrix. Therefore, for the time being, keep the soft menu key [VEC■] selected while using the matrix writer.

⬇ The [←WID] key is used to decrease the width of the columns in the spreadsheet. Press this key a couple of times to see the column width decrease in your matrix writer.

⬇ The [WID→] key is used to increase the width of the columns in the spreadsheet. Press this key a couple of times to see the column width increase in your matrix writer.

⬇ The [GO→] key, when selected, automatically selects the next cell to the right of the current cell when you press [ENTER]. This option is selected by default.

⬇ The [GO↓] key, when selected, automatically selects the next cell below the current cell when you press [ENTER].

Moving to the right vs. moving down in the matrix writer

> Activate the matrix writer and enter [3][SPC][5][SPC][2][ENTER][ENTER] with the [GO→■] key selected (default). Next, enter the same sequence of numbers with the [GO↓■] key selected to see the difference. In the first case you entered a vector of three elements. In the second case you entered a matrix of three rows and one column.

Activate the matrix writer again by using [↰][MTRW], and press [NXT] to check out the second soft key menu at the bottom of the display. It will show the keys:

[+ROW][-ROW][+COL][-COL][→STK][GOTO].

- The [+ROW] key will add a row full of zeros at the location of the selected cell of the spreadsheet.

- The [-ROW] key will delete the row corresponding to the selected cell of the spreadsheet.

- The [+COL] key will add a column full of zeros at the location of the selected cell of the spreadsheet.

- The [-COL] key will delete the column corresponding to the selected cell of the spreadsheet.

- The [→STK] key will place the contents of the selected cell on the stack.

- The [GOTO] key, when pressed, will request that the user indicate the number of the row and column where he or she wants to position the cursor.

Pressing [NXT] once more produces the last menu, which contains only one function [DEL] (delete).

- The function [DEL] will delete the contents of the selected cell and replace it with a zero.

To see these keys in action try the following exercise:

(1) Activate the matrix writer by using [←][MTRW]. Make sure the [VEC■] and [GO→■] keys are selected.

(2) Enter the following:

[1][SPC][2][SPC][3][ENTER]
[NXT][GOTO][2][SPC][1][OK][OK]
[4][SPC][5][SPC][6][ENTER]
[7][SPC][8][SPC][9][ENTER]

(3) Move the cursor up two positions by using [▲][▲]. Then press [-ROW]. The second row will disappear.

(4) Press [+ROW]. A row of three zeroes appears in the second row.

(5) Press [-COL]. The first column will disappear.

(6) Press [+COL]. A row of two zeroes appears in the first row.

(7) Press [GOTO][3][SPC][3][OK][OK] to move to position (3,3).

(8) Press [→STK]. This will place the contents of cell (3,3) on the stack, although you will not be able to see it yet.

(9) Press [NXT][DEL]. This should place a zero in location (3,3), however, with ROM 1.17b it does not seem to be working.

In summary, to enter a vector using the matrix writer, simply activate the writer ([←][MTRW]), and place the elements of the vector, separated by spaces [SPC] on the stack, then press [ENTER][ENTER]. Make sure that the [VEC■] and [GO→■] keys are selected.

Example:
[←][MTRW] [→][`] [ALPHA][←][X] [y^x][2] [▶] [SPC] [2] [SPC] [5][+/-][ENTER] [ENTER] produces:

['x^2' 2 -5]

Building a vector with →ARRY

The function [→ARRY], available in the menu [←][PRG][TYPE], can be used to build an array in the following way:

(1) Enter the n elements of the array in the order you want them to appear in the array (when read from left to right) into the stack.

(2) Enter n as the last entry.

(3) Use the keystroke sequence: [←][PRG][TYPE][→ARRY].

The function [→ARRY] (create an ARRaY - see below) takes the objects from stack levels n+1, n, n-1, ..., down to stack levels 3 and 2, and converts them into a vector of n elements. The object originally at stack level n+1 becomes the first element, the object originally at level n becomes the second element, and so on.

For example:

[5][ENTER][3][+/-][ENTER][→][`][ALPHA][←][M][ENTER][3][ENTER] [←][PRG][TYPE][→ARRY], produces the array:

[5 -3 'm'].

Identifying, extracting, and inserting elements of a vector

If you store a vector into a variable name, say A, you can identify elements of the vector by using A(i), where i is an integer number less than or equal to the vector size. For example, create the following array and store it in variable A: [-1. -2. -3. -4. -5.]:

[←][[ ]] [1] [+/-] [SPC][2] [+/-] [SPC][3] [+/-] [SPC][4] [+/-] [SPC][5] [+/-] [ENTER]
[→][`][ALPHA][A][ENTER] [STO▶]

To recall the third element of A, for example, you could use:

[VAR] [→][`] [A   ] [←][( )] [3] [ENTER] [→][EVAL] to recall the value A(3) to the stack. Result: -3.

You can operate with elements of the array by writing and evaluating algebraic expressions such as:

'A(2)+A(5)' [→][EVAL]   Result: -7.
'A(1)-A(4)' [→][EVAL]   Result: 3.
'A(3)*A(2)' [→][EVAL]   Result: 6.
'A(3)/A(5)' [→][EVAL]   Result: .6
'A(5)' [ENTER] [→][LN] [→][EVAL]   Result: (1.6094, 3.1416).
'A(3)' [ENTER] [←][e^x] [→][EVAL]   Result: 4.9787E-2
'√ (A(2)^2+A(4)^2)' [→][EVAL]   Result: 4.4721
Etcetera.

More complicated expressions involving elements of A can also be written, for example:]

[EQW][→][Σ] [ALPHA][←][J][▶][1][▶][5][▶][ALPHA][A][→][ALPHA][←][J][ENTER]
Result: 'Σ(j=1,5,A(j))'. [→][EVAL] Result: -15.

Note: On paper, you would write A(j) as A_j, and would refer to the latter result as

$$\sum_{j=1}^{5} A_j = -15 .$$

The vector A can also be referred to as an indexed variable because the name A represents not one, but many values identified by a sub-index.

You can change the value of an element of A, by storing a new value in that particular element. For example, if we want to change the contents of A(3) to read 4.5 instead of its current value of -3., use:

[4][.][5][ENTER][→]['][ALPHA][A][←] [( )] [3][ENTER][STO]

To verify that the change took place use: [→][A]. The result is now: [-1 -2 4.5 -4 -5].

If you have an array in the stack, you can extract a value from the array by entering an positive integer number indicating the position of the element you want to extract, and then using the function GET. This function is available by using [←][PRG][LIST][ELEM][GET]. For example, place array A in the stack, by using [→][A]. Then, enter [3][ENTER], and use [←][PRG][LIST][ELEM][GET]. The result is, of course, 4.5, the element we just stored in A(3). This operation, however, gets rid of the array in the stack.

To replace an element in an array located in the stack, enter the index of the element to be replaced and the value that will replace the element. Then, use the function PUT. This function is available through:
[←][PRG][LIST][ELEM][PUT]. For example, place array A in the stack, by using [VAR][→][A]. Then, enter [2][ENTER] [3][.][5][ENTER], and use [←][PRG][LIST][ELEM][PUT] to replace A(2) with 3.5. The result is the vector [-1. 3.5 4.5 -4. -5.].

To find the length of a vector you can use the function SIZE, available through [←][PRG][LIST][ELEM][SIZE]. For example, for vector A, use: [VAR][→][A], and then through [←][PRG][LIST][ELEM][SIZE]. The result is { 5. }.

To determine the position of the first occurrence of an element in an array enter the array in the stack followed by the element of interest, the press through [↩][PRG][LIST][ELEM][POS]. For example, using A, try the following: [VAR][→][A][2][+/-][ENTER] [↩][PRG][LIST] [ELEM][SIZE]. The result is { 5. }.

Arrays, vectors, and matrices

So far we have used the term vector to refer to a sequence of objects (numbers and algebraics) enclosed between square brackets, and the terms matrix to refer to a collection of objects (numbers and algebraics) arranged in rows and columns. Array is a generic term for any arrangement of objects whose elements that can be identified by one, two, or more indices. The number of indices necessary to identify elements in an array is known as the dimension of the array. We can say, for example, that a vector is a one-dimensional array, while a matrix is a two-dimensional array. Arrays are sometimes known as indexed variables.

Simple operations with vectors

Addition and subtraction operations with vectors require that the vectors involved in the operation be of the same length. For example:

Example of addition: [-1. 2. 5. 'a'][ENTER][2. 'B' -5. 1.][ENTER][+] will produce [1. '2.+B' 0. 'a+1'].

Example of subtraction: Press [→][UNDO] to get back the two arrays listed earlier. Now, press [-]. The result is [-3. '2.-B' 10. 'a-1'].

Note: If you cannot see all the elements of an array on the stack, you can always launch the matrix writer (as a matrix editor), by pressing [▼]. Then you can use the left- and right-arrow keys ([◄], [►]) to move about the vector.

To change the sign of a vector, simply use [+/-]. For example, [-1. 2] [+/-] produces [1. -2.].

Multiplication or division of a vector by a scalar (a simple number), is straightforward. For example:

Multiplication by a scalar: [-2. 3. 5.][ENTER][5][×]   Results in [-10 15 25].
Division by a scalar: [12. 6. -24.][ENTER][6][÷]   Results in [2. 1. -4.]

The VECTR menu

Multiplication of two vectors, and other vector operations, require us to use the VECTR menu under MTH. To access this menu use: [↩][MTH][VECTR]. The functions available are the following:

[ABS][DOT][CROSS][V→][→V2][→V3]

Press [NXT] to show the second menu:

$$[\text{RECT}\blacksquare][\text{CYLIN}][\text{SPHER}][\quad][\quad][\text{MTH}]$$

Press [NXT] to return to the first menu. The operation of these functions is described below:

✦ The function ABS, also available through [←][ABS], applied to a vector $A = [a_1\ a_2\ ...\ a_N]$ produces the Euclidean magnitude, or absolute value, of the vector, i.e., $|A| = (a_1^2 + a_2^2 + ... + a_N^2)^{1/2}$. For example:
[2. -3. 5.][ABS] results in 6.1644.

✦ The function DOT calculates the dot product (or internal product) of two vectors $A = [a_1\ a_2\ ...\ a_N]$, and $B = [b_1\ b_2\ ...\ b_N]$, written as $A\bullet B = a_1 \cdot b_1 + a_2 \cdot b_2 + ... + a_N \cdot b_N$. Notice that the dot product of two vectors produces a scalar. Also, $A\bullet A = a_1^2 + a_2^2 + ... + a_N^2 = |A|^2$. Examples:

[2. 3. -5.][ENTER][-1. 7. 4.][ENTER][DOT] produces -1.
[2. -3. 5.][ENTER][ENTER][DOT] produces 38 = $|A|^2$. Thus, $|A| = \sqrt{38} = 6.1644$.

✦ The function CROSS is defined only for vectors with three components and represents the cross, or external, product of two vectors $A = [a_1\ a_2\ a_3]$, and $B = [b_1\ b_2\ b_3]$. The result can be written as:

$$A \times B = [\ b_3 \cdot a_2 - b_2 \cdot a_3 \quad a_3 \cdot b_1 - b_3 \cdot a_1 \quad b_2 \cdot a_1 - a_2 \cdot b_1\ ].$$

Example, [-2. 3. 1][ENTER][5. 3. -2.][ENTER][CROSS] results in [-9. 1. -21.].

Note:[←][MATRICES][NXT][VECTR] also provide access to [DOT] and [CROSS] products for vectors.

✦ The function [V→] decomposes a vector of size n in stack level 1, placing its elements, from left to right, in stack levels n, n-1, ..., 1. (This only works with arrays whose elements are all numbers). For example, [2. 3. -5. 7.][ENTER][V→], produces:

```
4:              2.
3:              3.
2:             -5.
1:              7.
```

✦ The function [→V2] takes the terms in stack levels 2 and 1, and creates a two-dimensional vector. For example: [5][+/-][ENTER][3][ENTER][→V2] produces [-5. 3].

✦ The function [→V3] takes the terms in stack levels 3, 2 and 1, and creates a three-dimensional vector. For example: [1][ENTER][7][+/-][ENTER][2][ENTER][→V2] produces [1. -7. 2.].

Press [NXT] to access the second menu. The functions available in the second menu are:

✦ [RECT■]: This is the default coordinate system, the rectangular or Cartesian coordinate system (x,y,z).
✦ [CYLIN]: This is the polar coordinate system (r, θ, z).
✦ [SPHER]: This is the spherical coordinate system (ρ, θ, ϕ).

The functions RECT, CYLIN, and SPHER, are used to select a coordinate system for two- and three-dimensional vectors. These functions were introduced in Chapter 3 (Changing the angle mode and coordinate system). We will have more to say about using these functions in later sections.

Vectors as physical quantities

A vector in two- or three-dimensions represents a directed segment, i.e., a mathematical object characterized by a magnitude and a direction in the plane or in space. Vectors in the plane or in space can be used to represent physical quantities such as the position, displacement, velocity, and acceleration of a particle, angular velocity and acceleration or a rotating body, forces and moments.

Operations with vectors

Vectors in two- or three-dimensions are represented by arrows. The length of the arrow represents the magnitude of the vector, and the arrowhead indicates the direction of the vector. The figure below shows a vector A, and its negative -A. As you can see, the negative of a vector is a vector of the same magnitude, but with opposite sense.

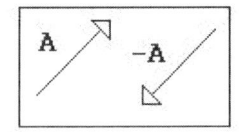

Vectors can be added and subtracted. To illustrate vector addition and subtraction refer to the figure below. Consider two vectors in the plane or in space, A and B, as shown in the figure below, item (a). There are two ways that you can construct the vector sum, A+B: (1) By attaching the origin of the vector B to the tip of vector A, as shown in the figure, item (b). In this case, the vector sum is the vector extending from the origin of vector A to the tip of vector B. (2) By showing the vectors A and B with a common origin, and completing the parallelogram resulting from drawing lines parallel to vectors A and B at the tips of vectors B and A, respectively, as shown in the figure below, item (c). The vector sum, also known as the resultant, is the diagonal of the parallelogram that starts at the common origin of vectors A and B.

Subtraction is accomplished by using the definition:

$$A - B = A + (-B).$$

In other words, subtracting vector B from vector A is equivalent to adding vectors A and (-B),. This operation is illustrated in the figure below, item (d).

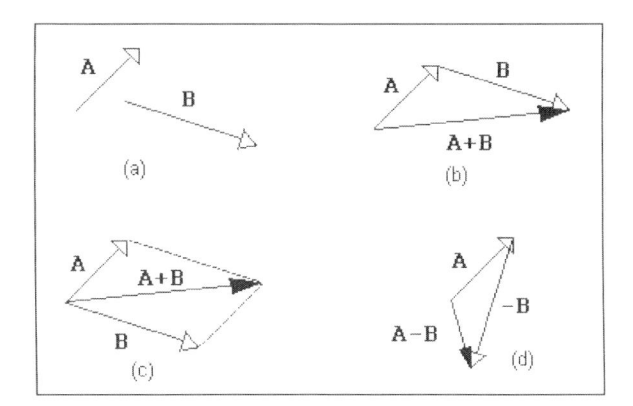

A vector A can be <u>multiplied by a scalar</u> c, resulting in a vector c·A, parallel and oriented in the same direction as A, and whose magnitude is c times that of A. The figure below illustrates the case in which c = 2. If the scalar c is negative, then the orientation of the vector c·A will be opposite that of A.

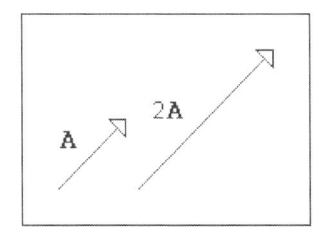

The <u>magnitude</u> of a vector A is represented by |A|. A <u>unit vector</u> in the direction of A is a vector of magnitude 1 parallel to A. The unit vector corresponding to a vector A is shown in the figure below, item (a). The unit vector along the direction of A is defined by

$$\mathbf{e}_A = \frac{\mathbf{A}}{|\mathbf{A}|}.$$

By definition, $|e_A|$ = 1.0.

The <u>projection of vector B onto vector A</u>, $P_{B/A}$, is shown in the figure below, item (b). If θ represents the angle between the two vectors, we see from the figure that $P_{B/A}$ = |B|·cos θ = $|e_A|$·|B|·cos θ, or

$$P_{B/A} = \frac{|\mathbf{A}| \cdot |\mathbf{B}| \cdot \cos\theta}{|\mathbf{A}|}.$$

We will define the <u>dot product, or internal product</u>, of two vectors A and B as the scalar quantity

$$A \bullet B = |A| \cdot |B| \cdot \cos \theta,$$

where θ is the angle between the vectors when they have a common origin. Notice that A•B = B•A.

$$\theta = \arccos\left(\frac{\mathbf{A} \cdot \mathbf{B}}{|\mathbf{A}| \cdot |\mathbf{B}|}\right)$$

From the definition of the dot product it follows that the <u>angle between two vectors</u> can be found from

From the same definition it follows that if the angle between the vectors A and B is $\theta = 90° = \pi/2$ rad, i.e., if A is <u>perpendicular (or normal)</u> to B (A⊥B), $\cos \theta = 0$, and A•B = 0. The reverse statement is also true, i.e., if A•B = 0, then A⊥B.

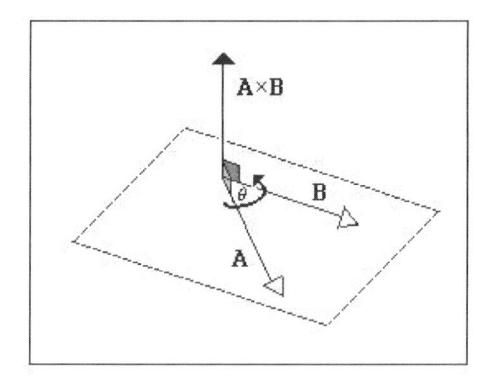

The <u>cross product, or vector product</u>, of two vectors in space is defined as the vector C = A×B, such that $|C| = |A| \cdot |B| \cdot \sin \theta$, where θ is the angle between the vectors, and A•C = 0 and B•C = 0 (i.e., C is perpendicular to both A and B). The cross product is illustrated in the figure below. Since there could be two orientations for a vector C perpendicular to both A and B, we need to refine the definition of C = A×B by indicating its orientation. The so-<u>called right-hand rule</u> indicates that if we were to curl the fingers of the right hand in the direction shown by the curved arrow in the figure (i.e., from A to B), the right hand thumb will point towards the orientation of C. Obviously, the order of the factors in a cross product affects the sign of the result, for the right-hand rule indicates that B×A = - C.

Three vectors A, B, and C, in space, having a common origin, determine a solid figure called a parallelepiped, as shown in the figure below. It can be proven that the <u>volume of the parallelepiped</u> is obtained through the expression A• (B×C). This expression is known as the <u>vector triple product</u>.

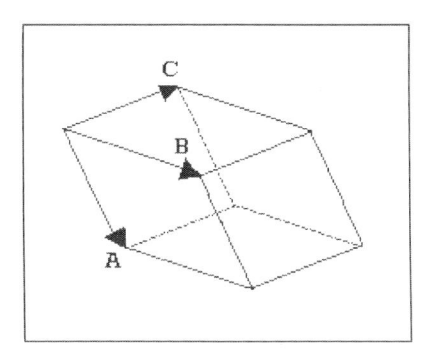

Vectors in Cartesian coordinates

The mathematical representation of a vector typically requires it to be referred to a specific coordinate system. Using a Cartesian coordinate system, we introduce the <u>unit vectors</u> i, j, and k, corresponding to the x-, y-, and z-directions, respectively. The unit vectors i, j, and k, are shown in the figure below. These unit vectors are such that

$$i \bullet j = i \bullet k = j \bullet k = 0,$$

and

$$i \times j = k, \quad j \times k = i, \quad k \times i = j, \quad i \times k = -j, \quad k \times j = -i, \quad j \times i = -k.$$

n terms of these unit vectors, any vector A can be written as

$$A = A_x \cdot i + A_y \cdot j + A_z \cdot k,$$

where the values A_x, A_y, and A_z, are called the <u>Cartesian components</u> of the vector A.

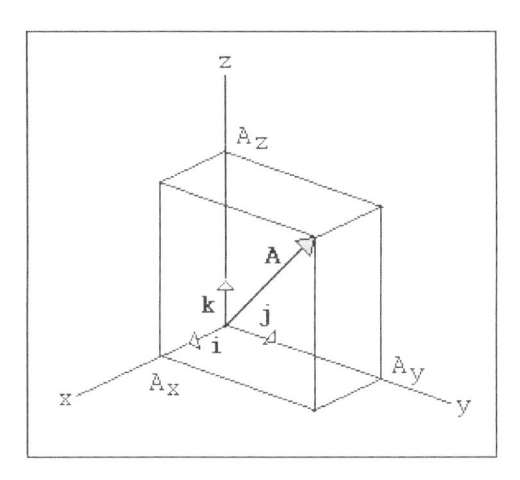

Using Cartesian components, therefore, we can also write $B = B_x \cdot i + B_y \cdot j + B_z \cdot k$, and define the following <u>vector operations:</u>

- negative of vector: $-A = -(A_x \cdot i + A_y \cdot j + A_z \cdot k) = -A_x \cdot i - A_y \cdot j - A_z \cdot k$.

- addition: $A+B = (A_x + B_x) \cdot i + (A_y + B_y) \cdot j + (A_z + B_z) \cdot k$.

- subtraction: $A-B = (A_x - B_x) \cdot i + (A_y - B_y) \cdot j + (A_z - B_z) \cdot k$.

- multiplication by a scalar, c: $c \cdot A = c \cdot (A_x \cdot i + A_y \cdot j + A_z \cdot k) = c \cdot A_x \cdot i + c \cdot A_y \cdot j + c \cdot A_z \cdot k$.

- dot product: $A \bullet B = (A_x \cdot i + A_y \cdot j + A_z \cdot k) \bullet (B_x \cdot i + B_y \cdot j + B_z \cdot k) = A_x \cdot B_x + A_y \cdot B_y + A_z \cdot B_z$.

- magnitude: $|A|^2 = A \bullet A = A_x^2 + A_y^2 + A_z^2$; $|A| = \sqrt{(A \bullet A)} = (A_x^2 + A_y^2 + A_z^2)^{1/2}$.

- unit vector: $e_A = A/|A| = (A_x \cdot i + A_y \cdot j + A_z \cdot k)/(A_x^2 + A_y^2 + A_z^2)^{1/2}$.

- cross product: $A \times B = (A_x \cdot i + A_y \cdot j + A_z \cdot k) \times (B_x \cdot i + B_y \cdot j + B_z \cdot k) =$
  $(B_z \cdot A_y - B_y \cdot A_z) \cdot i + (A_z \cdot B_x - B_z \cdot A_x) \cdot j + (B_y \cdot A_x - A_y \cdot B_x) \cdot k$.

Calculating 2×2 and 3×3 determinants

The calculation of a cross product can be simplified if the cross product is written as a determinant of a matrix of 3 rows and 3 columns (also referred to as a 3×3 matrix). A determinant is a number associated with a square matrix, i.e., a matrix with the same number of rows and columns. While there is a general rule to obtain the determinant of any square matrix, we concentrate our attention on 2×2 and 3×3 determinants. Next, we present a simple way to calculate the determinant for 2×2 and 3×3 matrices.

The elements of a matrix are identified with two sub-indices, the first representing the row and the second the column. Therefore, a 2×2 and a 3×3 matrix will be represented as:

$$\begin{bmatrix} a_{11} & a_{12} \\ a_{21} & a_{22} \end{bmatrix}, \quad \begin{bmatrix} a_{11} & a_{12} & a_{13} \\ a_{21} & a_{22} & a_{23} \\ a_{31} & a_{32} & a_{33} \end{bmatrix}.$$

The determinants corresponding to these matrices are represented by the same arrangement of elements enclosed between vertical lines, i.e.,

$$\begin{vmatrix} a_{11} & a_{12} \\ a_{21} & a_{22} \end{vmatrix}, \quad \begin{vmatrix} a_{11} & a_{12} & a_{13} \\ a_{21} & a_{22} & a_{23} \\ a_{31} & a_{32} & a_{33} \end{vmatrix}.$$

A 2×2 determinant is calculated by multiplying the elements in its diagonal and adding those products accompanied by the positive or negative sign as indicated in the diagram shown below.

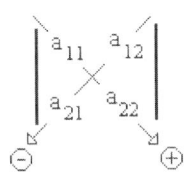

The 2×2 determinant is, therefore,

$$\begin{vmatrix} a_{11} & a_{12} \\ a_{21} & a_{22} \end{vmatrix} = a_{11} \cdot a_{22} - a_{12} \cdot a_{21}.$$

A 3×3 determinant is calculated by augmenting the determinant, an operation that consists on copying the first two columns of the determinant, and placing them to the right of column 3, as shown in the diagram below. The diagram also shows the elements to be multiplied with the corresponding sign to attach to their product, in a similar fashion as done earlier for a 2×2 determinant. After multiplication the results are added together to obtain the determinant.

Therefore, a 3×3 determinant produces the following result:

$$\begin{vmatrix} a_{11} & a_{12} & a_{13} \\ a_{21} & a_{22} & a_{23} \\ a_{31} & a_{32} & a_{33} \end{vmatrix} = a_{11} \cdot a_{22} \cdot a_{33} + a_{12} \cdot a_{23} \cdot a_{31} + a_{13} \cdot a_{21} \cdot a_{32}$$

$$- (a_{13} \cdot a_{22} \cdot a_{31} + a_{11} \cdot a_{23} \cdot a_{31} + a_{12} \cdot a_{21} \cdot a_{33}).$$

Determinants can be calculated in the HP 49 G calculator by using the keystroke sequence:

[↰][MTH][MATRX][NORM][NXT][DET],

or

[↵][MATRICES][OPER][DET].

For example, calculate the following 3×3 determinant:

[↵][[ ]]   Opens external set of brackets to indicate a matrix
[↵][[ ]] [3][SPC][5][SPC][2]   Enters first row of a 3×3 matrix
[▶] [↵][[ ]] [1][+/-][SPC][6][SPC][1]   Enters second row of matrix
[▶] [↵][[ ]] [4] [SPC][5][SPC][7][+/-]   Enters third row of matrix
[ENTER]   Enter matrix into the stack
[↵][MATRICES][OPER][DET]   Calculate the determinant

The result is -214.

Cross product as a determinant

The cross product A×B in Cartesian coordinates can be expressed as a determinant if the first row of the determinant consists of the unit vectors i, j, and k. The components of vector A and B constitute the second and third rows of the determinant, i.e.,

$$\mathbf{A} \times \mathbf{B} = \begin{vmatrix} \mathbf{i} & \mathbf{j} & \mathbf{k} \\ A_x & A_y & A_z \\ B_x & B_y & B_z \end{vmatrix}.$$

Evaluation of this determinant, as indicated above, will produce the result

$$A \times B = (B_z \cdot A_y - B_y \cdot A_z) \cdot i + (A_z \cdot B_x - B_z \cdot A_x) \cdot j + (B_y \cdot A_x - A_y \cdot B_x) \cdot k.$$

Operations with 2- and 3-dimensional vectors in the HP 49 G calculator

In this section we present examples of operations with 2- and 3-dimensional vectors using the features of the HP 49 G calculator. The examples are taken from applications in different physical sciences.

Example 1 - Position vector

The position vector of a particle is a vector that starts at the origin of a system of coordinates and ends at the particle's position. If the current position of the particle is P(x,y,z), then the position vector is

$$r = x \cdot i + y \cdot j + z \cdot k,$$

as illustrated in the figure below.

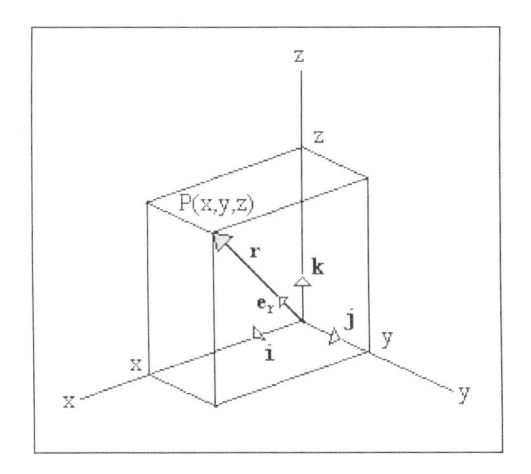

If a particle is located at point P(3, -2, 5), the position vector for this particle is r = 3i -2j+5k. In the HP 49 G calculator this position vector is written as [3 -2 5]. Press [ENTER] twice to keep two extra copies of the vector in the stack. To determine the magnitude of the position vector use [←][ABS]. The result is '√38', if the calculator is in the Exact mode. To find a numerical result use [→][→NUM]. This will produce a value of 6.164414. You can write |r| = 6.164414. You should have a copy of the position vector in stack level 2. With the magnitude in stack level 1, pressing [÷] will produce a unit vector in the direction of r, i.e., e_r = r/|r| = [0.4866 -0.3244 0.81111].

Example 2 - Center of mass of a system of discrete particles

Consider a system of discrete particles of mass m_i, located at position P_i (x_i, y_i, z_i), with i = 1, 2, 3, ..., n. We can write position vectors for each particle as $r_i = x_i \cdot i + y_i \cdot j + z_i \cdot k$. The center of mass of the system of particles will be located at a position r_{cm} defined by

$$\mathbf{r}_{cm} = \frac{\sum_{i=1}^{n} m_i \cdot \mathbf{r}_i}{\sum_{i=1}^{n} m_i}.$$

A system of four particles is illustrated in the figure below.

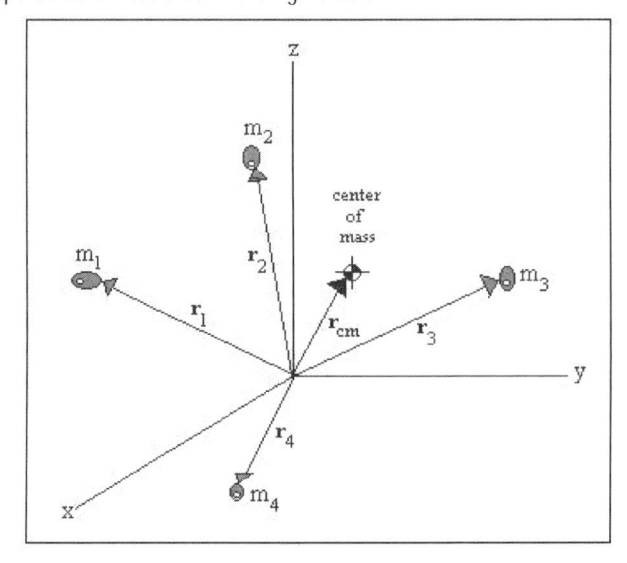

The following table shows the coordinates and masses of 5 particles. Determine their center of mass.

i	x_i	y_i	z_i	m_i
1	2	3	5	12
2	-1	6	4	15
3	3	-1	2	25
4	5	4	-7	10
5	5	3	2	30

First, enter a list of the masses {12 15 25 19 30} [ENTER][ENTER]. You will have two copies in the stack. Next, enter the five position vectors in a list: { [2 3 5] [-1 6 4] [3 -1 2] [5 4 -7] [5 3 2] } [ENTER]. User [×] [←][MTH][LIST][ΣLIST] to calculate Σ $m_i \cdot r_i$, i.e., the numerator in the formula for r_{cm}. The result is [284 231 160]. Press [▶] to swap stack levels 1 and 2. You should now have the list of masses in stack level 1. User [←][MTH][LIST][ΣLIST] to calculate Σ m_i, i.e., the denominator in the formula for r_{cm}. Finally, press [÷] to get r_{cm} = ['71/23' '231/92' '40/23'], in exact mode. Use [→][→NUM] to get the value, r_{cm} = [3.0869 2.5108 1.7391].

Example 3 - Resultant of forces

The figure below shows a vertical pole buried in the ground and supported by four cables EA, EB, EC, and ED. The magnitude of the tensions in each of those cables, as shown in the figure, are T_1 = 150 N, T_2 = 300 N, T_3 = 200 N, and T_4 = 150 N.

To find the vector resultant of all those forces you need to add the four tensions written out as vectors. The tension in cable i will be given by $\vec{T}_i = T_i \cdot e_i$, where e_i is a unit vector in the direction of the cable where the tension acts. (The tension vectors act so that the cables pull away from point E, where the cables are attached to the pole.)

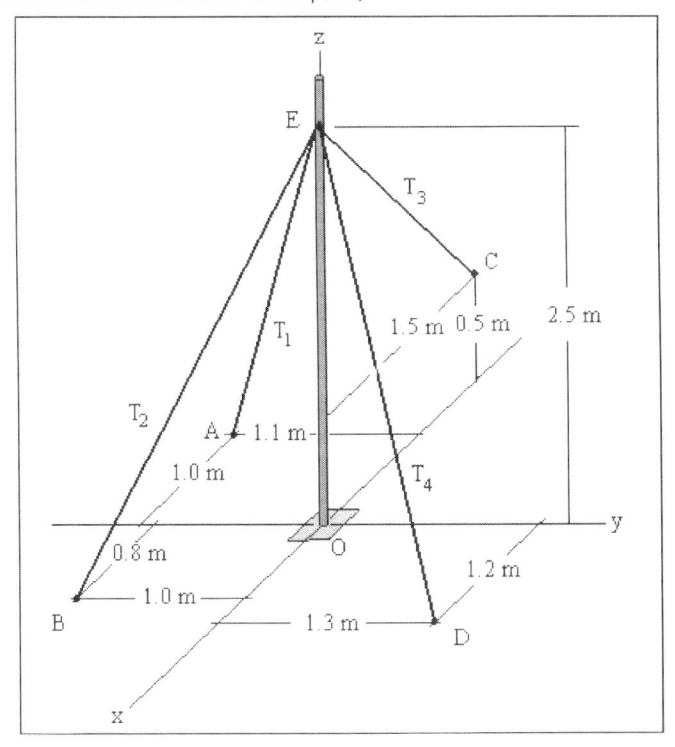

Writing out tension T_1

Tension T_1 acts along cable EA. To determine a unit vector along cable EA, you need to write the vector $r_{EA} = r_A - r_E$, where r_A is the position vector of point A (the tip of the vector r_{EA}), and r_A is the position vector of point A (the origin of the vector r_{EA}). The coordinates of these points, obtained from the figure, are A(-1.0 m, -1.1 m, 0) and E(0, 0, 2.5 m). Therefore, we can write r_E = (2.5k) m, and r_A = (-i -1.1j) m.

In the HP 49 G calculator we will enter r_E and r_A in this order: [-1 -1.1 0] [ENTER] [0 0 2.5] [ENTER]. To obtain, $r_{EA} = r_A - r_E$, just press [-], the result is [-1 -1.1 -2.5], or , r_{EA} = -i -1.1j-2.5k. To find the unit vector along which T_1 acts (i.e., along cable EA), press [ENTER] to get a second copy of the vector r_{EA}. Then, use [↰][ABS] to obtain the magnitude of this vector. ($|r_{EA}|$ = 2.909). Finally, press [÷] to get the unit vector, $e_{EA} = r_{EA}/|r_{EA}|$ = [-0.344 - 0.378 -0.860] = -0.344i - 0.378j -0.860k. (To verify that this is indeed a unit vector, press [ENTER] to keep an additional copy of the vector, then press [↰][ABS]. You should get as a result 1.000. Press [⇦] to drop this result from the stack.)

With the unit vector e_{EA} = [0.344 0.378 0.860] in stack level 1, enter the magnitude of T_1, i.e., T_1 = 150 N. [1][5][0][ENTER], and press [×] to multiply it with the unit vector.

The result is the vector [-51.57 -56.73 -128.93], or T_1 = (-51.57i -56.73j -128.93k) N.

Writing the vector that joins two points in space

The vector that joints points A and B, where $A(x_A, y_A, z_A)$ is the origin and point $B(x_B, y_B, z_B)$ the tip of the vector, is written as

$$r_{AB} = r_B - r_A = (x_B - x_A) \cdot i + (y_B - y_A) \cdot j + (z_B - z_A) \cdot k,$$

where $r_A = x_A \cdot i + y_A \cdot j + z_A \cdot k$ is the position vector of point A, and $r_B = x_B \cdot i + y_B \cdot j + z_B \cdot k$ is the position vector of point B. This operation is illustrated in the figure below.

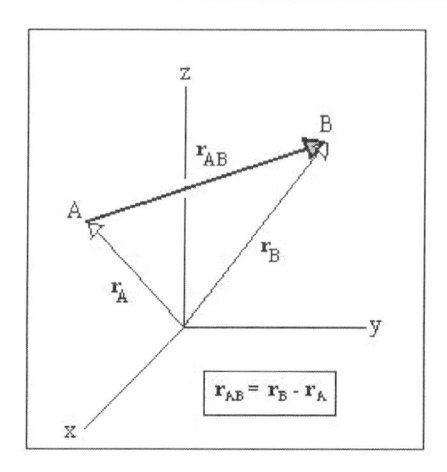

Relative position vector

The previous operation can also be interpreted as obtaining the relative position vector of point $B(x_B, y_B, z_B)$ with respect to point $A(x_A, y_A, z_A)$, and written as

$$r_{B/A} = r_B - r_A = (x_B - x_A) \cdot i + (y_B - y_A) \cdot j + (z_B - z_A) \cdot k.$$

Writing out tensions T_2, T_3, and T_4

Having developed a procedure to determine the unit vector along any of the cables, we can proceed to write out the vectors representing the tensions T_2, T_3, and T_4. First, we determine the coordinates of relevant points from the figure describing the problem. These points are: A(-1.0 m, -1.1m, 0) B(0.8 m, -1.0 m, 0.0), C(-1.5 m, 0.0, 0.5 m), D(1.2 m, 1.3m, 0.0), and E(0, 0, 2.5 m). The tensions, as vectors, will be written as $T_2 = T_2 \cdot e_{EB}$, $T_3 = T_3 \cdot e_{EC}$, and $T_4 = T_4 \cdot e_{ED}$. (Recall that T_2 = 300 N, T_3 = 200 N, and T_4 = 150 N.). Thus, we need to write out the vectors r_{EB}, r_{EC}, and r_{ED}, find their magnitudes, $|r_{EB}|$, $|r_{EC}|$, and $|r_{ED}|$, and calculate the unit vectors e_{EB} = $r_{EB}/|r_{EB}|$, $e_{EC} = r_{EC}/|r_{EC}|$, and $e_{ED} = r_{ED}/|r_{ED}|$, in order to obtain T_2, T_3, and T_4. Calculating each tension at a time, for T_2, along cable EB, we will follow the following steps:

[0.8 -1.0 0.0] [ENTER]   Enter coordinates of point B as position vector r_B
[0.0 0.0 2.5] [ENTER]   Enter coordinates of point E as position vector r_E
[-] [ENTER]   Calculate $r_{EB} = r_B - r_E$, and keep an extra copy.
[↰][ABS]   Calculates the magnitude $|r_{EB}|$.
[÷]   Calculates the unit vector e_{EB}.

[3][0][0][ENTER] [×]   Enter the magnitude T$_2$ and calculate T$_2$.

The result is the vector [85.44 -106.80 -267.01], or T$_2$ = (85.44i -106.80j -267.01k) N. The reader is invited to follow this procedure to obtain for cable EC, T$_3$ = (-120i -160k) N, and for cable ED, T$_4$ = (58.77i +63.67j -122.44k) N.

By the time you finish calculating the four tension vectors, your calculator's stack should look like this:

```
4:   [ -51.57 -56.73 -1...
3:   [ 85.44 -106.80 -2...
2:   [ -120.00 0.00 -16...
1:   [ 58.77 63.67 -122...
```

The stack holds tensions T$_1$, T$_2$, T$_3$, and T$_4$, in stack levels 4, 3, 2, and 1, respectively. To find the resultant of these four forces, namely, R = T$_1$ + T$_2$ + T$_3$ + T$_4$, we just need to press the [+] key three times to get
[-27.36 -99.86 -678.38], or R = (-27.36i -99.86j -678.38k) N.

Example 4 - Equation of a plane in space

The equation of a plane in space, in Cartesian coordinates, can be obtained given a point on the plane A(x$_A$,y$_A$,z$_A$), and a vector normal to the plane n = n$_x$ i+n$_y$ j+n$_z$ k. Let P(x,y,z) be a generic point on the plane of interest. We can form a relative position vector r$_{P/A}$ = r$_P$ - r$_A$ = (x-x$_A$)·i+(y-y$_A$)·j+(z-z$_A$)·k, which is contained in the plane, as shown in the figure below.

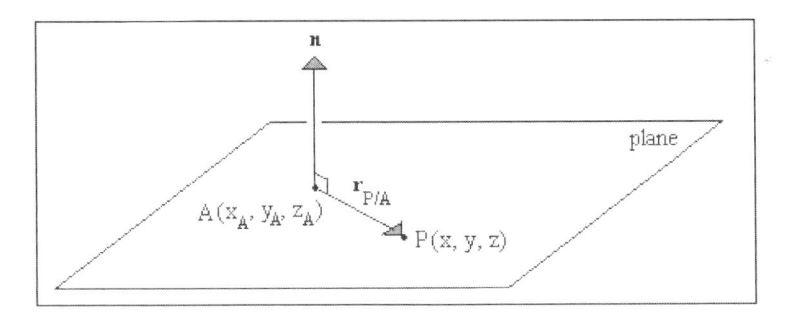

Because the vectors n and r$_{P/A}$ are perpendicular to each other, then, we can write n•r$_{P/A}$ = 0, or

$$(n_x i+n_y j+n_z k) \bullet ((x-x_A) \cdot i+(y-y_A) \cdot j+(z-z_A) \cdot k) = 0,$$

which results in the equation

$$n_x \cdot (x-x_A) + n_y \cdot (y-y_A) + n_z \cdot (z-z_A) = 0.$$

This result can be obtained using the HP 49 G calculator as follows:

[←][[ ]]   Open square brackets for vector n
[→]['] [ALPHA][←][N] [ALPHA][←][X] [▶] [SPC]   Enter 'nx'_
[→]['] [ALPHA][←][N] [ALPHA][←][Y] [▶] [SPC]   Enter 'ny'_

[→]['] [ALPHA][↵][N] [ALPHA][↵][Z] [ENTER]	Enter 'nz'_ Vector n is shown in stack
[↵][[ ]] [→]['] [ALPHA][↵][X] [▶] [SPC] [→]['] [ALPHA][↵][Y] [▶] [SPC] [→]['] [ALPHA][↵][Z] [ENTER]	Open square brackets for vector r_P Enter 'x'_ Enter 'y'_ Enter 'z'_ Vector r_P is shown in stack
[↵][[ ]] [→]['] [ALPHA][↵][X] [ALPHA][A] [▶] [SPC] [→]['] [ALPHA][↵][Y] [ALPHA][A] [▶] [SPC] [→]['] [ALPHA][↵][Z] [ALPHA][A] [ENTER]	Open square brackets for vector r_A Enter 'xA'_ Enter 'yA'_ Enter 'zA'_ Vector r_A is shown in stack
[-] [↵][MTH][VECTR][DOT]	Calculates $r_{P/A} = r_P - r_A$ Calculates $n \cdot r_{P/A}$

Let's try to find the equation of a plane if the normal vector is n = 3i + 2j −k, and the plane includes the point A(5, 0, -1). Using the HP 49 G calculator this is accomplished as follows:

[3 2 -1] [ENTER]	Enter n = 3i + 2j −k
['x' 'y' 'z'] [ENTER]	Enter generic point P(x,y,z) as a vector
[5 0 -1] [ENTER]	Enter point A(5, 0, -1) as a vector
[-]	Calculate $r_{P/A}$
[↵][MTH][VECTR][DOT]	Calculate $n \cdot r_{P/A}$

which results in '-1*(z+1)+2*y+3*(x-5)'. Use [→][ALG][EXPA] to obtain '3*x+(2*y-(z+16))'.

> The equation of the plane, of course, can be expanded, by hand, to read: 3x + 2y − z − 16 = 0.

To verify that the point A(5, 0, -1) satisfies the equation, first, press [ENTER] to keep a second copy of the expression available for future use. Then, enter the list

and use
{'x' 5 'y' 0 'z' -1}[ENTER],

[→][|][ENTER]

to replace the values in the expression. The result is ' 3*5+(2*0-(-1+16))'. Press [→][EVAL] to get 0.

Example 5 - Moment of a force

The figure below shows a force F acting on a point P in a rigid body. Suppose that the body is allowed to rotate about point O. Let r be the position vector of point P with respect to the point of rotation O, which we make coincide with the origin of our Cartesian coordinate system. The moment of the force F about point O is defined as M = r×F.

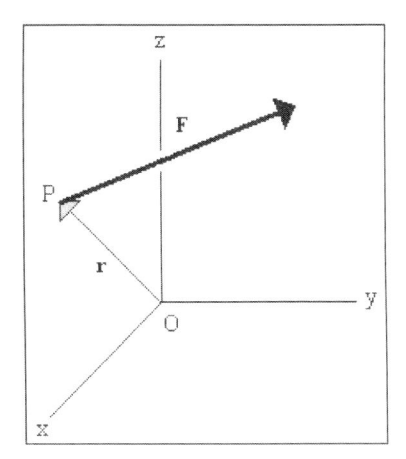

Referring to the figure of Example 3, if we want to calculate the moment of tension $T_1 = (-51.57i -56.73j -128.93k)$ N, about point O, we will use as the position vector r of the force's line of action through the pole, the vector $r_{OE} = (2.5k)$ m. The moment $M_1 = r_{OE} \times T_1$, can be calculated as follows:

[0. 0. 2.5][ENTER]   Enter r_{OE}
[-57.51 -56.73 -128.93][ENTER] Enter T_1
[↵][MTH][VECTR][CROSS]   Calculate $M_1 = r_{OE} \times T_1$

The result is [141.825 -143.775 0], or $M_1 = (141.825 i -143.775 j)$ m·N.

To find the magnitude of the moment, use [↵][ABS], thus, $|M_1| = M_1 = 201.954$ m·N.

> Note: Moments are vector quantities and obey all rules of vectors, i.e., they can be added, subtracted, multiplied by a scalar, undergo internal and external vector products. As an exercise, the reader may want to calculate the moments corresponding to the other tensions in Example 3, as well as the resultant moment from all four tensions.

Example 6 - Cartesian and polar representations of vectors in the x-y plane

A position vector in the x-y plane can be written simply as $r = x \cdot i + y \cdot j$. Let its magnitude be $r = |r|$. A unit vector along the direction of r is given by $e_r = r/r = (x/r) \cdot i + (y/r) \cdot j$. If we use polar coordinates, we recognize

$$x/r = \cos \theta, \text{ and } y/r = \sin \theta,$$

thus, we can write the unit vector as

$$e_r = r/r = \cos \theta \cdot i + \sin \theta \cdot j.$$

Thus, if we are given the magnitude, r, and the direction, q, of a vector (i.e., its polar coordinates (r,θ)), we can easily put together the vector as

$$r = r \cdot e_r = r \cdot (\cos \theta \cdot i + \sin \theta \cdot j).$$

This result is illustrated in the figure below.

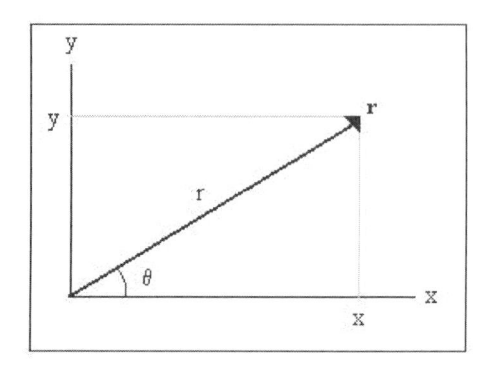

Polar to Cartesian

In the HP 49 G calculator entering the polar coordinate description of a vector is accomplished very easily by entering the vector [r ∠θ], where the symbol ∠ is obtained by using [ALPHA][→][6]. If the Cartesian (or rectangular) coordinate system is selected, then the calculator automatically produces the Cartesian representation of the vector. Make sure to check the top of the display to verify which angular units and coordinate system is currently selected. For example, for angular units set to degrees ([↰][PROG][NXT][MODES][ANGLE][DEG]), and coordinate systems set to rectangular ([↰][MTH][VECTR][NXT][RECT]), the top line of the display will start with DEG XYZ.

For example, given a vector with magnitude r = 6.0, and a direction q = 25°, enter:

[↰][[ ]]   Open square brackets for vector
[6][SPC]   Enter r, i.e., 6_
[ALPHA][→][6]   Enter the "angle" symbol, ∠
[2][5]   Enter angle value. Stack shows [6 ∠25].
[ENTER]   Converts vector to Cartesian components

The result is the vector [5.4378 2.5357], i.e., r = 5.4378·i+ 2.5357·j.

Cartesian to Polar

A quick way to obtain the magnitude and direction of a vector in the plane given its Cartesian components is to enter the vector as [x y], and use the conversion to polar coordinates: [↰][MTH][VECT][NXT] [CYLIN] to obtain the vector as [r ∠θ]. For example, to get the polar coordinates corresponding to r = 2·i+3·j, use
[2 3][ENTER] [↰][MTH][VECT][NXT][CYLIN], to get [3.61 ∠56.31], i.e., r = 3.61 and θ = 56.31°.

Example 7 - Planar motion of a rigid body

In the study of the planar motion of a rigid body the following equations are used to determine the velocity and acceleration of a point B (v_B, a_B) given the velocity and acceleration of a reference point A (v_A, a_A):

$$v_B = v_A + \omega_{AB} \times r_{B/A},$$

$$a_B = a_A + \alpha_{AB} \times r_{B/A} - \omega_{AB}^2 \cdot r_{B/A}.$$

The equations also use the angular velocity of the body connecting A and B, ω_{AB}, whose magnitude is ω_{AB}; the angular acceleration of the body connecting A and B, α_{AB}, and the relative position vector of point B with respect to point A, $r_{B/A}$. The following figure illustrates the calculation of relative velocity and acceleration in planar motion of a rigid body.

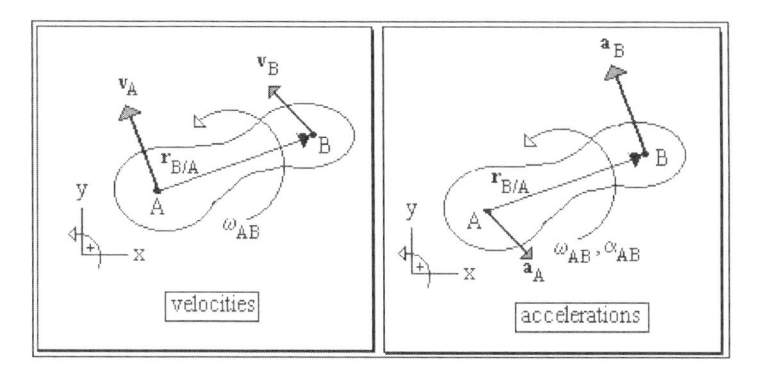

To present applications of this equation using the HP 49 G calculator, we use the data from the mechanism shown in the figure below.

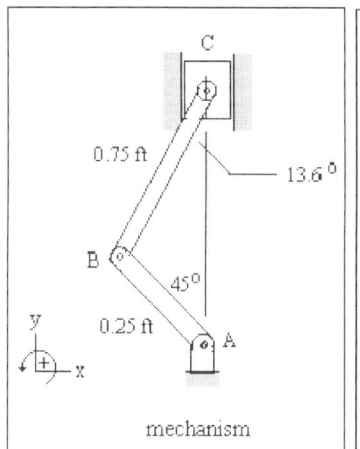

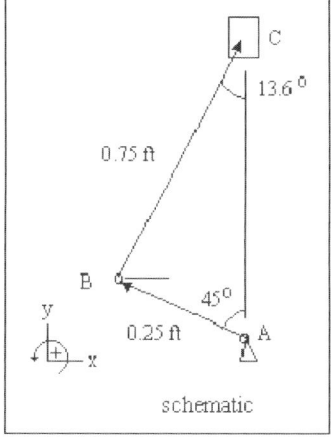

In this mechanism there are two bars AB and AC pin-connected at B. Bar AB is pin supported at A, and bar BC is attached through a pin to piston C. Piston C is allowed to move in the vertical direction only. At the instant shown the angular velocity and acceleration of bar AB are 10 rad/s and 20 rad/s^2 clockwise. You are asked to determine the angular velocity and acceleration of bar BC and the linear velocity and acceleration of piston C.

Angular velocity and acceleration

Angular velocities and accelerations in the x-y plane are represented as vectors in the z direction, i.e., normal to the x-y plane. These vectors are positive if the angular velocity or acceleration is counterclockwise. In general, thus we can write, $\omega_{AB} = \pm\omega_{AB} \cdot k$, and $\alpha_{AB} = \pm\alpha_{AB} \cdot k$. For the data in this problem we can write, therefore,

$$\omega_{AB} = (-10k) \text{ rad/s}, \alpha_{AB} = (-20k)\text{rad/s}^2, \omega_{BC} = (\omega_{BC} \cdot k) \text{ rad/s, and } \alpha_{BC} = (\alpha_{BC} \cdot k).$$

Relative position vector

The relative position vectors of interest in this problem are $r_{B/A}$ and $r_{C/B}$, which are obtained as follows:

- To obtain $r_{B/A}$, we can use HP 49 G vectors as follows (set coordinates to rectangular, angles to degrees): [←][[ ]] [.][2][5][SPC][ALPHA][→][6][1][3][5][ENTER]. The result will be displayed in rectangular coordinates as [-0.177 0.177]. Notice that for the xy coordinate system shown, the angle corresponding to vector $r_{B/A}$ is $\theta_{B/A} = 90° + 45° = 135°$. Thus, we can write

$$r_{B/A} = (-0.177i + 0.177j)\text{ft}.$$

- For $r_{C/B}$, the angle $\theta_{C/B} = 90° - 13.6° = 76.4°$. Thus, we can write: [←][[ ]] [.][7][5][SPC] [ALPHA][→][6][7][6][.][5][ENTER]. This result in the vector [0.175 0.729], or

$$r_{C/B} = (0.175i + 0.729j)\text{ft}.$$

Velocity

Since point A is fixed point, $v_A = 0$, and we can write $v_B = \omega_{AB} \times r_{B/A} = (-10k)$ rad/s × (-0.177i + 0.177j)ft. Using the HP 49 G calculator we would write:

[0 0 -10][ENTER][-0.177 0.177 0][ENTER] [←][MTH][VECTR][CROSS]

to get [-1.77 -1.77 0], or

$$\boxed{v_B = (-1.77i - 1.77j) \text{ ft/s}.}$$

Note: Radians are basically dimensionless units, thus rad·ft = ft, rad·m = m.

To find the velocity of point C, we use

$$v_C = v_B + \omega_{BC} \times r_{C/B} = (-1.77i - 1.77j) \text{ ft/s} + (\omega_{BC} \cdot k) \text{ rad/s} \times (0.175i + 0.729j)\text{ft}.$$

Using the HP 49 G calculator we follow these step (Recall that to get you use [ALPHA][→][V]):

[-1.77 -1.77 0][ENTER]	Enter v_B
[0 0 'ωBC'][ENTER]	Enter ω_{BC}
[0.175 0.729 0][ENTER]	Enter $r_{C/B}$
[↰][MTH][VECTR][CROSS]	Calculate $\omega_{BC} \times r_{C/B}$ =
[+]	Calculate $v_B + \omega_{BC} \times r_{C/B} = v_C$

The result is ['-1.77-0.729*ωBC' '-1.77+ωBC*.175' 0], or

$$v_C = ((-1.77-0.729 \cdot \omega_{BC}) \cdot i + (-1.77+0.175 \cdot \omega_{BC}) \cdot j) \text{ ft/s}.$$

The figure above shows that the piston C is forced to move in the vertical direction, thus, the velocity of point C can be written as

$$v_C = (v_C \cdot j) \text{ ft/s}.$$

Equating the two results presented immediately above for v_C we get:

$$(-1.77-0.729 \cdot \omega_{BC}) \cdot i + (-1.77+0.175 \cdot \omega_{BC}) \cdot j = 0 \cdot i + v_C \cdot j.$$

Since the x- and y-components of the two vectors in each side of the equal sign must be the same, we can write the system of equations:

$$-1.77-0.729 \cdot \omega_{BC} = 0$$
$$-1.77+ 0.175 \cdot \omega_{BC} = vC$$

Solution of equations - one at a time

The solution of this system of two linear equations in two unknowns (ω_{BC} and v_C) can be accomplished with the HP 49 G calculator as follows:

1. Decompose the vector left over from the previous calculation by using [↰][PRG][TYPE][OBJ→].

2. Press [⇐][⇐] to drop the contents in the first and second stack levels.

3. Press [▶] to swap new contents of levels 1 and 2 in the stack.

4. Enter [0][↦][=] to construct the equation: '-1.77-.729*ωBC=0'.

5. Enter the name of the variable to be solved for:

    [↦]['][ALPHA][↦][V][ALPHA][B][ALPHA][C][ENTER].

6. Use [↦][ALG][SOLVE] to obtain 'ωBC=-2.42798353909'.

7. Press [▶] to swap new contents of levels 1 and 2 in the stack once more.

8. Enter [↦]['][ALPHA][↰][V][ALPHA][C][ENTER] [↦][=] to construct the equation

    '-1.77+ωBC*.175=vC'

9. Press [▶] yet once more to swap new contents of levels 1 and 2 in the stack.

10. Use [↦][ALG][SUBST] to replace the value of ωBC into the equation built in step 8.

11. Press [↦][EVAL] to get '-2.198489711934=vC'.

> Thus, the solution of the system of equations is:
>
> $$\omega_{BC} = -2.42 \text{ rad/s, and } v_C = -2.198 \text{ ft/s}.$$
>
> The negative sign in ω_{BC} means that the angular velocity is clockwise. The negative sign in v_C means that point C is actually moving downwards in the vertical direction.

Acceleration

Again, because A is a fixed point, $a_A = 0$. Thus, the acceleration of point B is given by

$$a_B = \alpha_{AB} \times r_{B/A} - \omega_{AB}^2 \cdot r_{B/A} = (-20k)\text{rad/s}^2 \times (-0.177i + 0.177j)\text{ft} - (10 \text{ rad/s})^2 \times (-0.177i + 0.177j)\text{ft}$$

To use vectors in the HP 49G calculator we will re-write the known data as follows (no units will be used, it is understood that system of units is consistent):

$$\alpha_{AB} = [0\ 0\ -20]\ ;\ \omega_{AB} = 10;\ \text{and},\ r_{B/A} = [-0.177\ 0.177\ 0].$$

To calculate a_B from: $a_B = \alpha_{AB} \times r_{B/A} - \omega_{AB}^2 \cdot r_{B/A}$, use the following keystroke sequence:

[↵][[ ]] [.][1][7][7] [+/-][SPC] [.][1][7][7][SPC][0][ENTER]	Enter $r_{B/A}$
[ENTER]	Copy $r_{B/A}$
[1][0][↵][x²][+/-]	Enter $-\omega_{AB}^2$
[×]	Calculate $-\omega_{AB}^2 \cdot r_{B/A}$
[▶]	Place $r_{B/A}$ in level 1
[↵][[ ]] [0][SPC][0][SPC][2][0][+/-][ENTER]	Enter α_{AB}
[▶]	Place $r_{B/A}$ to level 1, α_{AB} to level 2.
[↵][MTH][VECTR][CROSS]	Calculate $\alpha_{AB} \times r_{B/A}$
[+]	Calculate $\alpha_{AB} \times r_{B/A} - \omega_{AB}^2 \cdot r_{B/A} = a_B$

The result is [21.240 -14.160 0.000], or

$$a_B = (21.24i - 14.16j)\text{ft/s}^2.$$

To calculate the acceleration of point C we use:

$a_C = a_B + \alpha_{BC} \times r_{C/B} - \omega_{BC}^2 \cdot r_{C/B} =$
$\quad (21.24i - 14.16j)\text{ft/s}^2 + (\alpha_{BC}k) \times (0.175i + 0.729j)\text{ft} - (-2.42 \text{ rad/s})^2 \cdot (0.175i + 0.729j)\text{ft}$

[21.24 -14.16 0][ENTER]	Enter a_B
[0.175 0.729 0]ENTER]	Enter $r_{C/B}$
[ENTER]	Copy $r_{C/B}$
[2][.][4][2][+/-][↵][x²][+/-]	Enter $-\omega_{BC}^2$
[×]	Calculate $-\omega_{BC}^2 \cdot r_{C/B}$
[▶]	Place $r_{C/B}$ in level 1
[0 0 'αBC'][ENTER]	Enter α_{BC}
[▶]	Place $r_{C/B}$ to level 1, α_{BC} to level 2.
[↵][MTH][VECTR][CROSS]	Calculate $\alpha_{BC} \times r_{C/B}$

[+]   Calculate $\alpha_{BC} \times r_{C/A} - \omega_{BC}^2 \cdot r_{C/A}$

[+] $a_B + \alpha_{AB} \times r_{B/A} - \omega_{AB}^2 \cdot r_{B/A} = a_C$

The result is ['21.24+(-1.02487-0.729*αBC)' '-14.16+(-4.2693156+αBC*.175)' 0], or

$$a_C = ((21.24 - 1.02487 - 0.729 \cdot \alpha_{BC}) \cdot i + (-14.16 + (-4.2693156 + 0.175 \cdot \alpha_{BC})) \cdot j) \text{ ft/s}^2.$$

Also, because the motion of point C is in the vertical direction, we can write

$$a_C = a_C \cdot j.$$

Equating the two expressions for the vector aC shown above, we get the following equations:
$$21.24 - 1.02487 - 0.729 \cdot \alpha_{BC} = 0,$$
$$-14.16 + (-4.2693156 + 0.175 \cdot \alpha_{BC}) = a_C.$$

Solution of a system of linear equations using matrices

The solution of this system of two linear equations in two unknowns (ω_{BC} and v_C) can be accomplished with the HP 49 G calculator as follows:

1. Decompose the vector left over from the previous calculation by using [↩][PRG][TYPE][OBJ→].

2. Press [⇐][⇐] to drop the contents in the first and second stack levels.

3. Press [→][EVAL] to simplify the equation to '.175*αBC-18.4293156'

4. Press [▶] to swap new contents of levels 1 and 2 in the stack.

5. Press [→][EVAL] to simplify the equation to '-(.729*αBC-20.21513)'

6. Enter [0][→][=] to construct the equation: '-(.729*αBC-20.21513)=0'.

7. Press [▶] to swap new contents of levels 1 and 2 in the stack.

8. Enter [→]['][ALPHA][↩][A][ALPHA][C][ENTER] [→][=] to construct the equation:

    '-1.77+ωBC*.175=aC'

9. At this point you'll have two equations in your stack. In order to solve them simultaneously, you need to re-write them to look like this:

    'aC-.175*αBC=18.4293156'
    '.729*αBC=20.21513'

Or, as a matrix equation as:

$$\begin{bmatrix} 1 & -0.175 \\ 0 & 0.729 \end{bmatrix} \begin{bmatrix} a_C \\ \alpha_{BC} \end{bmatrix} = \begin{bmatrix} 18.4293156 \\ 20.21513 \end{bmatrix}$$

> This matrix equation can be solved by first entering the right hand-side vector, then entering the matrix of coefficients, and, finally, dividing the two of them. This follows from writing the matrix equation as
>
> $A \cdot x = b$, thus, $x = A^{-1} \cdot b$ (or b/A).
>
> 10. Enter the right-hand side vector as [18.4293156 20.21513][ENTER].
>
> 11. Enter the matrix of coefficients [[1 -0.175] [0 0.729]][ENTER].
>
> 12. Divide the vector in stack level 2 by stack level 1, [÷]. The result is: [23.2821 27.7300], or
>
> $a_C = 23.2821$ ft/s^2, $\alpha_{BC} = 27.7300$ rad/s^2.

We have presented here two methods for solution of equations. These methods, and others, are presented in more detail in a different chapter.

Row vectors, column vectors, and lists

The vectors we have used so far in this chapter are presented as row vectors, i.e., their elements are written in a horizontal row enclosed in square brackets. This format is preferred for operations that are typical of vectors, such as dot and cross products. For other applications, for example pre-programmed statistical functions in the HP 49 G calculator, you may want to write your vectors as column vectors. A column vector is basically a matrix with one column and many rows. Thus, you can enter data for a column vector by enclosing each data value between square brackets within an external pair of square brackets, for example: [[1.2] [2.5] [3.2] [4.5] [6.2]][ENTER]. This data entry represents the column vector,

$$\begin{bmatrix} 1.2 \\ 2.5 \\ 3.2 \\ 4.5 \\ 6.2 \end{bmatrix}$$

You can verify this format in the matrix editor by pressing [▼]. The matrix editor will show the data occupying only the first row of the spreadsheet. Press [ENTER] to return to normal calculator display.

Transforming a row vector into a column vector

As an example, enter the vector [1.2 3.2 -4.5 7.2 3.1][ENTER]. Then, decompose the vector by using [↰][PRG][TYPE][OBJ→]. This operation will place the elements of the vector in stack levels 2 and above, and will also place a list { 5 } in stack level 1. The list indicates the size of the vector. (Note: You could re-build the vector at this point by pressing [→ARRY].) To generate a column vector using the elements of the original row vector, use the following keystrokes:

[1][+]   Creates the list {5 1} meaning 5 rows and 1
column
[←][PRG][TYPE][→ARRY]   Creates the column vector

Transforming a column vector into a row vector

Using the result from the previous exercise, use the following keystrokes:

[←][PRG][TYPE][OBJ→]   Decomposes the vector. Matrix size shown in
level 1
[OBJ→][⇔][→LIST]   Creates the list { 5 } in stack level 1
[→ ARRY]   Creates the row vector

The result is [1.2 3.2 -4.5 7.2 3.1].

An alternative approach using the original column vector: [[1.2] [2.5] [3.2] [4.5] [6.2]][ENTER], is to use:

[←][PRG][TYPE][OBJ→]   Decomposes the vector. Matrix size shown in
level 1
[←][MTH][LIST][REVLI]   Creates the list { 1 5 } in stack level 1
[←][PRG][TYPE][→ ARRY]   Creates the row vector

Notice that the result is now shown as [[1.2 3.2 -4.5 7.2 3.1]]. Although this format shows two sets of square brackets, the vector operations defined earlier apply to this alternate vector format.

Transforming a list into a vector

Enter the list {1.2 -5.2 3.5 2.7}. To create a row vector use the following keystrokes:

[←][PRG][TYPE][OBJ→]   Decomposes the list. List size shown in level 1
[1][→LIST]   Creates the list { 4 }
[→ ARRY]   Creates the row vector with a single set of
brackets

Alternatively, enter the list {1.2 -5.2 3.5 2.7}. Then use the following keystrokes:

[←][PRG][TYPE][OBJ→]   Decomposes the list. List size shown in level 1
[1] [ENTER] [▶] [2][→LIST]   Creates the list { 1 4 }
[→ ARRY]   Creates the row vector with double set of
brackets

To create a column vector, for the list under consideration, i.e., {1.2 -5.2 3.5 2.7}, use:

[←][PRG][TYPE][OBJ→]   Decomposes the list. List size shown in level 1
[1] [ENTER][2][→LIST]   Creates the list { 4 1 }
[→ ARRY]   Creates the column vector

Transforming a vector (or matrix) into a list

Enter the vector [1 2 3 4 5], then use [↰][MATRICES][OPER][AXL] (AXL stands for Array transformed to List), to get {1 2 3 4 5}.

10 Matrices and linear algebra

A matrix is a rectangular arrangement of numbers in rows and columns enclosed in brackets. Examples of matrices follow:

* A matrix with one row and three columns (a 1×3 matrix) is basically the same as a three-dimensional row vector, e.g.,

$$[5.\ 6.\ -2.]$$

* A matrix with one column and five rows (a 5×1 matrix) is the same as a column vector of five elements, e.g.,

$$\begin{bmatrix} -5 \\ 3 \\ 2.5 \\ 0 \\ 4 \end{bmatrix}$$

* The following is a matrix of two rows and four columns (a 2×4 matrix):

$$\begin{bmatrix} -5 & 2 & 3 & 0 \\ 11 & -7 & 0 & -1 \end{bmatrix}$$

* A matrix can have variables and algebraic expressions as their elements, for example:

$$\begin{bmatrix} a_{11}^2 - \lambda & b \\ x & a_{22}^2 - 2\mu \end{bmatrix}$$

* A matrix can have complex numbers as elements, for example:

$$\begin{bmatrix} -1+5i & 3 \\ 2i & \pi - 5i \\ 5 - i\sqrt{2} & e^{i\pi/2} \end{bmatrix}$$

Definitions

⬥ The <u>elements of a matrix</u> are referred to by using two sub-indices: the first one representing the row, and the second one the column where the element is located. A matrix A with n rows and m columns can be represented by

$$A = \begin{bmatrix} a_{11} & a_{12} & \Lambda & a_{1,m-1} & a_{1,m} \\ a_{21} & a_{22} & \Lambda & a_{2,m-1} & a_{2,m} \\ M & M & O & M & M \\ a_{n-1,1} & a_{n-1,2} & \Lambda & a_{n-1,m-1} & a_{n-1,m} \\ a_{n,1} & a_{n,2} & \Lambda & a_{n,m-1} & a_{n,m} \end{bmatrix}$$

Thus, a generic element of matrix A belonging in row i and column j will be written as $a_{i,j}$ or a_{ij}. The matrix A, itself, can be written in a simplified form as

$$A_{n \times m} = [a_{ij}].$$

⬥ A matrix having the same number of rows and columns is called a <u>square matrix</u>. The following is a 3×3 square matrix:

$$\begin{bmatrix} -2.5 & 4.2 & 2.0 \\ 0.3 & 1.9 & 2.8 \\ 2 & -0.1 & 0.5 \end{bmatrix}$$

⬥ The elements of a square matrix with equal sub-indices, i.e., a_{11}, a_{22}, ..., a_{nn}, belong to the matrix's <u>main diagonal</u>.

$$\begin{bmatrix} 12.5 & 0 & 0 \\ 0 & -9.2 & 0 \\ 0 & 0 & 0.75 \end{bmatrix}$$

⬥ A <u>diagonal matrix</u> is a square matrix having non-zero elements only in the main diagonal. An example of a 3×3 diagonal matrix is:

$$I_{3 \times 3} = \begin{bmatrix} 1 & 0 & 0 \\ 0 & 1 & 0 \\ 0 & 0 & 1 \end{bmatrix}$$

⬥ A diagonal matrix whose main diagonal elements are all equal to 1.0 is known as an <u>identity matrix,</u> because multiplying I by any matrix results in the same matrix, i.e., $I \cdot A = A \cdot I = A$. The identity matrix is typically given the symbol I. The following is a 3×3 identity matrix:

Matrices as tensors and the Kronecker's delta function

A sub-indexed variable, such as those used to identify a matrix, is also referred to as a <u>tensor</u>. The number of sub-indices determines the <u>order of the tensor</u>. Thus, a vector (see Chapter 9) is a first-order tensor, and a matrix is a second order tensor. A scalar value is referred to as a zero-th order tensor.

The <u>Kronecker's delta function</u>, δ_{ij}, is a tensor function defined as $\delta_{ij} = 1.0$, if $i = j$, and $\delta_{ij} = 0$, if $i \neq j$.

Using the Kronecker's delta function, therefore, an n×n <u>identity matrix</u> can be written as

$$I_{n \times n} = [\delta_{ij}].$$

- A <u>tridiagonal matrix</u> is a matrix having non-zero elements in the main diagonal and the upper and lower diagonals adjacent to the main diagonal. Tridiagonal matrices typically arise from numerical solution of partial differential equations, and, more often than not, the terms in the diagonals off the main diagonal are the same. An example of a 5×5 tridiagonal matrix follows:

$$\begin{bmatrix} -2.5 & 4 & 0 & 0 & 0 \\ 2 & -3.5 & 2 & 0 & 0 \\ 0 & -2 & 6.5 & -2 & 0 \\ 0 & 0 & 3 & -4 & 3 \\ 0 & 0 & 0 & 0 & 5 \end{bmatrix}$$

Matrix operations

- The <u>transpose of a matrix</u> results from exchanging rows for columns and columns for rows. Therefore, given the matrix $A_{n \times m} = [a_{ij}]$, of n rows and m columns, its transpose matrix is $A^T_{n \times m} = [a^T_{ij}]$, of m rows and n columns, such that $a^T_{ij} = a_{ji}$, (i = 1,2, ..., n; j = 1,2, ..., m).

- Consider the matrices $A_{n \times m} = [a_{ij}]$, and $B_{n \times m} = [b_{ij}]$, and $C_{n \times m} = [c_{ij}]$. The operations <u>of addition, subtraction, and multiplication by as scalar</u>, are defined as:

- Addition: $C_{n \times m} = A_{n \times m} + B_{n \times m}$, implies $c_{ij} = a_{ij} + b_{ij}$.
- Subtraction: $C_{n \times m} = A_{n \times m} + B_{n \times m}$, implies $c_{ij} = a_{ij} + b_{ij}$.
- Multiplication by a scalar, k: $C_{n \times m} = k \cdot A_{n \times m}$, implies $c_{ij} = k \cdot a_{ij}$.

- <u>Matrix multiplication</u> requires that the number of rows of the first matrix be equal to the number of columns of the second matrix. In other words, the only matrix multiplication allowed is such that,

$$A_{n\times m} \cdot B_{m\times p} = C_{n\times p},$$

with the elements of the matrix product given by

$$c_{ij} = \sum_{k=1}^{m} a_{ik} \cdot b_{kj}, \quad (i = 1, 2, ..., n;\ j = 1, 2, ..., p).$$

Schematically, the calculation of element cij of a matrix product, $C_{n\times p} = A_{n\times m} \cdot B_{m\times p}$, is shown below:

Thus, element c_{ij} of the product results from the summation:

$$c_{ij} = a_{i1} \cdot b_{1j} + a_{i2} \cdot b_{2j} + ... + a_{ik} \cdot b_{kj} + ... + a_{i,m-1} \cdot b_{m-1,j} + a_{i,m} \cdot b_{m,j}.$$

which is the term-by-term multiplication of the elements of row i from A and column j from B which then are added together.

Note: Matrix multiplication is, in general, non-commutative, i.e., A·B ≠ B·A. In fact, if one of these products exist, the other may not even be defined. The only case in which both A·B and B·A are defined is when both A and B are square matrices of the same order.

Einstein's summation convention for tensor algebra

When developing his general theory of relativity, Albert Einstein was faced with the daunting task of writing huge amounts of tensor summations. He figured out that he did not need to write the summation symbol, Σ, with its associated indices, if he used the convention that, whenever two indices were repeated in an expression, the summation over all possible values of the repeating index was implicitly expressed.

Thus, the equation for the generic term of a matrix multiplication, expressed above as a summation, can be simplified to read

$$c_{ij} = a_{ik} \cdot b_{kj}, \quad (i = 1, 2, ..., n;\ j = 1, 2, ..., p).$$

Because the index k is repeated in the expression, the summation of all the products indicated by the expression is implicit over the repeating index, k = 1, 2, ..., m.

The dot or internal product of two vectors of the same dimension (see Chapter 9), a = [a₁ a₂ ... aₙ] and b = [b₁ b₂ ... bₙ], can be expressed, using Einstein's summation convention, as

$$a \bullet b = a_i \cdot b_i, \text{ or } a \bullet b = a_k \cdot b_k, \text{ or even } a \bullet b = a_r \cdot b_r.$$

The repeating index in this, or in the previous, expression is referred to as a dummy index and can be replaced by any letter, as long as we are aware of the range of values over which the summation is implicit.

✤ The <u>inverse</u> of a square matrix $A_{n \times n}$, referred to as A^{-1}, is defined in terms of matrix multiplication and the n×n identity matrix, I, as

$$A \cdot A^{-1} = A^{-1} \cdot A = I.$$

Entering matrices in the HP 49 G stack

Using the matrix editor

As with the case of vectors, discussed in Chapter 9, matrices can be entered into the stack by using the matrix editor. For example, to enter the matrix:

$$\begin{bmatrix} -2.5 & 4.2 & 2.0 \\ 0.3 & 1.9 & 2.8 \\ 2 & -0.1 & 0.5 \end{bmatrix},$$

first, start the matrix writer by using [↰][MTRW]. Make sure that the option [GO→■] is selected. Then use the following keystrokes:

[2][.][5][+/-][SPC][4][.][2][SPC][2][ENTER]   Enter first row.
[▼][◄][◄][◄]   Move cursor to first element of second
row.
[.][3][SPC][1][.][9][SPC][2][.][8][ENTER]   Enter second row. Cursor moves to 3ʳᵈ
row.
[2][SPC][.][1][+/-][SPC][.][5][ENTER]   Enter third row. Cursor moves to 4ᵗʰ row.
[ENTER]   Enters matrix into stack, quits matrix
writer.

If you have selected the textbook display option (using [MODE][DISPLAY] and checking off ✓Textbook), the matrix will look like the one shown above. Otherwise, the display will show:

```
3:
2:
1: [[-2.5 4.2 2 ]
    [.3 1.9 2.8 ]
    [2 -.1 .5 ]]
```

Note: Details on the use of the matrix writer were presented in Chapter 9.

Typing in the matrix directly into the stack

The same result can be achieved by entering the following directly into the stack:

[←][[ ]]   Opens outer set of square brackets.
[←][[ ]] [2][.][5][+/-][SPC][4][.][2][SPC][2][▶]   Type in first row of matrix as a vector.
[←][[ ]] [.][3][SPC][1][.][9][SPC][2][.][8][▶]   Type in second row of matrix as a vector.
[←][[ ]] [2][SPC][.][1][+/-][SPC][.][5]   Type in third row of matrix as a vector.
[ENTER]   Enter matrix into the stack.

Examples of matrix operations

In this section we will define a number of variables in which matrices of different dimensions will be stored. Then, we will utilize those matrices to check the results of simple matrix operations.

To get started, therefore, enter the following matrices and store them in the names suggested. Notice that the names of the variables correspond to an upper case letter of the alphabet followed by two numbers. The numbers represent the number of rows and columns that the matrix has. This way we can purposely select some particular matrices to illustrate matrix operations that are and that are not allowed. Thus, proceed to store the following variables:

$$A11 = [[3]], B11 = [[-2]], C11 = [[5]]$$

$$A12 = [[-5 \quad 6]], B12 = [[3 \quad -2]], C12 = [[-10 \quad 20]].$$

$$A13 = [[1 \quad -2 \quad 6]], B13 = [[0 \quad 3 \quad -4]], C13 = [[5 \quad 3 \quad -10]].$$

$$A21 = \begin{bmatrix} -7 \\ 3 \end{bmatrix}, B21 = \begin{bmatrix} 3 \\ 5 \end{bmatrix}, C21 = \begin{bmatrix} -2 \\ 2 \end{bmatrix}.$$

$$A22 = \begin{bmatrix} -3 & 0 \\ 4 & -6 \end{bmatrix}, B22 = \begin{bmatrix} 5 & -2 \\ 5 & 4 \end{bmatrix}, C22 = \begin{bmatrix} -1 & 4 \\ 8 & 2 \end{bmatrix}.$$

$$A23 = \begin{bmatrix} 8 & 0 & -1 \\ 5 & -2 & 3 \end{bmatrix}, B23 = \begin{bmatrix} 1 & 0 & 1 \\ 0 & 1 & -1 \end{bmatrix}, C23 = \begin{bmatrix} 2 & -3 & -5 \\ 6 & 4 & -2 \end{bmatrix}.$$

$$A31 = \begin{bmatrix} -10 \\ 2 \\ 5 \end{bmatrix}, B31 = \begin{bmatrix} 3 \\ -7 \\ -2 \end{bmatrix}, C31 = \begin{bmatrix} 0 \\ 2 \\ 6 \end{bmatrix}.$$

$$A32 = \begin{bmatrix} 1 & 0 \\ 1 & 2 \\ 5 & 2 \end{bmatrix}, B32 = \begin{bmatrix} 9 & 2 \\ 3 & 0 \\ 6 & -5 \end{bmatrix}, C32 = \begin{bmatrix} 5 & 8 \\ 6 & -7 \\ -3 & -2 \end{bmatrix}.$$

$$A33 = \begin{bmatrix} 2 & -1 & 5 \\ 0 & 2 & 1 \\ -7 & 2 & -5 \end{bmatrix}, B33 = \begin{bmatrix} 3 & 1 & 2 \\ 0 & 5 & 2 \\ -4 & 2 & 1 \end{bmatrix}, C33 = \begin{bmatrix} 2 & 1 & 2 \\ 3 & -7 & 0 \\ 2 & 1 & 4 \end{bmatrix}.$$

Once these variables have been entered, try the following exercises [fill the missing solutions by hand]:

Addition and subtraction of 1×1 matrices (i.e., scalars)

[A11][B11][+],   i.e., A11 + B11 = [1].
[A11][C11][+],   i.e., A11 + C11 = [8].
[A11][B11][C11][+][+] i.e., A11 + B11 + C11 = [6].
[A11][B11][-]   i.e., A11 - B11 = [5].
[A11][C11][-]   i.e., A11 - C11 = [-2].
[B11][C11][-]   i.e., B11 - C11 = [-7].
[A11][B11][C11][-][-] i.e., A11 - (B11 - C11) = [10].

Addition and subtraction of 1×2 matrices (i.e., two-dimensional row vectors)

[A12][B12][+],   i.e., A12 + B12 =
[A12][C12][+],   i.e., A12 + C12 =
[A12][B12][C12][+][+] i.e., A12 + B12 + C12 =
[A12][B12][-]   i.e., A12 - B12 =
[A12][C12][-]   i.e., A12 - C12 =
[B12][C12][-]   i.e., B12 - C12 =
[A12][B12][C12][-][-] i.e., A12 - (B12 - C12) =

Addition and subtraction of 2×1 matrices (i.e., two dimensional column vectors)

[A21][B21][+],   i.e., A21 + B21 =
[A21][C21][+],   i.e., A21 + C21 =
[A21][B21][C21][+][+] i.e., A21 + B21 + C21 =
[A21][B21][-]   i.e., A21 - B21 =
[A21][C21][-]   i.e., A21 - C21 =
[B21][C21][-]   i.e., B21 - C21 =
[A21][B21][C21][-][-] i.e., A21 - (B21 - C21) =

Addition and subtraction of 2×2 matrices (i.e., two dimensional column vectors)

[A22][B22][+],   i.e., A22 + B22 =
[A22][C22][+],   i.e., A22 + C22 =
[A22][B22][C22][+][+] i.e., A22 + B22 + C22 =
[A22][B22][-]   i.e., A22 - B22 =
[A22][C22][-]   i.e., A22 - C22 =
[B22][C22][-]   i.e., B22 - C22 =
[A22][B22][C22][-][-] i.e., A22 - (B22 - C22) =

Addition and subtraction of 1×3 matrices (i.e., three dimensional column vectors)

[A13][B13][+],   i.e., A13 + B13 =
[A13][C13][+],   i.e., A13 + C13 =
[A13][B13][C13][+][+]   i.e., A13 + B13 + C13 =
[A13][B13][-]   i.e., A13 - B13 =
[A13][C13][-]   i.e., A13 - C13 =
[B13][C13][-]   i.e., B13 - C13 =
[A13][B13][C13][-][-]   i.e., A13 - (B13 - C13) =

Addition and subtraction of 2×3 matrices (i.e., two dimensional column vectors)

[A23][B23][+],   i.e., A23 + B23 =
[A23][C23][+],   i.e., A23 + C23 =
[A23][B23][C23][+][+]   i.e., A23 + B23 + C23 =
[A23][B23][-]   i.e., A23 - B23 =
[A23][C23][-]   i.e., A23 - C23 =
[B23][C23][-]   i.e., B23 - C23 =
[A23][B23][C23][-][-]   i.e., A23 - (B23 - C23) =

Addition and subtraction of 3×1 matrices (i.e., two dimensional column vectors)

[A31][B31][+],   i.e., A31 + B31 =
[A31][C31][+],   i.e., A31 + C31 =
[A31][B31][C31][+][+]   i.e., A31 + B31 + C31 =
[A31][B31][-]   i.e., A31 - B31 =
[A31][C31][-]   i.e., A31 - C31 =
[B31][C31][-]   i.e., B31 - C31 =
[A31][B31][C31][-][-]i.e., A31 - (B31 - C31) =

Addition and subtraction of 3×2 matrices (i.e., two dimensional column vectors)

[A32][B32][+],   i.e., A32 + B32 =
[A32][C32][+],   i.e., A32 + C32 =
[A32][B32][C32][+][+]   i.e., A32 + B32 + C32 =
[A32][B32][-]   i.e., A32 - B32 =
[A32][C32][-]   i.e., A32 - C32 =
[B32][C32][-]   i.e., B32 - C32 =
[A32][B32][C32][-][-]   i.e., A32 - (B32 - C32) =

Addition and subtraction of 3×3 matrices (i.e., two dimensional column vectors)

[A33][B33][+],   i.e., A33 + B33 =
[A33][C33][+],   i.e., A33 + C33 =
[A33][B33][C33][+][+]   i.e., A33 + B33 + C33 =
[A33][B33][-]   i.e., A33 - B33 =
[A33][C33][-]   i.e., A33 - C33 =
[B33][C33][-]   i.e., B33 - C33 =
[A33][B33][C33][-][-]   i.e., A33 - (B33 - C33) =

Notes:

1. The subtraction A - B can be interpreted as A + (-B).

2. Addition is commutative, i.e., A + B = B + A, and associative, i.e., A+B+C = A+(B+C) = (A+B)+C.

3. Addition and subtraction can only be performed between matrices of the same dimensions. Verify this by trying [A23][A21][+]. You will get an error message: `<!> + Error: Invalid Dimension`

Multiplication by a scalar

[2][ENTER][A11][×]   i.e., 2·A11 =
[3][+/-][ENTER][A12][×]   i.e., -3·A12 =
[1][+/-][ENTER][A21][×]   i.e., -1·A21 =
[5][ENTER][A22][×]   i.e., 5·A22 =
[2][+/-][ENTER][A13][×]   i.e., -2·A13 =
[1][0][ENTER][A31][×]   i.e., 10·A31 =
[5][+/-][ENTER][A23][×]   i.e., -5·A23 =
[2][ENTER][A32][×]   i.e., 2·A32 =
[1][.][5][ENTER][A33][×]   i.e., 1.5·A33 =

Matrix multiplication

[A11][B11][×]   i.e., A11·B11 =
[A11][B11][×]   i.e., B11·A11 =
[A12][B21][×]   i.e., A12·B21 =
[B21][A12][×]   i.e., B21·A12 =
[A12][B22][×]   i.e., A12·B22 =
[A21][B12][×]   i.e., A21·B12 =
[B12][A21][×]   i.e., B12·A21 =
[A22][B21][×]   i.e., A22·B21 =
[A22][B22][×]   i.e., A22·B22 =
[B22][A22][×]   i.e., B22·A22 =
[A13][B31][×]   i.e., A13·B31 =
[B31][A13][×]   i.e., B31·A13 =
[A13][B32][×]   i.e., A13·B32 =
[A13][B33][×]   i.e., A11·B11 =
[A23][B31][×]   i.e., A23·B31 =
[A23][B32][×]   i.e., A23·B32 =
[A32][B23][×]   i.e., A32·B23 =
[A23][B33][×]   i.e., A23·B33 =
[A33][B31][×]   i.e., A33·B31 =
[A33][B32][×]   i.e., A23·B31 =
[A33][B33][×]   i.e., A23·B31 =
[B33][A33][×]   i.e., B33·A33 =

> Note: Some of the examples above illustrate the fact that $A_{m \times n} \cdot B_{n \times m} \neq B_{n \times m} \cdot A_{m \times n}$. That is, multiplication, in general, is not commutative.

[A12][B21][C12][×][×]	i.e., A12·(B21·C12) = [[-150 300]]
[A12][B21][×][C12][×]	i.e., (A12·B21)·C12 = [[-150 300]]
[A22][B23][C32][×][×]	i.e., A22·(B21·C12) = [[-6 -18] [-46 54]]
[A22][B23][×][C32][×]	i.e., (A22·B21)·C12 = [[-6 -18] [-46 54]]
[A32][B23][C32][×][×]	i.e., A32·(B23·C3) = [[2 6][20 -4][28 20]]
[A32][B23][×][C32][×]	i.e., (A32·B23)·C3 = [[2 6][20 -4][28 20]]

> Note: The examples above illustrate the fact that multiplication is associative:
> $A_{m \times n} \cdot B_{n \times m} \cdot C_{m \times p} = A_{m \times n} \cdot (B_{n \times m} \cdot C_{m \times p}) = (A_{m \times n} \cdot B_{n \times m}) \cdot C_{m \times p}$.

Inverse matrices

Inverse matrices exists only for square matrices. In the HP 49 G calculator, inverse matrices are calculated by using the [1/x] key. Notice that the standard notation for the inverse matrix of A is A^{-1}, and not 1/A, even though we will be using the [1/x] key to calculate inverses. Try the following exercises:

[A11][1/x]	i.e., $(A11)^{-1}$ = [['1/3']]
[B11][1/x]	i.e., $(B11)^{-1}$ = [['-1/2']]
[C11][1/x]	i.e., $(C11)^{-1}$ = [['1/5']]
[A22][1/x]	i.e., $(A22)^{-1}$ = [['-1/3' 0] ['-2/9' '-1/6']]
[B22][1/x]	i.e., $(B22)^{-1}$ = [['2/15' '1/15'] ['-1/6' '1/6']]
[C22][1/x]	i.e., $(C22)^{-1}$ = [['-1/17' '2/17'] ['4/17' '1/34']]
[A33][1/x]	i.e., $(A33)^{-1}$ = [['-37/164' '-1/164' '-11/164'] ['-1/164' '31/164' '13/164'] ['-11/164' '13/164' '-21/164']]
[B33][1/x]	i.e., $(B33)^{-1}$ = [['-67/248' '15/124' '29/248'] ['23/124' '-7/62' '3/124'] ['107/248' '-11/124' '-13/248']]
[C33][1/x]	i.e., $(C33)^{-1}$ = [['20/317' '-6/317' '-45/317'] ['-8/317' '-61/317' '18/317'] ['29/317' '23/317' '14/317']]

> Note: Some matrices has no inverse, for example, matrices with one row or column whose elements are all zeroes, or a matrix with a row or column being proportional to another row or column, respectively.
>
> Try the following examples:
>
> [[1 2 3] [0 0 0] [-2 5 2]][ENTER][1/x]
> [[1 2 3] [2 4 6] [5 2 -1]][ENTER][1/x]

Verifying properties of inverse matrices

[A22][1/x][A22][×]	i.e., $(A22)^{-1} \cdot A22 = I_{2\times 2}$
[A22][A22][1/x][×]	i.e., $A22 \cdot (A22)^{-1} = I_{2\times 2}$
[B22][1/x][B22][×]	i.e., $(B22)^{-1} \cdot B22 = I_{2\times 2}$
[B22][B22][1/x][×]	i.e., $B22 \cdot (B22)^{-1} = I_{2\times 2}$
[C22][1/x][C22][×]	i.e., $(C22)^{-1} \cdot C22 = I_{2\times 2}$
[C22][C22][1/x][×]	i.e., $C22 \cdot (C22)^{-1} = I_{2\times 2}$
[A33][1/x][A33][×]	i.e., $(A33)^{-1} \cdot A33 = I_{3\times 3}$
[A33][A33][1/x][×]	i.e., $A33 \cdot (A33)^{-1} = I_{3\times 3}$
[B33][1/x][B33][×]	i.e., $(B33)^{-1} \cdot B33 = I_{3\times 3}$
[B33][B33][1/x][×]	i.e., $B33 \cdot (B33)^{-1} = I_{3\times 3}$
[C33][1/x][C33][×]	i.e., $(C33)^{-1} \cdot C33 = I_{3\times 3}$
[C33][C33][1/x][×]	i.e., $C33 \cdot (C33)^{-1} = I_{3\times 3}$

Creating matrices using calculator functions

Some matrices can be created by using the calculator functions available in either

[←][MTH][MATRX][MAKE] (Let's call it the MAKE menu)

or in

[←][MATRICES][CREAT] (Let's call the CREATE menu).

Examine these two menus by pressing [NXT] to get acquainted with the functions they contain. As you can see from exploring these menus, they both have the same functions GET, GETI, PUT, PUTI, SUB, REPL, RDM, RANM, HILBE, VANDE, IDN, CON, →DIAG, and DIAG→. The CREATE menu includes the ROW and COL menus, that are also available under [←][MTH][MATRX]. The MAKE menu includes the functions SIZE, that the CREATE menu does not include. Basically, however, both menus, MAKE and CREATE, provide the user with the same set of functions. In the examples that follow, we will show how to access functions through use of the matrix MAKE menu. At the end of this section we present a table with the keystroke required to obtain the same functions with the CREATE menu.

Functions GET and PUT

The functions GET, GETI, PUT, and PUTI, operate with matrices in a similar manner as with lists or vectors, i.e., you need to provide the location of the element that you want to GET or PUT. However, while in lists and vectors we only needed one index to identify an element, in matrices we need a list of two indices { row column } to identify matrix elements. Examples of the use of GET and PUT follow.

Let's use the matrix we just entered to use the GET and PUT functions. Use the following keystrokes to extract element a_{23} from the matrix:

[ENTER]   Keeps an additional copy of the matrix available.
[←][{][2][SPC][3][ENTER]   List showing location of an element to get, i.e.,{2 3}.
[←][MTH][MATRX][MAKE][NXT][GET]   Copies element in row 2, column 3, into stack.
([←][MATRICES][CREAT][NXT][GET])

Notice that the matrix, from which the element is extracted, disappears. (The copy we made earlier is maintained).

Suppose that we want to place the value 'π' into element a$_{31}$ of the matrix. Use the following keystrokes:

[⇦][ENTER]   Drop value from level 1, make a copy of the matrix.
[←][{][3][SPC][1][ENTER]   List showing location of element to replace, {3 1}.
[→]['][←][π][ENTER]   Place 'π' in stack level 1.

[←][MTH][MATRX][MAKE][NXT][PUT]   Places value 'π' in location {3 1} of the matrix at 3:

The result is:

```
3:
2:
1: [[-2.5  4.2  2  ]
    [ .3   1.9  2.8]
    ['π'  -.1  .5 ]]
```

Functions GETI and PUTI

The functions PUTI and GETI are used in programs since they keep track of an index for repeated application of the PUT and GET functions. The index list in matrices varies by columns first. For example, using the matrix currently in the stack, enter:

[←][{][2][SPC][2][ENTER]   List showing location of an element to get, i.e.,{2 2}.
[←][MTH][MATRX][MAKE][NXT][GETI]   Copies element in row 2, column 3, into stack.

The result is now:

```
3: [[-2.5  4.2  2  ]
    [ .3   1.9  2.8]
    ['π'  -.1  .5 ]]
2:                   {2. 3.}
1:                       1.9
```

Thus, we have in level 1 the element extracted from the matrix. Level 2 contains a list of indices with the index list used {2 2} with the column index increased by one, i.e., {2 3}. The original matrix is still available in level 3.

Now, suppose that you want to insert the value [2] in element {3 1}, use:

[⇦][⇦][ENTER]   Drop value from level 1, make a copy of the matrix.
[↵][{][3][SPC][1][ENTER]   List showing location of element to replace, {3 1}.
[2][ENTER]   Place a 2 in stack level 1.
[↵][MTH][MATRX][MAKE][NXT][PUTI]   Places a 2 in location {3 1} of the matrix at 3:

The result is :

```
3:
2:   [[-2.5 4.2 2 ]
      [.3 1.9 2.8 ]
      [2 -.1 .5 ]]
1:              {3. 2.}
```

Thus, the 2 was replaced in position {3 1}, and the index list was increased by 1 (by column first). The matrix is in level 2, and the incremented index list is in level 1.

The function SIZE

The function SIZE provides a list showing the number of rows and columns of the matrix in stack level 1. The matrix is deleted when applying this function. To try this function, use the matrix currently in the stack. Press [⇦][ENTER] to delete the list {3 2} and make an additional copy of the matrix. Then, press:

[↵][MTH][MATRX][MAKE][SIZE]

The result is {3. 3.}, as expected. Of course, for small matrices, we can determine their size by eye. This function is useful for larger matrices, and for programming purposes.

The function TRN

The function TRN produces the transpose of a real matrix or the conjugate transpose of a complex matrix. Let's use the matrix still in the stack to try this function. Press [⇦][ENTER] to delete the list {3 2} and make an additional copy of the matrix. Then, press:

[↵][MTH][MATRX][MAKE][TRN]

The result is

```
2:
1:   [[-2.5 .3 2 ]
      [4.2 1.9 -.1 ]
      [2 2.8 .5 ]]
```

Note: The calculator also includes the function TRAN ([↵][MATRICES][OPER][NXT][NXT][TRAN]) to calculate the transpose.

The function CON

The function generates a matrix with constant elements. A list of two elements, corresponding to the number of row and columns of the matrix to be generated, should be placed first in the stack. Next, enter the constant value to fill the matrix. For example, use:

[←][{][3][SPC][2][ENTER]   List showing number of rows and columns.
[5][.][2][ENTER]   Enter constant value to load in matrix.

[←][MTH][MATRX][MAKE][CON]   Create the constant matrix.

The result is:

```
3:
2:
1: [[ 5 5 5 ]
    [ 5 5 5 ]]
```

The function IDN

The function IDN (IDeNtity matrix) creates an identity matrix given its order. Recall that an identity matrix has to be a square matrix, therefore, only one value is required to describe it completely. For example, to create a 4×4 identity matrix enter:

[4][ENTER]   Order or size of identity matrix to be generated.
[←][MTH][MATRX][MAKE][IDN].   Generate identity matrix.

The result is:

```
1: [[ 1 0 0 0 ]
    [ 0 1 0 0 ]
    [ 0 0 1 0 ]
    [ 0 0 0 1 ]]
```

The function RDM

The function RDM (Re-DiMensioning) is used to re-write vectors and matrices as matrices and vectors. The input to the function consists of the original vector or matrix followed by a list of a single number, if converting to a vector, or two numbers, if converting to a matrix. In the former case the number represents the vector's dimension, in the latter the number of rows and columns of the matrix. There is nothing better than examples to understand the operation of this function. Here they are:

Re-dimensioning a vector into a matrix

Enter the vector [1 2 3 4 5 6][ENTER], and the re-dimensioning list, {2 3}[ENTER]. Then press [←][MTH][MATRX][MAKE][RDN]. to get the matrix:

```
1:[[ 1 2 3 ]
   [ 4 5 6 ]]
```

Re-dimensioning a matrix into another matrix

Now, enter {3 2}[ENTER] and [←][MTH][MATRX][MAKE][IDN] to get the re-dimensioned matrix:

```
1:[[ 1 2 ]
   [ 3 4 ]
   [ 5 6 ]]
```

Re-dimensioning a matrix into a vector

Enter { 6 }[ENTER] and [←][MTH][MATRX][MAKE][IDN] to get the re-dimensioned vector:

```
1:     [ 1 2 3 4 5 6]
```

Note: the function RDM provides a more direct and efficient way to transform lists to arrays and vice versa, than that provided at the end of chapter 9.

The function RANM

The function RANM (RANdom Matrix) will generate a matrix with random integer elements given a list with the number of rows and columns (i.e., the dimensions of the matrix). For example, enter
{ 2 3 }[ENTER], followed by [←][MTH][MATRX][MAKE][RANM] to generate a random matrix. Of course, the results you will get in your calculator, very likely, will be different than what I got in mine, which were:

```
1: [[ -5 -2 2 ]
    [ 0 1 3 ]]
```

The function SUB

The function SUB extracts a sub-matrix from an existing matrix, provided you indicate the initial and final position of the sub-matrix. For example, if we want to extract elements a_{12}, a_{13}, a_{22}, and a_{23} from the last result, as a 2×2 sub-matrix, enter {1 2}[ENTER] { 2 3 } [ENTER]. Then, enter

[←][MTH][MATRX][MAKE][NXT][SUB]

The result is

```
1: [[ -2  2 ]
    [  1  3 ]]
```

The function REPL

The function REPL replaces or inserts a sub-matrix into a larger one. The input for this function is the matrix where the replacement will take place, the location where the replacement begins, and the matrix to be inserted. For example, keeping the matrix that we inherited from the previous example, enter the matrix: [[1 2 3] [4 5 6] [7 8 9]] [ENTER]. Next, press [▶], to exchange stack levels 1 and 2. Next, enter { 2 2 }[ENTER] to indicate that the sub-matrix insertion starts at position (2,2). Press [▶] once more to exchange stack levels 1 and 2. Now, press [←][MTH][MATRX][MAKE][NXT][REPL], to obtain:

```
1: [[ 1  2  3 ]
    [ 4 -2  2 ]
    [ 7  1  3 ]]
```

The function →DIAG

The function →DIAG takes the main diagonal of a square matrix of dimensions n×n, and creates a vector of dimension n containing the elements of the main diagonal. For example, for the matrix remaining from the previous exercise, when you press [←][MTH][MATRX][MAKE][NXT][NXT][→DIAG], you obtain:

```
1:          [ 1 -2  3 ]
```

The function DIAG→

The function DIAG→ takes a vector and a list of matrix dimensions { rows columns }, and creates a diagonal matrix with the main diagonal replaced with the proper vector elements. For example, enter
[1 -1 2 3][ENTER], { 3 3 }, and then [←][MTH][MATRX][MAKE][NXT][NXT][DIAG→], you obtain:

```
1:  [[ 1  0  0 ]
     [ 0 -1  0 ]
     [ 0  0  2 ]]
```

Notice that although the vector has dimension 4, the dimensions of the matrix to be created were only {3 3}, therefore, the calculator took only the first three elements in the vector to create the main diagonal of the 3×3 matrix requested.

Another example of application of the DIAG→ function follows:

Enter [1 2 3 4 5][ENTER] {3 2 } [ENTER] [↵][MTH][MATRX][MAKE][NXT][NXT][DIAG→] to get:

```
1:    [[ 1  0 ]
       [ 0  2 ]
       [ 0  0 ]]
```

In this case a 3 2 matrix was to be created using as main diagonal elements as many elements as possible form the vector [1 2 3 4 5]. The main diagonal, for a rectangular matrix, starts at position (1,1) and moves on to position (2,2), (3,3), etc. until either the number of rows or columns is exhausted. In this case, the number of columns (2) was exhausted before the number of rows (3), so the main diagonal included only the elements in positions (1,1) and (2,2). Thus, only the first two elements of the vector were required to form the main diagonal.

The function VANDERMONDE

The function VANDERMONDE generates the Vandermonde matrix of dimension n based on a given list of input data. The dimension n is, of course, the length of the list. If the input list consists of objects

$$\{x_1\ x_2\ ...\ x_n\},$$

then, a Vandermonde matrix in the HP 49 G is a matrix made of the following elements:

$$\begin{bmatrix} 1 & x_1 & x_1^2 & x_1^3 & ... & x_1^{n-1} \\ 1 & x_2 & x_2^2 & x_2^3 & ... & x_2^{n-1} \\ 1 & x_3 & x_3^2 & x_3^3 & ... & x_3^{n-1} \\ . & . & . & . & & . \\ . & . & . & . & & . \\ 1 & x_n & x_n^2 & x_n^3 & ... & x_n^{n-1} \end{bmatrix}$$

For example, enter { 1 2 3 4}[ENTER], and then [↵][MTH][MATRX][MAKE][NXT][NXT][VANDE] ,to get

```
1:    [[ 1  1   1   1 ]
       [ 1  2   4   8 ]
       [ 1  3   9  27 ]
       [ 1  4  16  64 ]]
```

So, you wonder, what is the utility of the Vandermonde matrix? One practical application, which we will be utilizing in a future chapter, is the fact that the Vandermonde matrix corresponding to a list $\{x_1\ x_2\ ...\ x_n\}$, which is in turn associated to a list $\{y_1\ y_2\ ...\ y_n\}$, can be used to determine a polynomial fitting of the form

$$y = b_0 + b_1 \cdot x + b_2 \cdot x^2 + ... + b_{n-1} \cdot x^{n-1},$$

to the data sets (x,y).

The function HILBERT

The function HILBERT creates the Hilbert matrix corresponding to a dimension n. By definition, the n×n Hilbert matrix is $H_n = [h_{jk}]$, so that

$$h_{jk} = \frac{1}{j+k-1}.$$

The Hilbert matrix has application in numerical curve fitting by the method of linear squares.

As an example, try [3] [↰][MTH][MATRX][MAKE][NXT][NXT][HILBE] to get:

```
1:   [[ 1   '1/2' '1/3' ]
      ['1/2' '1/3' '1/...']
      ['1/3' '1/4' '1/...']]
```

Using the matrix CREATE menu

As indicated earlier, all the calls to the functions in this section have been through the matrix MAKE menu, i.e., through [↰][MTH][MATRX][MAKE]. The following table shows the keystrokes to access the same functions, but using the matrix CREATE menu instead of the matrix MAKE menu.

Function	Matrix CREATE menu
IDN	[↰][MATRICES][CREAT][IDN]
CON	[↰][MATRICES][CREAT][CON]
→DIAG	[↰][MATRICES][CREAT][→DIAG]
DIAG→	[↰][MATRICES][CREAT][DIAG→]
GET	[↰][MATRICES][CREAT][NXT]
GETI	[↰][MATRICES][CREAT][NXT][GET]
HILBERT	[↰][MATRICES][CREAT][NXT][GETI]
PUT	[↰][MATRICES][CREAT][NXT][PUT]
PUTI	[↰][MATRICES][CREAT][NXT][PUTI]
RANM	[↰][MATRICES][CREAT][NXT][RANM]
RDM	[↰][MATRICES][CREAT][NXT][NXT][RDM]
REPL	[↰][MATRICES][CREAT][NXT][NXT][REPL]
SUB	[↰][MATRICES][CREAT][NXT][NXT][SUB]
VANDERMONDE	[↰][MATRICES][CREAT][NXT][NXT][VANDE]

A program to build a matrix out of a number of lists

Lists represent columns of the matrix

The program [CRMTC] allows you to put together a p×n matrix (i.e., p rows, n columns) out of n lists of p elements each. To use this program, enter the n lists in the order that you want them as columns of the matrix, enter the value of n, and press [CRMT].

To create the program enter the following keystrokes:

Keystroke sequence:	Produces:
[→][<< >>]	<<
[→][→][SPC][ALPHA][↵][N]	→ n
[→][<< >>]	<<
[↵][PRG][BRCH][FOR][FOR]	FOR
[ALPHA][↵][J]	j
[↵][PRG][TYPE][OBJ->]	OBJ→
[->ARR]	→ARRY
[↵][PRG][BRCH][IF][IF]	IF
[ALPHA][↵][J][SPC]	j
[ALPHA][↵][N]	n
[↵][PRG][TEST][<]	<
[↵][PRG][BRCH][IF][THEN]	THEN
[ALPHA][↵][J][SPC][1][+]	j 1 +
[↵][PRG][STACK][NXT][ROLL]	ROLL
[↵][PRG][BRCH][IF][END]	END
[↵][PRG][BRCH][FOR][NEXT]	NEXT
[↵][PRG][BRCH][IF][IF]	IF
[ALPHA][↵][N][SPC][1]	n 1
[↵][PRG][TEST][>]	>
[↵][PRG][BRCH][IF][THEN]	THEN
[1][SPC]	1
[ALPHA][↵][N][SPC][1][-]	n 1 -
[↵][PRG][BRCH][FOR][FOR]	FOR
[ALPHA][↵][J][SPC]	j
[ALPHA][↵][J][SPC][1][+]	j 1 +
[↵][PRG][STACK][NXT][ROLL]	ROLL
[↵][PRG][BRCH][FOR][NEXT]	NEXT
[↵][PRG][BRCH][IF][END]	END
[ALPHA][↵][N][SPC]	n
[↵][MTH][MATRX][COL][COL→]	COL→
[ENTER]	Program is displayed in level 1

To save the program:

[→]['][ALPHA][ALPHA][C][R][M][T][C] [ALPHA] [STO▶]

Note: if you save this program in your HOME directory it will be available from any other sub-directory you use.

To see the contents of the program use [r→][CRMTC]. The program listing is the following:

```
<< → n << DUP 'n' STO 1 SWAP FOR j OBJ→ →ARRY IF j n < THEN j 1 + ROLL
END NEXT IF n 1 > THEN 1 n 1 - FOR j j 1 + ROLL NEXT END n COL→ >> >>
```

As an example, try the following exercise:

{1 2 3 4} [ENTER] { 1 4 9 16} [ENTER] {1 8 27 64 }[ENTER] 3 [ENTER] [CRMTC]

The result is:

```
1:    [[ 1  1  1  ]
       [ 2  4  8  ]
       [ 3  9  27 ]
       [ 4  16 64 ]]
```

Lists represent rows of the matrix

The previous program can be easily modified to create a matrix when the input lists will become the rows of the resulting matrix. The only change to be performed is to change COL→ for ROW→ in the program listing. To perform this change use:

[r→][CRMTC]   List program CRMTC in stack
[▼][r→][▼][▲][◄][◄][◄][⇐][⇐][⇐]   Moves to end of program, delete COL
[ALPHA][ALPHA][R][O][W][ALPHA][ENTER]   Type in ROW, enter program

To store the program use

[r→]['][ALPHA][ALPHA][C][R][M][T][R] [ALPHA] [STO▶]

As an example, try the following exercise:

{1 2 3 4} [ENTER] { 1 4 9 16} [ENTER] {1 8 27 64 }[ENTER] 3 [ENTER] [CRMTR]

The result is:

```
1:    [[1 2 3 4]
       [ 1 4 9 16 ]
       [ 1 8 27 64 ]]
```

I developed these programs in a HP 48 GX calculator for statistical applications, specifically to create the statistical matrix ΣDAT. Examples of the use of these program are shown in a latter chapters.

Manipulating matrices by columns

The HP 49 G provides a menu with functions for manipulating matrices by operating in their columns. This menu is accessed by either [←][MTH][MATRX][COL], or by [←][MATRICES][CREAT][COL]. The functions provided by the COL menu are:

[→COL][COL→][COL+][COL-][CSWP][MATRX or CREAT]

The operation of these functions is presented below.

The function →COL

The function →COL takes a matrix in stack level 1, and decomposes it into vectors corresponding to its columns. The first column occupies the highest stack level after decomposition, and stack level 1 is occupied by the number of columns of the original matrix. The matrix does not survive decomposition, i.e., it is no longer available in the stack. For example, enter the matrix: [[2 -1 3] [3 5 4] [4 2 -7]]
[ENTER] , and press [←][MTH][MATRX][COL][→COL] to produce:

```
4:              [ 2  3  4 ]
3:              [ -1  5  2]
2:              [ 3  4  -7 ]
1:                        3.
```

The function COL→

The function COL→ has the opposite effect of the function →COL, i.e., given n vectors of the same length in stack levels n+1, n, n-1,...,2, and the number n in stack level 1, the function COL→ puts together the matrix by placing the vectors as columns in the resulting matrix. Since we already have the right set up in the stack from the previous exercise, just press [←][MTH][MATRX][COL][COL→] to produce:

```
1: [[ 2 -1  3 ]
    [ 3  5  4]
    [ 4  2 -7 ]
```

The function COL+

The function COL+ adds a vector, placed in stack level 2, at column n, where n is placed in stack level 1, of the matrix placed in stack level 3. For example, if we wanted to add the column [-1 0 1], after column 2, to the matrix we inherited from the previous exercise, use the following: [-1 0 1][ENTER][2][ENTER], then press [←][MTH][MATRX][COL][COL+] to obtain:

```
1: [[ 2 -1 -1  3 ]
    [ 3  0  5  4]
    [ 4  1  2 -7 ]
```

The function COL-

The function COL- extracts the column whose index is placed in stack level 1 from the matrix occupying stack level 2. The extracted column is then shown in stack level 1, as a vector, and the reduced matrix occupies stack level 2. For example, using the expression currently in the stack, enter [3], and [←][MTH][MATRX][COL][COL-] to obtain:

```
2:  [[ 2 -1  3 ]
     [ 3  0  4 ]
     [ 4  1 -7 ]]
1:           [-1 5 2]
```

The function CSWP

The function CSWP (Column SWaP) lets you swap the columns whose indices are listed in stack levels 1 and 2. For example, using the result from the previous exercise, first press [⇦] to drop the vector in stack level 1. Next, enter the values [1][ENTER][2][ENTER], and [←][MTH][MATRX][COL][CSWP] to obtain:

```
1:  [[ -1  2  3 ]
     [  0  3  4]
     [  1  4 -7 ]]
```

As you can see, the columns that originally occupied positions 1 and 2 have been swapped. Swapping of columns, and of rows, is commonly used when solving systems of linear equations with matrices. Details of these operations will be given in a later section.

Manipulating matrices by rows

The HP 49 G provides a menu with functions for manipulating matrices by operating in their rows. This menu is accessed by either [←][MTH][MATRX][ROW], or by [←][MATRICES][CREAT][ROW]. The functions provided by the ROW menu are:

[→ROW][ROW→][ROW+][ROW-][RCI][RCIJ]

Pressing [NXT]:

[RSWP][   ][   ][   ][   ][MATRX or CREAT]

The operation of these functions is presented below.

The function →ROW

The function →ROW takes a matrix in stack level 1, and decomposes it into vectors corresponding to its rows. The first row occupies the highest stack level after decomposition, and stack level 1 is occupied by the number of rows of the original matrix. The matrix does not survive decomposition, i.e., it is no longer available in the stack. For example, enter the matrix: [[2 -1 3] [3 5 4] [4 2 -7]]
[ENTER] , and press [←][MTH][MATRX][ROW][→ROW] to produce:

```
4:            [ 2 -1 3 ]
3:            [ 3  5 4]
2:            [ 4  2 -7 ]
1:                      3.
```

The function ROW→

The function ROW→ has the opposite effect of the function →ROW, i.e., given n vectors of the same length in stack levels n+1, n, n-1,...,2, and the number n in stack level 1, the function ROW→ puts together the matrix by placing the vectors as rows in the resulting matrix. Since we already have the right set up in the stack from the previous exercise, just press [←][MTH][MATRX][ROW][ROW→] to produce:

```
1: [[ 2 -1 3 ]
    [ 3  5 4]
    [ 4  2 -7 ]
```

The function ROW+

The function ROW+ adds a vector, placed in stack level 2, at row n, where n is placed in stack level 1, of the matrix placed in stack level 3. For example, if we wanted to add the row [-1 0 1], after row 2, to the matrix we inherited from the previous exercise, use the following: [-1 0 1][ENTER][2][ENTER], then press [←][MTH][MATRX][ROW][ROW+] to obtain:

```
1: [[ 2 -1 3 ]
    [ -1 0 1 ]
    [ 3  5 4 ]
    [ 4  2 -7 ]]
```

The function ROW-

The function ROW- extracts the row whose index is placed in stack level 1 from the matrix occupying stack level 2. The extracted row is then shown in stack level 1, as a vector, and the reduced matrix occupies stack level 2. For example, using the expression currently in the stack, enter [3], and [←][MTH][MATRX][ROW][ROW-] to obtain:

```
2:  [[  2 -1  3 ]
     [ -1  0  1]
     [  4  2 -7 ]]
1:              [3  5  4]
```

The function RSWP

The function RSWP (Row SWaP) lets you swap the rows whose indices are listed in stack levels 1 and 2. For example, using the result from the previous exercise, first press [◁] to drop the vector in stack level 1. Next, enter the values [1][ENTER][2][ENTER], and [↰][MTH][MATRX][ROW][NXT][RSWP] to obtain:

```
1:  [[ -1  0  1 ]
     [  2 -1  3]
     [  4  2 -7 ]]
```

As you can see, the rows that originally occupied positions 1 and 2 have been swapped. Swapping of rows, and of rows, is commonly used when solving systems of linear equations with matrices. Details of these operations will be given in a later section.

The function RCI

The function RCI stands for multiplying Row I by a Constant value and replace the resulting row at location I. For example, with the matrix left over from the previous exercise, enter [2][ENTER][1][ENTER] (or, you can enter [2][SPC][1][ENTER]). The number 2 is the constant that will multiply row 1. Then, the resulting row is replaced instead of the current value of row 1. Press [↰][MTH][MATRX][ROW][RCI] to get:

```
1:  [[ -2  0  2 ]
     [  2 -1  3]
     [  4  2 -7 ]]
```

The function RCIJ

The function RCIJ stands for taking Row I and multiplying it by a constant C and then add that multiplied row to row J, replacing row J with the resulting sum. For example, with the result from the previous example readily available, enter [2][+/-][ENTER][2][ENTER][3] (or, use [2][+/-][SPC][2][SPC][3]).
Then, press [↰][MTH][MATRX][ROW][RCI] to get:

```
1:  [[ -2  0  2 ]
     [  2 -1  3]
     [  4  2 -7 ]]
```

In this operation we first multiplied row 2 by -2, then added this result (-2 2 -6) to row 3 (0 4 -13), and replaced the sum for row 3.

The operation used by RCIJ is common in row operations in the solution of linear systems using matrices. More detailed in their use in linear solutions is presented later in this chapter.

Symmetric and anti-symmetric matrices

A square matrix, $A_{n \times n} = [a_{ij}]$, is said to be symmetric if $a_{ij} = a_{ji}$, for $i \neq j$, i.e., $A^T = A$. Also, a square matrix, $A_{n \times n} = [a_{ij}]$, is said to be anti-symmetric if $a_{ij} = -a_{ji}$, for $i \neq j$, i.e., $A^T = -A$.

The matrix A below is symmetric, while the matrix C below is anti-symmetric:

$$A = \begin{bmatrix} -2 & 3 & -5 \\ 3 & 6 & 4 \\ -5 & 4 & 2 \end{bmatrix}, \quad C = \begin{bmatrix} 12 & 1.3 & 15 \\ -1.3 & -2 & -2.4 \\ -15 & 2.4 & 5 \end{bmatrix}$$

Any square matrix, $B_{n \times n} = [b_{ij}]$ can be written as the sum of a symmetric $B'_{n \times n} = [b'_{ij}]$ and an anti-symmetric $B''_{n \times n} = [b''_{ij}]$ matrices. Because

$$b_{ij} = \tfrac{1}{2} \cdot (b_{ij} + b_{ji}) + \tfrac{1}{2} \cdot (b_{ij} - b_{ji}),$$

we can write

$$b'_{ij} = \tfrac{1}{2} \cdot (b_{ij} + b_{ji}), \text{ and } b''_{ij} = \tfrac{1}{2} \cdot (b_{ij} - b_{ji}).$$

Therefore,

$$B'_{n \times n} = \tfrac{1}{2} \cdot (B_{n \times n} + B^T_{n \times n}), \text{ and } B''_{n \times n} = \tfrac{1}{2} \cdot (B_{n \times n} - B^T_{n \times n}),$$

where $B^T_{n \times n} = [b^T_{ij}] = [b_{ji}]$ is the transpose of matrix B.

For example, take the matrix B33 defined earlier, and use the following keystrokes to find its symmetric and anti-symmetric components:

[VAR][B33][ENTER]   Enter matrix B33 and create a copy
[←][MATRICES][OPER][NXT][NXT][TRAN]   Calculate transpose of B33
[+][2][÷]   Add B33 and its transpose and divide by 2

The result is a symmetric matrix.

[VAR][B33][ENTER]   Enter matrix B33 and create a copy
[←][MATRICES][OPER][NXT][NXT][TRAN]   Calculate transpose of B33
[+][2][÷]   Subtract the transpose from B33, divide by 2

The result is an anti-symmetric matrix.

[+]   Recovers original matrix B33.

Matrices and solution of linear equation systems

A system of n linear equations in m variables can be written as

$$a_{11}x_1 + a_{12}x_2 + a_{13}x_3 + \ldots + a_{1,m-1}x_{m-1} + a_{1,m}x_m = b_1,$$
$$a_{21}x_1 + a_{22}x_2 + a_{23}x_3 + \ldots + a_{2,m-1}x_{m-1} + a_{2,m}x_m = b_2,$$
$$a_{31}x_1 + a_{32}x_2 + a_{33}x_3 + \ldots + a_{3,m-1}x_{m-1} + a_{3,m}x_m = b_3,$$
$$\vdots$$
$$a_{n-1,1}x_1 + a_{n-1,2}x_2 + a_{n-1,3}x_3 + \ldots + a_{n-1,m-1}x_{m-1} + a_{n-1,m}x_m = b_{n-1},$$
$$a_{n1}x_1 + a_{n2}x_2 + a_{n3}x_3 + \ldots + a_{n,m-1}x_{m-1} + a_{n,m}x_m = b_n.$$

This system of linear equations can be written as a matrix equation,

$$A_{n \times m} \cdot x_{m \times 1} = b_{n \times 1},$$

if we define the following matrices:

Solution to a system of linear equations using the numerical solver (NUM.SLV)

There are many ways to solve a system of linear equations. The calculator provides an input screen for solving linear equations through the keystroke sequence [↱][NUM.SLV][▼][▼][▼][OK]. The linear system solver screen looks as:

```
###### SOLVE SYSTEM A·X=B ######
A:
B:
X:
```

To solve the linear system A·x = b, enter the matrix A, as [[...][....]] in the A: field. Enter the vector b in the B: field. When the X: field is highlighted, press [SOLVE]. If a solution is available, the solution vector x will be shown in the X: field. The solution is also copied to stack level 1. Some examples follow.

Example 1 - A system with more unknowns than equations

The system of linear equations

$$2x_1 + 3x_2 - 5x_3 = -10,$$
$$x_1 - 3x_2 + 8x_3 = 85,$$

can be written as the matrix equation A·x = b, if

$$\mathbf{A} = \begin{bmatrix} 2 & 3 & -5 \\ 1 & -3 & 8 \end{bmatrix}, \quad \mathbf{x} = \begin{bmatrix} x_1 \\ x_2 \\ x_3 \end{bmatrix}, \quad \text{and} \quad \mathbf{b} = \begin{bmatrix} -10 \\ 85 \end{bmatrix}.$$

This system has more unknowns than equations, therefore, it is not uniquely determined. We can visualize the meaning of this statement by realizing that each of the linear equations represents a plane in the three-dimensional Cartesian coordinate system (x_1, x_2, x_3). The solution to the system of equations shown above will be the intersection of two planes in space. We know, however, that the intersection of two (non-parallel) planes is a straight line, and not a single point. Therefore, there is more than one point that satisfy the system. In that sense, the system is not uniquely determined.

Let's use the numerical solver to attempt a solution to this system of equations:

[→][NUM.SLV][▼][▼][▼][OK]　　　　Launch numerical solver for linear system
[↵][[ ]] [↵][[ ]] [2][SPC][3][SPC][5][+/-][▶]　Enter the first row of matrix A
[↵][[ ]] [1][SPC][3][+/-][SPC][8][OK]　　Finish entering matrix A
[↵][[ ]] [1][0][+/-][SPC][8][5][OK]　　　Enter the vector b
[SOLVE]　　　　　　　　　　　　Obtain a solution

The result is

```
###### SOLVE SYSTEM A·X=B ######
A: [[ 2. 3. -5. ] [ 1...
B: [ -10. 85. ]
X: [ 15.3731343284 2. ...
```

To see the solution in the stack press [ON] or [ENTER]. The solution is

$$x = [15.3731343284 \quad 2.46268656716 \quad 9.6268657164].$$

To check that the solution is correct, try the following:

[→][EVAL]　　　　　　　　　　Remove the tag from the result
[↵][[ ]] [↵][[ ]] [2][SPC][3][SPC][5][+/-][▶]　Enter the first row of matrix A
[↵][[ ]] [1][SPC][3][+/-][SPC][8][ENTER]　Finish entering matrix A
[▶][×]　　　　　　　　　　　　Exchange stack levels 1 and 2, multiply A·x

The result is [-9.99999999999 85.], close enough to [-10 85], the original vector b.

Try also this:

[↵][CMD][OK][ENTER]　　　　　　To recover the matrix A
[↵][[ ]] [1][5][SPC][→]['][1][0][÷][3] [▶][1][0][ENTER] Enter a new x = [15 '10/3' 10]
[×]　　　　　　　　　　　　　Calculate A·x

The result is [-10 85]. Therefore, x = [15 '10/3' 10] is also a solution to the system, confirming our observation that a system with more unknowns than equations is not uniquely determined.

So, how does the calculator came up with the solution x = [15.37... 2.46... 9.62...] shown earlier? Actually, the calculator minimizes the distance from a point, which will constitute the solution, to each of the planes represented by the equations in the linear system. The calculator uses a least-square method, i.e., minimizes the sum of the squares of those distances or errors.

Example 2 - A system with more equations than unknowns

The system of linear equations

$$x_1 + 3x_2 = 15,$$
$$2x_1 - 5x_2 = 5,$$
$$-x_1 + x_2 = 22,$$

can be written as the matrix equation A·x = b, if

$$\mathbf{A} = \begin{bmatrix} 1 & 3 \\ 2 & -5 \\ -1 & 1 \end{bmatrix}, \quad \mathbf{x} = \begin{bmatrix} x_1 \\ x_2 \end{bmatrix}, \quad and \quad \mathbf{b} = \begin{bmatrix} 15 \\ 5 \\ 22 \end{bmatrix}.$$

This system has more equations than unknowns. The system does not have a single solution. Each of the linear equations in the system presented above represents a straight line in a two-dimensional Cartesian coordinate system (x_1, x_2). Unless two of the three equations in the system represent the same equation, the three lines will have two different intersection points. For that reason, the solution is not unique. Some numerical algorithms can be used to force a solution to the system by minimizing the distance from the presumptive solution point to each of the lines in the system. Such is the approach followed by the HP 49 G numerical solver.

Let's use the numerical solver to attempt a solution to this system of equations:

[↱][NUM.SLV][▼][▼][▼][OK]	Launch numerical solver for linear system
[↰][[ ]] [↰][[ ]] [1][SPC][3][▶]	Enter the first row of matrix A
[↰][[ ]] [2][SPC][5][+/-][▶]	Enter the first row of matrix A
[↰][[ ]] [1][+/-][SPC][1][OK]	Finish entering matrix A
[↰][[ ]] [1][5][SPC][5][SPC][2][2][OK]	Enter the vector b
[SOLVE]	Obtain a solution

The result is

```
###### SOLVE SYSTEM A·X=B ######
A: [[ 1. 3.] [ 2. -5....
B: [ 15. 5. 22. ]
X: [ 3.02054794521 1. ...
```

To see the solution in the stack press [ON] or [ENTER]. The solution is

x = [15.3731343284   2.46268656716   9.6268657164].

To check that the solution is correct, try the following:

[↱][EVAL]	Remove the tag from the result
[↰][[ ]] [↰][[ ]] [1][SPC][3][▶]	Enter the first row of matrix A
[↰][[ ]] [2][SPC][5][+/-][▶]	Enter the first row of matrix A
[↰][[ ]] [1][+/-][SPC][1][ENTER]	Finish entering matrix A
[▶][×]	Exchange stack levels 1 and 2, multiply A·x

The result is [8.6917... -3.4109... -1.1301...], which is not equal to [15 5 22], the original vector b. The "solution" is simply the point that is closest to the three lines represented by the three equations in the system.

Example 3 - A system with equal number of equations and unknowns

The system of linear equations

$$2x_1 + 3x_2 - 5x_3 = 13,$$
$$x_1 - 3x_2 + 8x_3 = -13,$$
$$2x_1 - 2x_2 + 4x_3 = -6,$$

can be written as the matrix equation A·x = b, if

$$\mathbf{A} = \begin{bmatrix} 2 & 3 & -5 \\ 1 & -3 & 8 \\ 2 & -2 & 4 \end{bmatrix}, \quad \mathbf{x} = \begin{bmatrix} x_1 \\ x_2 \\ x_3 \end{bmatrix}, \quad and \quad \mathbf{b} = \begin{bmatrix} 13 \\ -13 \\ -6 \end{bmatrix}.$$

This system has the same number of equations as of unknowns. In general, there should be a unique solution to the system. The solution will be the point of intersection of the three planes in the coordinate system (x_1, x_2, x_3).

Let's use the numerical solver to attempt a solution to this system of equations:

[↱][NUM.SLV][▼][▼][▼][OK]   Launch numerical solver for linear system
[↰][[ ]] [↰][[ ]] [2][SPC][3][SPC][5][+/-][▶]   Enter the first row of matrix A
[↰][[ ]] [1][SPC][3][+/-][SPC][8][▶]   Enter the second row of matrix A
[↰][[ ]] [2][SPC][2][+/-][SPC][4][OK]   Finish entering matrix A
[↰][[ ]] [1][3][SPC][1][3][+/-][SPC][6][+/-][OK] Enter the vector b
[SOLVE]   Obtain a solution

The result is

```
###### SOLVE SYSTEM A·X=B ######
A: [[ 2. 3. -5. ] [ 1...
B: [ 13. -13. -6. ]
X: [ 1. 2. -1.]
```

To see the solution in the stack press [ON] or [ENTER]. The solution is x = [1. 2. -=1.].

To check that the solution is correct, try the following:

[↱][EVAL]   Remove the tag from the result
[↰][[ ]] [↰][[ ]] [2][SPC][3][SPC][5][+/-][▶]   Enter the first row of matrix A
[↰][[ ]] [1][SPC][3][+/-][SPC][8][▶]   Enter the second row of matrix A
[↰][[ ]] [2][SPC][2][+/-][SPC][4][OK]   Finish entering matrix A
[▶][×]   Exchange stack levels 1 and 2, multiply A·x

The result is [13. -13. -6.], the original vector b.

Direct solution of a linear system in the stack

If you have a linear system of equations such that its coefficient matrix, A, is square, you can solve the system directly in the stack by dividing the vector b by the coefficient matrix A. In other words, you are calculating x = b/A. This last equation will make any mathematician's back hairs stand up on end, for it is the most improper equation you can write in matrix algebra since matrix division is an operation that is not defined. Still, that is what the calculator produces, b/A. Try this approach using the linear system from the last example:

[↵][[ ]] [1][3][SPC][1][3][+/-][SPC][6][+/-][ENTER]   Enter the vector b
[↵][[ ]] [↵][[ ]] [2][SPC][3][SPC][5][+/-][▶]   Enter the first row of matrix A
[↵][[ ]] [1][SPC][3][+/-][SPC][8][▶]   Enter the second row of matrix A
[↵][[ ]] [2][SPC][2][+/-][SPC][4][ENTER]   Finish entering matrix A
[÷]   Calculate x = b/A.

The solution is again [1 2 -1].

Solution using the inverse matrix

The proper way to write the solution to the system A·x = b is x = A^{-1}· b. This results from multiplying the first equation by A^{-1}, i.e., A^{-1}·A·x = A^{-1}·b. By definition, A^{-1}·A = I, thus we write I·x = A^{-1}·b. Also, I·x = x, thus, we have,

$$x = A^{-1} \cdot b.$$

For the example used earlier we have:

[↵][[ ]] [↵][[ ]] [2][SPC][3][SPC][5][+/-][▶]   Enter the first row of matrix A
[↵][[ ]] [1][SPC][3][+/-][SPC][8][▶]   Enter the second row of matrix A
[↵][[ ]] [2][SPC][2][+/-][SPC][4][ENTER]   Finish entering matrix A
[1/x]   Calculate the inverse A^{-1}.
[↵][[ ]] [1][3][SPC][1][3][+/-][SPC][6][+/-][ENTER]   Enter the vector b
[×]   Calculate x = A^{-1}· b.

The result is, once more, [1. 2. -1].

Characterizing a matrix

In the previous section we mentioned that, under certain conditions, a system of n linear equations with n unknowns may not have a unique solution. In order to determine when such situations occur, we can use certain measures or norms that characterize a matrix. Some of those measures are the determinant and the rank of the matrix. These and other ways of characterizing a matrix are presented in this section.

The matrix NORM menu

The matrix NORM menu is accessed through the keystroke sequence [↵][MTH][MATRX][NORM]. This menu contains the following functions:

[ABS][SNRM][RNRM][CNRM][SRAD][COND]

Pressing [NXT]:
[RANK][DET][TRACE][TRAN][   ][MATRX].

These functions are described following. Because many of these functions use concepts of matrix theory, such as singular values, rank, etc., we will include short descriptions of these concepts intermingled with the description of functions.

The function ABS

The function ABS calculates what is known as the Frobenius norm of a matrix. For a matrix $A_{m \times n} = [a_{ij}]$, the Frobenius norm of the matrix is defined as

$$\| \mathbf{A} \|_F = \sqrt{\sum_{i=1}^{n} \sum_{j=1}^{m} | a_{ij} |^2 }.$$

If the matrix under consideration in a row vector or a column vector, then the Frobenius norm, $\|A\|_F$, is simply the vector's magnitude. The function ABS is accessible directly in the keyboard as [←][ABS] ([÷] key).

Try the following exercises (using the matrices stored earlier for matrix operations):

[A11] [←][ABS]   The result is 3.
[A12] [←][ABS]   The result is '√61'.
[A13] [←][ABS]   The result is '√41'.
[A22] [←][ABS]   The result is '√61'.
[A23] [←][ABS]   The result is '√103'.
[A33] [←][ABS]   The result is '√118'.

Matrix decomposition

To understand the operation of the function SNRM, presented below, we need to introduce the concept of matrix decomposition. Basically, matrix decomposition involves the determination of two or more matrices that, when multiplied in a certain order (and, perhaps, with some matrix inversion or transposition thrown in), produce the original matrix. So, you will see later examples of what is called LU decomposition, where a square matrix A is written as A = L·U, where L is a lower-triangular matrix, and U is an upper-triangular matrix. (A lower-triangular matrix is such that elements above and to the right of the main diagonal are zero, while an upper-triangular matrix is such that elements below and to the left of the main diagonal are zero.) An example of LU decomposition is shown below:

$$\mathbf{A} = \begin{bmatrix} 2 & 4 & -2 \\ 4 & 9 & -3 \\ -2 & -3 & 7 \end{bmatrix} = \begin{bmatrix} 1 & 0 & 0 \\ 2 & 1 & 0 \\ -1 & 1 & 1 \end{bmatrix} \cdot \begin{bmatrix} 2 & 4 & -2 \\ 0 & 1 & 1 \\ 0 & 0 & 4 \end{bmatrix} = \mathbf{L} \cdot \mathbf{U}$$

Singular value decomposition and rank

The Singular Value Decomposition (SVD) is such that a rectangular matrix $A_{m \times n}$ is written as

$$A_{m \times n} = U_{m \times m} \cdot S_{m \times n} \cdot V^T_{n \times n},$$

Where U and V are orthogonal matrices, and S is a diagonal matrix. The diagonal elements of S are called the <u>singular values</u> of A and are usually ordered so that $s_i \geq s_{i+1}$, for i = 1, 2, ..., n-1. The columns [u_j] of U and [v_j] of V are the corresponding <u>singular vectors</u>. (<u>Orthogonal matrices</u> are such that $U \cdot U^T = I$).

The <u>rank</u> of a matrix can be determined from its SVD by counting the number of non-singular values. Examples of SVD will be presented in a subsequent section.

The function SNRM

The function SNRM calculates the Spectral NoRM of a matrix, which is defined as the matrix's largest singular value. This value corresponds also to the Euclidean norm of the matrix. For example,

[A23] [↵][MTH][MATRX][NORM][SNRM]   produces the value 9.53341026006.

Row norm and column norm of a matrix

The row norm of a matrix is calculated by taking the sums of the absolute values of all elements in each row, and then, selecting the maximum of these sums. The column norm of a matrix is calculated by taking the sums of the absolute values of all elements in each column, and then, selecting the maximum of these sums.

The functions RNRM and CNRM

The function RNRM returns the Row NoRM of a matrix, while the function CNRM returns the Column NoRM of a matrix. Examples,

[A23] [↵][MTH][MATRX][NORM][RNRM]   produces the value 10.
[A23] [↵][MTH][MATRX][NORM][CNRM]   produces the value 13.

Eigenvalues and eigenvectors of a matrix

The eigenvalues of a square matrix result from the matrix equation $A \cdot x = \lambda \cdot x$. The values of λ that satisfy the equation are known as the eigenvalues of the matrix A. The values of x that result from the equation for each value of λ are known as the eigenvectors of the matrix. Further details on calculating eigenvalues and eigenvectors are presented later in the chapter.

The function SRAD

The function SRAD determines the Spectral RADius of a matrix, defined as the largest of the absolute values of its eigenvalues. For example,

[A33] [↰][MTH][MATRX][NORM][SRAD]   produces the value 8.97201587219.

Determinants, singular matrices, and conditions numbers

In Chapter 9 we introduce the concept of determinants for square matrices of dimensions 2×2 and 3×3. Calculation of determinants for matrices of higher order is described later in this section. It will be enough at this point to indicate that it is possible to calculate the determinant of any square matrix. If the determinant of matrix A, written det A, is zero, then the matrix A is said to be singular. Otherwise, it is non-singular. Singular matrices do not have an inverse.

The condition number of a square non-singular matrix is defined as the product of the matrix norm times the norm of its inverse, i.e.,
$$\text{cond}(A) = ||A|| \cdot ||A^{-1}||.$$

We will choose as the matrix norm, $||A||$, the maximum of its row norm (RNRM) and column norm (CNRM). The condition number of a singular matrix is infinity. The condition number of a non-singular matrix is a measure of how close the matrix is to being singular. The larger the value of the condition number, the closer it is to singularity.

The function COND

The function COND determines the condition number of a matrix. Try the following exercise:

[A33][ENTER][ENTER][ENTER][ENTER]   Place matrix A33 in stack and make 4 copies of it.
[↰][MTH][MATRX][NORM][COND]   Produces the value cond(A33) = 3.58536585366.
[◁][RNRM]   Produces the value RNRM(A33) = 12
[▶][CNRM]   Produces the value CNRM(A33) = 12

Since RNRM(A33) = CNRM(A33) = 12, then we take $||A33|| = 12$.

[◁][◁][▶][1/x][RNRM]   Swap levels 1 & 2, obtain RNRM(A33^{-1}) = '49/164'
[▶][1/x][CNRM]   Produces the value CNRM(A33^{-1}) = '49/164'

As before, since RNRM(A33⁻¹) = CNRM(A33⁻¹) = '49/164', then we take ||A33⁻¹|| = '49/164'.

[1][2][×]   [→][→NUM]   Multiply ||A33⁻¹|| and ||A33||.

The result of the latter multiplication is 3. 58536585366. Thus, we have ||A33⁻¹|| · ||A33|| = cond(A33).

The rank of a matrix

The rank of a square matrix is the maximum number of linearly independent rows or columns that the matrix contains. Suppose that you write a square matrix $A_{n \times n}$ as $A = [c_1 \; c_2 \; ... \; c_n]$, where c_i (i = 1, 2, ..., n) are vectors representing the columns of the matrix A, then, if any of those columns, say c_k, can be written as

$$c_k = \sum_{j \neq k, j \in \{1,2,...,n\}} d_j \cdot c_j,$$

where the values d_j are constant, we say that c_k is linearly dependent on the columns included in the summation. (Notice that the values of j include any value in the set {1, 2, ..., n}, in any combination, as long as j≠k.) If the expression shown above cannot be written for any of the column vectors then we say that all the columns are linearly independent. A similar definition for the linear independence of rows can be developed by writing the matrix as a column of row vectors.

Thus, if we find that rank(A) = n, then the matrix has an inverse and it is a non-singular matrix. If, on the other hand, rank(A) < n, then the matrix is singular and no inverse exist.

The function RANK

The function RANK determines the rank of a square matrix. Try the following examples:

[A33][←][MTH][MATRX][NORM][NXT][RANK]   Produces the value rank(A33) = 3.

Try finding the rank for the matrix:

[[1 2 3][2 4 6][5 -2 1]][ENTER]
[←][MTH][MATRX][NORM][NXT][RANK]

You will find that the rank is 2. That is because the second row [2 4 6] is equal to the first row [1 2 3] multiplied by 2, thus, row two is linearly dependent of row 1 and the maximum number of linearly independent rows is 2. You can check that the maximum number of linearly independent columns is 3. The rank being the maximum number of linearly independent rows or columns becomes 2 for this case.

The determinant of a matrix

In chapter 9 we presented a way to calculate the determinant of 2×2 and 3×3 matrices. For square matrices of higher order determinants can be calculated by using smaller order determinant called cofactors. The general idea is to "expand" a determinant of a n×n matrix (also referred to as a n×n determinant) into a sum of the cofactors, which are (n-1)×(n-1) determinants, multiplied by the elements of a single row or column, with alternating positive and negative signs. This "expansion" is then carried to the next (lower) level, with cofactors of order (n-2)×(n-2), and so on, until we are left only with a long sum of 2×2 determinants. The 2×2 determinants are then calculated through the method presented in chapter 9.

The method of calculating a determinant by cofactor expansion is very inefficient in the sense that it involves a number of operations that grows very fast as the size of the determinant increases. A more efficient method, and the one preferred in numerical applications, is to use a result from Gaussian elimination. The method of Gaussian elimination is used to solve systems of linear equations. Details of this method are presented in a later part of this chapter.

To refer to the determinant of a matrix A, we write det(A). As mentioned in chapter 9, the determinant of a matrix can also be written as the elements of the matrix enclosed between vertical bars, for example, given

$$\mathbf{A} = \begin{bmatrix} 2 & -1 \\ 5 & 7 \end{bmatrix}, \quad \text{then} \quad \det(\mathbf{A}) = \begin{vmatrix} 2 & -1 \\ 5 & 7 \end{vmatrix} = 2 \cdot 7 - (-1) \cdot 5 = 19.$$

Note: A singular matrix has a determinant equal to zero.

The function DET

The function DET calculates the determinant of a square matrix. For example,

[A33] [←][MTH][MATRX][NORM][NXT][DET] produces the value det(A33) = 164.

Properties of determinants

The calculation of determinants by hand typically requires simplifying the determinant to reduce the number of operations in its calculation. Manipulation of the determinant includes operations such as multiplying or dividing a row or column by a constant, exchanging rows or columns, or replacing a row or column by a linear combination of other rows or columns. Whenever one of these operations take place, it is necessary to modify the expression for the determinant to ensure that its value does not change. Some of the rules of determinant manipulation are the following:

(1) Multiplying or dividing a row or column in a determinant by a constant is equivalent to multiplying or dividing the determinant by that constant. For example, consider the determinant used in the previous example

$$\det(\mathbf{A}) = \begin{vmatrix} 2 & -1 \\ 5 & 7 \end{vmatrix} = 19.$$

If we multiply any row by 2, the determinant gets multiplied by 2. For example (check these with your
calculator):

$$\begin{vmatrix} 4 & -2 \\ 5 & 7 \end{vmatrix} = 2 \cdot 19 = 38, \quad \begin{vmatrix} 2 & -1 \\ 10 & 14 \end{vmatrix} = 38, \quad \begin{vmatrix} 4 & -1 \\ 10 & 7 \end{vmatrix} = 38, \quad \begin{vmatrix} 2 & -2 \\ 5 & 14 \end{vmatrix} = 38, \quad \begin{vmatrix} 4 & -2 \\ 10 & 14 \end{vmatrix} = 2 \cdot 2 \cdot 19 = 76.$$

(2) Switching any two rows or columns produces a change of sign in the determinant. For example, for the same case presented earlier we have:

$$\begin{vmatrix} 5 & 7 \\ 2 & -1 \end{vmatrix} = -19, \quad \begin{vmatrix} -1 & 2 \\ 7 & 5 \end{vmatrix} = -19.$$

(3) A row (or column) in a determinant can be replaced by a linear combination of rows (or columns) without changing the value of the determinant. For example, referring to det(A) above, we will replace the second row by the linear combination resulting from multiplying the first row by 2 and adding the second row multiplied by -1, i.e., 2·[5 7] +(-1)· [2 -1] = [8 15], i.e.,

$$\begin{vmatrix} 5 & 7 \\ 8 & 15 \end{vmatrix} = 5 \cdot 15 - 7 \cdot 8 = 19.$$

Of course, with the HP 49 G calculator you no longer need to calculate determinants by hand. Still, to understand their calculation and some of the matrix elimination methods to be presented later, you need to keep in mind these rules.

Example: In the following example, we use determinant operations to simplify its calculation. First, we divide the first row by 5, the second by 3, and the third by 2.

$$\begin{vmatrix} 5 & 10 & -20 \\ 3 & -12 & 9 \\ 8 & 6 & -2 \end{vmatrix} = 5 \cdot 3 \cdot 2 \cdot \begin{vmatrix} 1 & 2 & -4 \\ 1 & -4 & 3 \\ 4 & 3 & -1 \end{vmatrix}$$

Next, we replace row 2 with (row 1 - row 2), and row 3 with (4·row1-row3):

$$5 \cdot 3 \cdot 2 \cdot \begin{vmatrix} 1 & 2 & -4 \\ 1 & -4 & 3 \\ 4 & 3 & -1 \end{vmatrix} = 5 \cdot 3 \cdot 2 \cdot \begin{vmatrix} 1 & 2 & -4 \\ 0 & -6 & -7 \\ 0 & 5 & -15 \end{vmatrix}$$

The next step in the simplification is to divide the second row by -6 and the third row by 5, to get

$$5 \cdot 3 \cdot 2 \cdot \begin{vmatrix} 1 & 2 & -4 \\ 0 & 6 & -7 \\ 0 & 5 & -15 \end{vmatrix} = 5 \cdot 3 \cdot 2 \cdot (-6) \cdot 5 \cdot \begin{vmatrix} 1 & 2 & -4 \\ 0 & 1 & -7/6 \\ 0 & 1 & -3 \end{vmatrix}$$

Finally, we replace the third row with (row 2 - row 3), i.e.,

$$5 \cdot 3 \cdot 2 \cdot (-6) \cdot 5 \cdot \begin{vmatrix} 1 & 2 & -4 \\ 0 & 1 & -7/6 \\ 0 & 1 & -3 \end{vmatrix} = 5 \cdot 3 \cdot 2 \cdot (-6) \cdot 5 \cdot \begin{vmatrix} 1 & 2 & -4 \\ 0 & 1 & -7/6 \\ 0 & 0 & 11/6 \end{vmatrix}$$

Now we use the method for calculating 3×3 determinants presented in chapter 9, i.e.,

$$5 \cdot 3 \cdot 2 \cdot (-6) \cdot 5 \cdot \begin{vmatrix} 1 & 2 & -4 \\ 0 & 1 & -7/6 \\ 0 & 0 & 11/6 \end{vmatrix} = 5 \cdot 3 \cdot 2 \cdot (-6) \cdot 5 \cdot \begin{vmatrix} 1 & 2 & -4 \\ 0 & 1 & -7/6 \\ 0 & 0 & 11/6 \end{vmatrix} \begin{matrix} 1 & 2 \\ 0 & 1 \\ 0 & 0 \end{matrix} =$$

$$= 5 \cdot 3 \cdot 2 \cdot (-6) \cdot 5 \cdot (1 \cdot 1 \cdot 11/6 + 0 + 0 - (0+0+0)) = -1650.$$

You can check, using the calculator's function DET, that indeed, det(A) = -1650.

Cramer's rule for solving systems of linear equations

Cramer's rule for solving systems of n linear equations with n unknowns, consists in forming the matrix equation, A·x=b, as presented earlier in the chapter, and calculating the determinant Δ = det(A). After that, for each unknown x_i, the matrix A_i is formed consisting of the matrix A with column i replaced by the components of vector b. The determinant corresponding to A_i is called Δ_i = det(A_i). The unknown x_i is then calculated as $x_i = \Delta_i/\Delta$.

For example, we will use Cramer's rule to determine the solution to the following system of linear equations:

$$X + 2Y+3Z+4R = 4,$$
$$-X+5Y+2Z+7R = 11,$$
$$4X+2Y-Z+6R=2,$$
$$2X+Y-4Z+7R = 9.$$

To write the system as a matrix equation we use:

$$\mathbf{A} = \begin{bmatrix} 1 & 2 & 3 & 4 \\ -1 & 5 & 2 & 7 \\ 4 & 2 & -1 & 6 \\ 2 & 1 & -4 & 7 \end{bmatrix}, \quad \mathbf{x} = \begin{bmatrix} X \\ Y \\ Z \\ R \end{bmatrix}, \quad \text{and} \quad \mathbf{b} = \begin{bmatrix} 4 \\ 11 \\ 2 \\ 9 \end{bmatrix}.$$

Next, we form the matrix A_X, calculate the determinants Δ, Δ_X, and solve fore X as follows:

$$\Delta = \det(\mathbf{A}) = \begin{vmatrix} 1 & 2 & 3 & 4 \\ -1 & 5 & 2 & 7 \\ 4 & 2 & -1 & 6 \\ 2 & 1 & -4 & 7 \end{vmatrix} = -258,$$

$$\mathbf{A}_X = \begin{bmatrix} 4 & 2 & 3 & 4 \\ 11 & 5 & 2 & 7 \\ 2 & 2 & -1 & 6 \\ 9 & 1 & -4 & 7 \end{bmatrix}, \quad \Delta_X = \det(A_X) = 516, \quad X = \frac{\Delta_X}{\Delta} = \frac{516}{-258} = -12.$$

You can use your calculator to form the matrices A_Y, A_Z, and A_R, and calculate the determinants Δ_Y, Δ_Z, and Δ_R, previous to calculating the other unknowns Y, Z, and R. The solution is: Y = , Z = , R =.

The function TRACE

The function TRACE calculates the trace of square matrix, defined as the sum of the elements in its main diagonal, or

$$tr(\mathbf{A}_{n \times n}) = \sum_{i=1}^{n} a_{ii}.$$

Using Einstein's repeated index convention we can write simply, $tr(A_{n \times n}) = a_{ii}$.

For example, [A33] [←][MTH][MATRX][NORM][NXT][DET] produces tr(A33) = -8.

The function TRAN

The function TRAN returns the transpose of a real or the conjugate transpose of a complex matrix. TRAN is equivalent to TRN.

Additional matrix operations

The matrix OPER menu

the functions in the [←][MTH][MATRX][NORM] menu are accessible through [←][MATRICES][OPER] menu , which also includes the functions AXL, AXM, HADAMARD, LSQ, MAD, RSD, and SIZE. The function SIZE was presented earlier when we discussed the matrix MAKE menu. We discussed the functions AXL, AXM, HADAMARD, LSQ, MAD, and RSD, below.

The function AXL

The function AXL converts an array (matrix) into a list, and vice versa. For example,

[[3 4 5] [2 -1 0]] [←][MATRICES][OPER][AXL] produces {{ 3 4 5} {2 -1 0}}
{{ 1 5 4} {2 3 1}} [←][MATRICES][OPER][AXL] produces [[3 4 5] [2 -1 0]]

Note: the latter operation is similar to that of the program CRMTR presented earlier in this chapter.

The function AXM

The function AXM converts an array containing integer or fraction elements into its corresponding decimal, or approximate, form. For example,

[['1/2' '2/3'] [1 '-3/4']] [←][MATRICES][OPER][AXM] produces [[.5 .66666666666][1. -.75]].

The function HADAMARD

The function HADAMARD perform a term by term multiplication of two matrices of the same order. This is called the Hadamard product of two matrices. For example, calculate the Hadamard product of the following two matrices:

[[1 2][2 -1]] [ENTER] [[0 -1][5 8]] [ENTER] [←][MATRICES][OPER][NXT][HADAM]

The result is the matrix [[0 -2] [10 -8]].

The function LSQ

The LSQ function returns the minimum-norm least-square solution of a linear system Ax = b.

- If A is a square matrix and A is non-singular, LSQ returns the exact solution to the linear system.
- If A has less than full row rank (underdetermined system of equations), LSQ returns the solution with the minimum Euclidean length out of an infinity number of solutions.
- If A has less than full column rank (over-determined system of equations), LSQ returns the "solution" with the minimum residual value ε = A·x - b. The system of equations may not have a solution, therefore, the value returned is not a real solution to the system, just the one with the smallest residual.
- The function LSQ takes as input the vector b in stack level 2, and the matrix A in stack level 1.

Try the following examples:

Example 1 - Square matrix:

[14 18 15][ENTER][[1 2 3][4 -5 8][7 1 2]][ENTER] [↤][MATRICES][OPER][NXT][LSQ] results in [1. 2. 3.]. You can check that this is the exact solution by multiplying:

[[1 2 3][4 -5 8][7 1 2]][ENTER][1 2 3][×] to recover [14 18 15].

Example 2 - Underdetermined equation system:

[14 18][ENTER][[1 2 3][7 1 2]][ENTER] [↤][MATRICES][OPER][NXT][LSQ] results in [1.46892655367 1.90960451977 2.90395480226], which is a solution to the system of equations [[1 2 3][7 1 2]][[x][y][z]] = [14 18].

To verify that it is a solution, type [[1 2 3] [7 1 2]][ENTER][▶][×] to get [14. 18.].

The solution returned earlier is the one with the smallest Euclidean length, i.e., the smallest distance to the two planes represented by the equations in the system.

Example 3 - Over-determined equation system:

[14 18 15] [ENTER][[1 2][4 -5][7 1]][ENTER] [↤][MATRICES][OPER][NXT][LSQ] results in [2.8041958042 -0.538461538462].

Let's check what results from multiplying the A matrix times the "solution" found above. Type [[1 2][4 -5][7 1]][ENTER][▶][×] to get [1.72727272728 13.9090909091 19.0909090909].

The error ε = A·x - b, can be calculated by entering [14 18 15][-]. The result is [-12.272727272727 -4.090909090909 4.090909090909]. The "solution" found above represents a point which is close to two of the three straight lines represented by the equation system, as illustrated in the figure below.

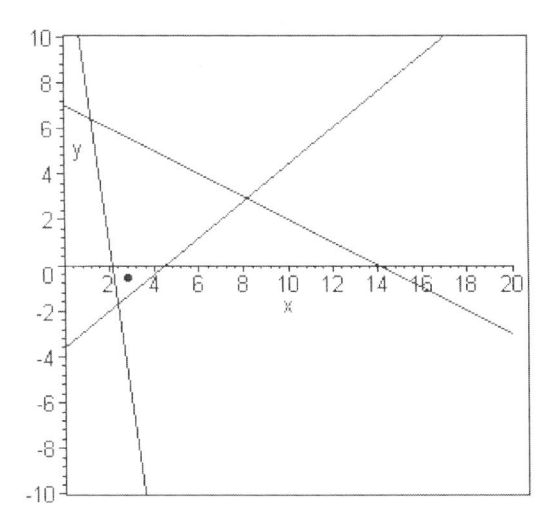

The function MAD

The function MAD is used to generate a number of properties of a square matrix, such as the determinant, the formal inverse (i.e., in exact mode), and other information related to what is known as the characteristic polynomial or characteristic equation of the matrix. The idea of characteristic polynomial of a matrix is presented in more detail in the section on eigenvalues and eigenvectors, therefore, we postpone discussion of this function until then.

The function RSD

The function RSD calculates the ReSiDuals or errors in the solution of the matrix equation $A \cdot x = b$, representing a system of n linear equations in n unknowns. We can think of solving this system as solving the matrix equation:

$$f(x) = b - A \cdot x = 0.$$

Suppose that, through a numerical method, we produce as a first approximation the solution $x_{(0)}$. Evaluating $f(x_{(0)}) = b - A \cdot x_{(0)} = \varepsilon \neq 0$. Thus, ε is a vector of residuals of the function for the vector $x = x_{(0)}$.

To use the function RSD you need to have the terms b, A, and $x_{(0)}$, in stack levels 3, 2, and 1, respectively. The vector returned is

$$\varepsilon = b - A \cdot x_{(0)}.$$

For example, using A = [[2 -1][0 2]], $x_{(0)}$ = [1.8 2.7], and b = [1 6], we can find the vector of residuals as follows:
[1 6] [ENTER] [[2 -1][0 2]] [ENTER] [1.8 2.7] [ENTER] [↰][MATRICES][OPER][RSD]

The result is $\varepsilon = b - A \cdot x_{(0)} = [\,0.1\ 0.6\,]$.

Note: If we let the vector $\Delta x = x - x_{(0)}$, represent the correction in the values of $x_{(0)}$, we can write a new matrix equation for Δx, namely $A \cdot \Delta x = \varepsilon$. Solving for Δx we can find the actual solution of the original system as $x = x_{(0)} + \Delta x$.

The function LCXM

The function LCXM can be used to generate matrices such that the element a_{ij} is a function of i and j. The input to this function consists of two integers, n and m, representing the number of rows and columns of the matrix to be generated, and a program that takes i and j as input. The numbers n, m, and the program occupy stack levels 3, 2, and 1, respectively. The function LCXM is accessible through the command catalog [CAT].

For example, to generate a 2×3 matrix whose elements are given by $a_{ij} = (i+j)^2$, use:

[2][ENTER][3][ENTER] << → i j << '(i+j)^2' EVAL >> [ENTER]
[CAT][ALPHA][ALPHA][L][C][X][ALPHA][OK]

To generate: [[4 9 16][9 16 25]].

Gaussian and Gauss-Jordan elimination

Gaussian elimination is a procedure by which the square matrix of coefficients belonging to a system of n linear equations in n unknowns is reduced to an upper-triangular matrix (echelon form) through a series of row operations. This procedure is known as forward elimination. The reduction of the coefficient matrix to an upper-triangular form allows for the solution of all n unknowns, utilizing only one equation at a time, in a procedure known as backward substitution.

Gaussian elimination using a system of equations

To illustrate the Gaussian elimination procedure we will use the following system of 3 equations in 3 unknowns:

$$2X + 4Y + 6Z = 14,$$
$$3X - 2Y + Z = -3,$$
$$4X + 2Y - Z = -4.$$

Forward elimination

First, we divide the first equation by 2, to get

$$X + 2Y + 3Z = 7,$$
$$3X - 2Y + Z = -3,$$
$$4X + 2Y - Z = -4.$$

Next, we replace the second equation by (equation 1 - 3*equation 2), and the third by (equation 1 - 4*equation 1), to get

$$X + 2Y + 3Z = 7,$$
$$-8Y - 8Z = -24,$$
$$-6Y - 13Z = -32.$$

Next, divide the second equation by -8, to get

$$X + 2Y + 3Z = 7,$$
$$Y + Z = 3,$$
$$-6Y - 13Z = -32.$$

Next, replace the third equation with (equation 2 + 6*equation 3), to get

$$X + 2Y + 3Z = 7,$$
$$Y + Z = 3,$$
$$-7Z = -14.$$

Backward substitution

The system of equations is now in an upper-triangular form, allowing us to solve for Z first, as

$$Z = -14/-7 = 2.$$

We then replace this value into equation 2 to solve for Y, as

$$Y = 3 - Z = 3 - 2 = 1.$$

Finally, we replace the values of Y and Z into equation 1 to solve for X as

$$X = 7-2Y-3Z = 7-2(1)-3(2) = 7-2-6 = -1.$$

The solution is, therefore, $\boxed{X = -1, Y = 1, Z = 2.}$

Gaussian elimination using matrices

The system of equations used in the example above, i.e.,

$$2X + 4Y + 6Z = 14,$$
$$3X - 2Y + Z = -3,$$
$$4X + 2Y - Z = -4.$$

Can be written as a matrix equation A·x = b, if we use:

$$\mathbf{A} = \begin{pmatrix} 2 & 4 & 6 \\ 3 & -2 & 1 \\ 4 & 2 & -1 \end{pmatrix}, \quad \mathbf{x} = \begin{bmatrix} X \\ Y \\ Z \end{bmatrix}, \quad \mathbf{b} = \begin{bmatrix} 14 \\ -3 \\ -4 \end{bmatrix}.$$

To obtain a solution to the system matrix equation using Gaussian elimination, we first create what is known as the <u>augmented matrix</u> corresponding to A, i.e.,

$$\mathbf{A}_{aug} = \begin{pmatrix} 2 & 4 & 6 & | & 14 \\ 3 & -2 & 1 & | & -3 \\ 4 & 2 & -1 & | & -4 \end{pmatrix}.$$

The matrix $\mathbf{A}_{aug}$ is nothing more than the original matrix A with a new row, corresponding to the elements of the vector b, added (i.e., augmented) to the right of the rightmost column of A.

Once the augmented matrix is put together, we can proceed to perform row operations on it that will reduce the original A matrix into an upper-triangular matrix. In your calculator, use the following keystrokes:

[[2 4 6 14][3 -2 1 -3][4 2 -1 -4]][ENTER][ENTER]   Enter augmented matrix, make extra copy
[↵][`][ALPHA][ALPHA][A][A][U][G][ALPHA][STO▶]   Save augmented matrix in variable AAUG

This step is not necessary, except as an insurance that you have an extra copy of the augmented matrix saved in case you make a mistake in the forward elimination procedure that we are about to undertake.

With a copy of the augmented matrix in the stack, press [↵][MTH][MATRX][ROW] to activate the ROW operation menu. Next, perform the following row operations on your augmented matrix.

[2][1/x][1][RCI]　　　　　　　　　Multiply row 1 by ½
[3][+/-][SPC][1][SPC][2][RCIJ]　　　Multiply row 1 by -3 add it to row 2, replacing it.
[4][+/-][SPC][1][SPC][3][RCIJ]　　　Multiply row 1 by -4 add it to row 3, replacing it.
[8][+/-][1/x][2][RCI]　　　　　　　Multiply row 2 by -1/8.
[6][SPC][2][SPC][3][RCIJ]　　　　　Multiply row 2 by 6 add it to row 3, replacing it.

If you were performing these operations by hand, you would write the following:

$$\mathbf{A}_{aug} = \begin{pmatrix} 2 & 4 & 6 & | & 14 \\ 3 & -2 & 1 & | & -3 \\ 4 & 2 & -1 & | & -4 \end{pmatrix} \cong \begin{pmatrix} 1 & 2 & 3 & | & 7 \\ 3 & -2 & 1 & | & -3 \\ 4 & 2 & -1 & | & -4 \end{pmatrix} \cong \begin{pmatrix} 1 & 2 & 3 & | & 7 \\ 0 & -8 & -8 & | & -24 \\ 0 & -6 & -13 & | & -32 \end{pmatrix},$$

$$\mathbf{A}_{aug} = \begin{pmatrix} 1 & 2 & 3 & | & 7 \\ 0 & 1 & 1 & | & 3 \\ 0 & -6 & -13 & | & -32 \end{pmatrix} \cong \begin{pmatrix} 1 & 2 & 3 & | & 7 \\ 0 & 1 & 1 & | & 3 \\ 0 & 0 & -7 & | & -14 \end{pmatrix}$$

The symbol ≅ (" is equivalent to") indicates that what follows is equivalent to the previous matrix with some row (or column) operations involved.

The resulting matrix is upper-triangular, and equivalent to the set of equations

$$X + 2Y + 3Z = 7,$$
$$Y + Z = 3,$$
$$-7Z = -14,$$

which can now be solved, one equation at a time, by backward substitution, as in the previous example.

Gauss-Jordan elimination using matrices

Gauss-Jordan elimination consists in continuing the row operations in the upper-triangular matrix resulting from the forward elimination process until an identity matrix results in place of the original A matrix. For example, for the case we just presented, we can continue the row operations as follows:

[7][+/-][1/x][3][RCI]　　　　　　　Multiply row 3 by -1/7.
[1][+/-][SPC][3][SPC][2][RCIJ]　　　Multiply row 3 by -1, add it to row 2, replacing it.
[3][+/-][SPC][3][SPC][1][RCIJ]　　　Multiply row 3 by -3, add it to row 1, replacing it.
[2][+/-][SPC][2][SPC][1][RCIJ]　　　Multiply row 2 by -2, add it to row 1, replacing it.

Writing the result by hand will result in

$$\mathbf{A}_{aug} = \begin{pmatrix} 1 & 2 & 3 & | & 7 \\ 0 & 1 & 1 & | & 3 \\ 0 & 0 & -7 & | & -14 \end{pmatrix} \cong \begin{pmatrix} 1 & 2 & 3 & | & 7 \\ 0 & 1 & 1 & | & 3 \\ 0 & 0 & 1 & | & 2 \end{pmatrix} \cong \begin{pmatrix} 1 & 2 & 3 & | & 7 \\ 0 & 1 & 1 & | & 1 \\ 0 & 0 & 1 & | & 2 \end{pmatrix} \cong \begin{pmatrix} 1 & 2 & 0 & | & 1 \\ 0 & 1 & 0 & | & 1 \\ 0 & 0 & 1 & | & 2 \end{pmatrix} \cong \begin{pmatrix} 1 & 0 & 0 & | & -1 \\ 0 & 1 & 0 & | & 1 \\ 0 & 0 & 1 & | & 2 \end{pmatrix}$$

The final result is equivalent to the equations: X = -1, Y = 1, Z = 2, which is the solution to the original system of equations.

Pivoting

If you look carefully at the row operations in the examples shown above, you will notice that many of those operations divide a row by its corresponding element in the main diagonal. This element is called a pivot element, or simply, a pivot. In many situations it is possible that the pivot element become zero, in which case we can not divide the row by its pivot. Also, to improve the numerical solution of a system of equations using Gaussian or Gauss-Jordan elimination, it is recommended that the pivot be the element with the largest absolute value in a given column. This operation is called partial pivoting. To follow this recommendation is it often necessary to exchange rows in the augmented matrix while performing a Gaussian or Gauss-Jordan elimination.

While performing pivoting in a matrix elimination procedure, you can improve the numerical solution even more by selecting as the pivot the element with the largest absolute value in the column and row of interest. This operation may require exchanging not only rows, but also columns, in some pivoting operations. When row and column exchanges are allowed in pivoting, the procedure is known as full pivoting.

When exchanging rows and columns in partial or full pivoting, it is necessary to keep track of the exchanges because the order of the unknowns in the solution is altered by those exchanges. One way to keep track of column exchanges in partial or full pivoting mode, is to create a permutation matrix $P = I_{n \times n}$, at the beginning of the procedure. Any row or column exchange required in the augmented matrix A_{aug} is also registered as a row or column exchange, respectively, in the permutation matrix. When the solution is achieved, then, we multiply the permutation matrix by the unknown vector x to obtain the order of the unknowns in the solution. In other words, the final solution is given by $P \cdot x = b'$, where b' is the last column of the augmented matrix after the solution has been found.

Example of Gauss-Jordan elimination with full pivoting

Let's illustrate full pivoting with an example. Solve the following system of equations using full pivoting and the Gauss-Jordan elimination procedure:

$$X + 2Y + 3Z = 2,$$
$$2X + 3Z = -1,$$
$$8X + 16Y - Z = 41.$$

The augmented matrix and the permutation matrix are as follows:

$$\mathbf{A}_{aug} = \begin{bmatrix} 1 & 2 & 3 & 2 \\ 2 & 0 & 3 & -1 \\ 8 & 16 & -1 & 41 \end{bmatrix}, \quad \mathbf{P} = \begin{bmatrix} 1 & 0 & 0 \\ 0 & 1 & 0 \\ 0 & 0 & 1 \end{bmatrix}.$$

Enter the augmented matrix in variable AAUG, then press [↱][AAUG] to get a copy in the stack. First, I recommend you change the display font option to small by using:

[MODE][DISP][▼][▼][✓CHK] (make sure a check is placed in the Stack: _ small option) [OK][OK].

We want to keep the CSWP (Column Swap) command readily available, for which we use:

[CAT][ALPHA][ALPHA][C][S][ALPHA] (find CSWP), [OK]

You'll get an error message, press [ON], and ignore the message.

Next, get the ROW menu available by pressing: [↤][MATRICES][CREAT][ROW].

Now we are ready to start the Gauss-Jordan elimination with full pivoting. We will need to keep track of the permutation matrix by hand, so take your notebook and write the P matrix shown above.

First, we check the pivot a_{11}. We notice that the element with the largest absolute value in the first row and first column is the value of a_{31} = 8. Since we want this number to be the pivot, then we exchange rows 1 and 3, by using: [1][SPC][3][NXT][RSWP]. The augmented matrix and the permutation matrix now are:

1	3	NXT	RSWP	NXT

$$\begin{bmatrix} 8 & 16 & -1 & 41 \\ 2 & 0 & 3 & -1 \\ 1 & 2 & 3 & 2 \end{bmatrix} \quad \begin{bmatrix} 0 & 0 & 1 \\ 0 & 1 & 0 \\ 0 & 0 & 1 \end{bmatrix}$$

Checking the pivot at position (1,1) we now find that 16 is a better pivot than 8, thus, we perform a column swap as follows:

1	2	CAT	OK

$$\begin{bmatrix} 16 & 8 & -1 & 41 \\ 0 & 2 & 3 & -1 \\ 2 & 1 & 3 & 2 \end{bmatrix} \quad \begin{bmatrix} 0 & 0 & 1 \\ 1 & 0 & 0 \\ 0 & 1 & 0 \end{bmatrix}$$

Now we have the largest possible value in position (1,1), i.e., we performed full pivoting at (1,1). Next, we proceed to divide by the pivot:

16	1/x	1	RCI

$$\begin{bmatrix} 1 & 1/2 & -1/16 & 41/16 \\ 0 & 2 & 3 & -1 \\ 2 & 1 & 3 & 2 \end{bmatrix} \quad \begin{bmatrix} 0 & 0 & 1 \\ 1 & 0 & 0 \\ 0 & 1 & 0 \end{bmatrix}$$

The next step is to eliminate the 2 from position (3,2) by using:

| 2 | +/- | SPC | 1 | SPC | 3 | RCIJ |

$$\begin{bmatrix} 1 & 1/2 & -1/16 & 41/16 \\ 0 & 2 & 3 & -1 \\ 0 & 0 & 25/8 & -25/8 \end{bmatrix} \quad \begin{bmatrix} 0 & 0 & 1 \\ 1 & 0 & 0 \\ 0 & 1 & 0 \end{bmatrix}$$

Having filled up with zeros the elements of column 1 below the pivot, now we proceed to check the pivot at position (2,2). We find that the number 3 in position (2,3) will be a better pivot, thus, we exchange columns 2 and 3 by using:

| 2 | 3 | CAT | OK |

$$\begin{bmatrix} 1 & -1/16 & 1/2 & 41/16 \\ 0 & 3 & 2 & -1 \\ 0 & 25/8 & 0 & -25/8 \end{bmatrix} \quad \begin{bmatrix} 0 & 1 & 0 \\ 1 & 0 & 0 \\ 0 & 0 & 1 \end{bmatrix}$$

Checking the pivot at position (2,2), we now find that the value of 25/8, at position (3,2), is larger than 3. Thus, we exchange rows 2 and 3 by using:

| 2 | 3 | NXT | RSWP | NXT |

$$\begin{bmatrix} 1 & -1/16 & 1/2 & 41/16 \\ 0 & 25/8 & 0 & -25/8 \\ 0 & 3 & 2 & -1 \end{bmatrix} \quad \begin{bmatrix} 0 & 1 & 0 \\ 0 & 0 & 1 \\ 1 & 0 & 0 \end{bmatrix}$$

Now, we are ready to divide row 2 by the pivot 25/8, by using (make sure that you enter [↱]['][2][5][÷][8][▶] to produce the value '25/8' in the stack):

| '8/25 ' | SPC | 2 | RCI |

$$\begin{bmatrix} 1 & -1/16 & 1/2 & 41/16 \\ 0 & 1 & 0 & -1 \\ 0 & 3 & 2 & -1 \end{bmatrix} \quad \begin{bmatrix} 0 & 1 & 0 \\ 0 & 0 & 1 \\ 1 & 0 & 0 \end{bmatrix}$$

Next, we eliminate the 3 from position (3,2) by using:

| 3 | +/- | SPC | 2 | SPC | 3 | RCIJ |

$$\begin{bmatrix} 1 & -1/16 & 1/2 & 41/16 \\ 0 & 1 & 0 & -1 \\ 0 & 0 & 2 & 2 \end{bmatrix} \quad \begin{bmatrix} 0 & 1 & 0 \\ 0 & 0 & 1 \\ 1 & 0 & 0 \end{bmatrix}$$

Having filled with zeroes the position below the pivot, we proceed to check the pivot at position (3,3). The current value of 2 is larger than ½ or 0, thus, we keep it unchanged. We do divide the whole third row by 2 to convert the pivot to 1, by using:

| 2 | 1/x | 3 | RCI |

$$\begin{bmatrix} 1 & -1/16 & 1/2 & 41/16 \\ 0 & 1 & 0 & -1 \\ 0 & 0 & 1 & 1 \end{bmatrix} \quad \begin{bmatrix} 0 & 1 & 0 \\ 0 & 0 & 1 \\ 1 & 0 & 0 \end{bmatrix}$$

Next, we proceed to eliminate the ½ in position (1,3) by using:

| 1/2' | SPC | 3 | SPC | 1 | RCIJ |

$$\begin{bmatrix} 1 & -1/16 & 0 & 33/16 \\ 0 & 1 & 0 & -1 \\ 0 & 0 & 1 & 1 \end{bmatrix} \quad \begin{bmatrix} 0 & 1 & 0 \\ 0 & 0 & 1 \\ 1 & 0 & 0 \end{bmatrix}$$

Finally, we eliminate the -1/16 from position (1,2) by using:

| '1/16' | SPC | 2 | SPC | 1 | RCIJ |

$$\begin{bmatrix} 1 & 0 & 0 & 2 \\ 0 & 1 & 0 & -1 \\ 0 & 0 & 1 & 1 \end{bmatrix} \quad \begin{bmatrix} 0 & 1 & 0 \\ 0 & 0 & 1 \\ 1 & 0 & 0 \end{bmatrix}$$

We now have an identity matrix in the portion of the augmented matrix corresponding to the original coefficient matrix A, thus we can proceed to obtain the solution while accounting for the row and column exchanges coded in the permutation matrix P. We identify the unknown vector x, the modified independent vector b' and the permutation matrix P as:

$$\mathbf{x} = \begin{bmatrix} X \\ Y \\ Z \end{bmatrix}, \quad \mathbf{b'} = \begin{bmatrix} 2 \\ -1 \\ 1 \end{bmatrix}, \quad \mathbf{P} = \begin{bmatrix} 0 & 1 & 0 \\ 0 & 0 & 1 \\ 1 & 0 & 0 \end{bmatrix}.$$

The solution is given by P·x=b', or

$$\begin{bmatrix} 0 & 1 & 0 \\ 0 & 0 & 1 \\ 1 & 0 & 0 \end{bmatrix} \cdot \begin{bmatrix} X \\ Y \\ Z \end{bmatrix} = \begin{bmatrix} 2 \\ -1 \\ 1 \end{bmatrix}.$$

Which results in

$$\begin{bmatrix} Y \\ Z \\ X \end{bmatrix} = \begin{bmatrix} 2 \\ -1 \\ 1 \end{bmatrix}.$$

Step-by-step calculator procedure for solving linear systems

The example we just worked is, of course, the step-by-step, user-driven procedure to use full pivoting for Gauss-Jordan elimination solution of linear equation systems. You can see the step-by-step procedure used by the calculator to solve a system of equations, without user intervention, by setting the step-by-step option in the calculator's CAS, as follows:

[MODE][CAS] [▼][▼][▼][✓CHK] [OK][OK].

Then, for this particular example, use:

[2 -1 41][ENTER][[1 2 3][2 0 3][8 16 -1]] [ENTER] [÷]

The calculator shows an augmented matrix consisting of the coefficients matrix A and the identity matrix I, while, at the same time, showing the next procedure to calculate:

$$\begin{array}{c} L2 = L2\text{-}2 \cdot L1 \\ \begin{bmatrix} 1 & 2 & 3 & 1 & 0 & 0 \\ 2 & 0 & 3 & 0 & 1 & 0 \\ 8 & 16 & -1 & 0 & 0 & 1 \end{bmatrix} \end{array}$$

L2 = L2-2·L1 stands for "replace row 2 (L2) with the operation L2 - 2·L1. If we had done this operation by hand, it would have corresponded to: [2][+/-][SPC][1][SPC][2][RCIJ]. Press [OK], and follow the operations in your calculator's screen. You will see the following operations performed:

L3=L3-8·L1, L1 = 2·L1--1·L2, L1=25·L--3·L3, L2 = 25·L2-3·L3,

and finally a message indicating "Reduction result" showing:

$$\begin{array}{c} \text{Reduction result} \\ \begin{bmatrix} 50 & 0 & 0 & -24 & 25 & 3 \\ 0 & -100 & 0 & -26 & 25 & -3 \\ 0 & 0 & -25 & -8 & 0 & 1 \end{bmatrix} \end{array}$$

When you press [OK], the calculator returns the final result [1 2 -1].

What the calculator showed was not exactly a Gauss-Jordan elimination with full pivoting, but a way to calculate the inverse of a matrix by performing a Gauss-Jordan elimination, without pivoting. This procedure for calculating the inverse is based on the augmented matrix $(A_{aug})_{n \times n}$ =

[A $_{n\times n}$ |I$_{n\times n}$]. The calculator showed you the steps up to the point in which the left-hand half of the augmented matrix has been converted to a diagonal matrix. From there, the next step is to divide each row by the corresponding main diagonal pivot, which, for the result shown above would have looked as:

$$\begin{bmatrix} 1 & 0 & 0 & -24/50 & 1/2 & 3/50 \\ 0 & 1 & 0 & 26/100 & -1/4 & 3/100 \\ 0 & 0 & 1 & 8/25 & 0 & -1/25 \end{bmatrix}$$

In other words, the calculator has transformed (A$_{aug}$)$_{n\times n}$ = [A $_{n\times n}$ |I$_{n\times n}$], into [I |A^{-1}]. To calculate the solution to the linear system, the calculator then multiplies the inverse matrix times the independent vector b, i.e., x = A$^{-1}\cdot$b.

Of course, if you uncheck the step-by-step option in the calculator's CAS, and you try this example again, you will just get the solution as a result.

Solving multiple set of equations with the same coefficient matrix

Suppose that you want to solve the following three sets of equations:

```
X +2Y+3Z = 14,      2X +4Y+6Z = 9,      2X +4Y+6Z = -2,
3X -2Y+ Z = 2,      3X -2Y+ Z = -5,     3X -2Y+ Z = 2,
4X +2Y -Z = 5,      4X +2Y -Z = 19,     4X +2Y -Z = 12.
```

We can write the three systems of equations as a single matrix equation: A·X = B, where

$$\mathbf{A} = \begin{bmatrix} 1 & 2 & 3 \\ 3 & -2 & 1 \\ 4 & 2 & -1 \end{bmatrix}, \quad \mathbf{X} = \begin{bmatrix} X_{(1)} & X_{(2)} & X_{(3)} \\ Y_{(1)} & Y_{(2)} & Y_{(3)} \\ Z_{(1)} & Z_{(2)} & Z_{(3)} \end{bmatrix}, \quad \mathbf{B} = \begin{bmatrix} 14 & 9 & -2 \\ 2 & -5 & 2 \\ 5 & 19 & 12 \end{bmatrix}.$$

The sub-indices in the variable names X, Y, and Z, determine to which equation system they refer to. To solve this expanded system we use the following procedure:

[[14 9 -2][2 -5 2][5 19 12]][ENTER] [[1 2 3][3 -2 1][4 2 -1]][ENTER][÷]

The result of this operation, after showing the various intermediate steps, is:

$$\mathbf{X} = \begin{bmatrix} 1 & 2 & 2 \\ 2 & 5 & 1 \\ 3 & -1 & -2 \end{bmatrix}.$$

Calculating the inverse matrix step-by-step

As mentioned before, the calculation of an inverse matrix can be considered as calculating the solution to the augmented system [A | I]. For example, for the matrix A used in the previous example, we would write this augmented matrix as

$$\mathbf{A}_{aug(I)} = \begin{bmatrix} 1 & 2 & 3 & | & 1 & 0 & 0 \\ 3 & -2 & 1 & | & 0 & 1 & 0 \\ 4 & 2 & -1 & | & 0 & 0 & 1 \end{bmatrix}.$$

To see the intermediate steps in calculating and inverse, just enter the matrix A from above, and press [1/x], while keeping the step-by-step option active in the calculator's CAS. Use the following:

[[1 2 3] [3 -2 1] [4 2 -1]] [ENTER][1/x].

After going through the different steps, the solution returned is:

$$\mathbf{A}^{-1} = \begin{bmatrix} 9/29 & -2/29 & 8/29 \\ -7/29 & 8/29 & -3/29 \\ 17/29 & -7/29 & -1/29 \end{bmatrix}.$$

Inverse matrices and determinants

Notice that all the elements in the inverse matrix calculated above are divided by the value 29. If you calculate the determinant of the matrix A, by using:

[[1 2 3] [3 -2 1] [4 2 -1]] [↵][MTH][MATRIX][NORM][NXT][DET],

you get det(A) = 56 = 2·29.

We could write, A^{-1} = C/det(A), where C is the matrix

$$\mathbf{C} = \begin{bmatrix} 18 & -4 & 16 \\ -14 & 16 & -6 \\ 34 & -14 & -2 \end{bmatrix}.$$

The result $(A^{-1})_{n \times n} = C_{n \times n} / det(A_{n \times n})$, is a general result that applies to any non-singular matrix A. A general form for the elements of C can be written based on the Gauss-Jordan algorithm.

The fact that I want to point out, however, is that, based on the equation A^{-1} = C/det(A), the inverse matrix, A^{-1}, is not defined if det(A) = 0.
Thus, the condition det(A) = 0 defines also a singular matrix.

Solution to linear systems using calculator functions

The simplest way to solve a system of linear equations, A·x = b, in the calculator is to enter b, enter A, and then use the division function [÷]. If the system of linear equations is over-determined or under-determined, a "solution" can be produced by using the function LSQ (Least-SQuares). The calculator, however, offers other possibilities for solving linear systems of equations by using the functions included in the matrices' LIN S (LINear Solutions) menu accessible through [↰][MATRICES][LIN S]. The functions included are:

[LINSO][REF][rref][RREF].

The function LINSOLVE

The function LINSOLVE takes an array of equations from stack level 2 and a vector containing the names of the unknowns and produces the solution to the linear system. For example, enter the following:

'X-2*Y+Z=-8' [ENTER] '2*X+Y-2*Z=6' [ENTER] '5*X-2*Y+Z=-12' [ENTER] [3][ENTER]
[↰][PRG][TYPE][→ARRY]['X' 'Y' 'Z'][ENTER] [↰][MATRICES][LIN S][LINSO]

to produce the solution: ['X=-1' 'Y=2' 'Z = -3'].

Thus, the function LINSOLVE works with symbolic expressions. The other functions in this menu, REF, rref, and RREF, work with the augmented matrix.

Functions REF, rref, RREF

The upper triangular form to which the augmented matrix is reduced during the forward elimination part of a Gaussian elimination procedure is known as an "echelon" form. <u>The function REF</u> (Reduce to Echelon Form) produces such a matrix given the augmented matrix in stack level 1.

Consider the augmented matrix,

$$\mathbf{A}_{aug} = \begin{bmatrix} 1 & -2 & 1 & | & 0 \\ 2 & 1 & -2 & | & -3 \\ 5 & -2 & 1 & | & 12 \end{bmatrix}.$$

Representing a linear system of equations, A·x = b, where

$$A = [[1 -2\ 1][2\ 1 -2][5 -2\ 1]],$$

and

$$b = [[0][-3][12]].$$

Enter the augmented matrix, and save it into variable AAUG:

[[1 -2 1 0][2 1 -2 -3][5 -2 1 12]][ENTER] [→]['][ALPHA][ALPHA][A][A][U][G][ALPHA][STO▶].

Press [AAUG] to bring the augmented matrix back to stack level 1, then press:

[←][MATRICES][LIN S][REF].

The result is the upper triangular (echelon form) matrix of coefficients resulting from the forward elimination step in a Gaussian elimination procedure:

$$\begin{bmatrix} 1 & 2 & 3 & | & 2 \\ 0 & 1 & '3/4' & | & '5/4' \\ 0 & 0 & 1 & | & -1 \end{bmatrix}.$$

The diagonal matrix that results from a Gauss-Jordan elimination is called a row-reduced echelon form. The function RREF (Row-Reduced Echelon Form) produces a list of the pivots and an equivalent matrix in row-reduced echelon form so that the matrix of coefficients is reduced to a diagonal matrix.

As an example, press [AAUG] to bring the augmented matrix back to stack level 1, then press:

[←][MATRICES][LIN S][RREF].

The result is the list of pivots and the diagonal matrix of coefficients resulting from the a Gauss Jordan elimination without pivoting:

Pivots: { 25 2. 2 1.}

$$\begin{bmatrix} 50 & 0 & 0 & | & 50 \\ 0 & -100 & 0 & | & -200 \\ 0 & 0 & -25 & | & 25 \end{bmatrix}.$$

A row-reduced echelon form for an augmented matrix can be obtained by using the function rref. The results of this function call is to produce the row-reduced echelon form so that the matrix of coefficients is reduced to an identity matrix. The extra column in the augmented matrix will contain the solution to the system of equations.

For example, press [AAUG] to bring the augmented matrix back to stack level 1, then press:

[←][MATRICES][LIN S][RREF].

The result is the diagonal matrix of coefficients resulting from the a Gauss-Jordan elimination with full pivot:

$$\begin{bmatrix} 1 & 0 & 0 & | & 1 \\ 0 & 1 & 0 & | & 2 \\ 0 & 0 & 1 & | & -1 \end{bmatrix}.$$

Eigenvalues and eigenvectors

Given a square matrix A, we can write the eigenvalue equation

$$A \cdot x = \lambda x,$$

where the values of λ that satisfy the equation are known as the <u>eigenvalues of matrix A</u>. For each value of λ, we can find, from the same equation, values of x that satisfy the eigenvalue equation. These values of x are known as the <u>eigenvectors of matrix A</u>.

The eigenvalues equation can be written also as

$$(A - \lambda I)x = 0.$$

This equation will have a non-trivial solution only if the matrix $(A - \lambda I)$ is singular, i.e., if

$$\det(A - \lambda I) = 0.$$

The last equation generates an algebraic equation involving a polynomial of order n for a square matrix $A_{n \times n}$. The resulting equation is known as the characteristic polynomial of matrix A. Solving the characteristic polynomial produces the eigenvalues of the matrix.

The HP 49 G calculator provides a number of functions to produce information regarding the eigenvalues and eigenvectors of a square matrix. Some of these functions are located under the menu [←][MATRICES][EIGEN].

The function PCAR

The function PCAR generates the characteristic polynomial of a square matrix using the contents of variable VX (a CAS reserved variable) as the unknown in the polynomial. For example,

[[1 5 -3][2 -1 4][3 5 2]] [←][MATRICES][EIGEN][PCAR] produces 'X^3+-2*X^2+-22*X+21'.

Using the variable λ to represent eigenvalues, this characteristic polynomial is to be interpreted as

$$\lambda^3 - 2\lambda^2 - 22\lambda + 21 = 0.$$

The function EGVL

The function EGVL (EiGenVaLues) produces the eigenvalues of a square matrix. For example, enter:

[[2 3] [2 -2]] [ENTER] [←][MATRICES][EIGEN][EGVL]

to produce the eigenvalues λ = ['-√10' '√10'].

Note: In some cases, you may not be able to find an 'exact' solution to the characteristic polynomial, and you will get an empty list as a result when using the function EGVL. If that were to happen to you, change the calculation mode to Approx in the CAS, i.e., [MODE][CAS] [▼][▼][✓CHK][OK][OK], and repeat the calculation.

For example, in exact mode, [[1 -2 1][2 -1 2][5 -2 1]] [ENTER] [←][MATRICES][EIGEN][EGVL] produces { }. Change mode to Approx and repeat the entry, to get [(-0.5849, 1.5923), (-0.5849, -1.5923), (4.1698,0.)].

The function EGV

The function EGV (EiGenValues and eigenvectors) produces the eigenvalues and eigenvectors of a square matrix. The eigenvectors are returned as the columns of a matrix in stack level 2, while the corresponding eigenvalues are the components of a vector returned in stack level 1.

For example, enter:

[[2 -1 1][-1 5 3][1 3 4]][ENTER] [←][MATRICES][EIGEN][EGV]

The result is

```
2: [[ 1.00  1.00 -0.03…
    [0.79 -0.51  1.00…
    [-0.91 0.65  0.84…
1:  [0.29  3.26  7.54 ]
```

Or, $\lambda_1 = 0.29$, $x_1 = [\ 0.79\ -0.91\ 0.29]^T$,
$\lambda_2 = 3.26$, $x_2 = [-0.51,\ 0.65,\ 3.26]^T$, and
$\lambda_3 = 7.54$, $x_1 = [-0.03,\ 1.00,\ 0.84]^T$.

Note: A symmetric matrix produces all real eigenvalues, and its eigenvectors are mutually perpendicular. For the example just worked out, you can check that $x_1 \bullet x_2 = 0$, $x_1 \bullet x_3 = 0$, and $x_2 \bullet x_3 = 0$.

The function JORDAN

The function JORDAN produces four outputs given a square matrix, namely:

The minimum polynomial of matrix A (stack level 4)
The characteristic polynomial of matrix A (stack level 3)
A list with the eigenvectors corresponding to each eigenvalue of matrix A (stack level 2)
A vector with the eigenvectors of matrix A (stack level 4)

For example, try the exercise:

[[4 1 -2] [1 2 -1][-2 -1 0]] [←][MATRICES][EIGEN][JORDAN].

The output is the following:

```
4:    'X^3+-6*x^2+2*X+8'
3:    'X^3+-6*x^2+2*X+8'
2: {-0.94: [4.88 2.94 13.52] 1.59: [-0.19 0.41 -0.02] 5.35: [-7.69 -3.35 3.50]}
1: [-0.94 1.59 5.35]
```

The function MAD

This function, although not available in the EIGEN menu, also provides information somewhat related to the eigenvalues of a matrix. The function MAD is available through

[←][MATRICES][OPER][NXT][MAD].

This function generate a number of properties of a square matrix, namely:
the determinant (stack level 4)
the formal inverse (stack level 3),
in stack level 2, the matrix coefficients of the polynomial p(x) defined by

$$(x \cdot I - A) \cdot p(x) = m(x) \cdot I,$$

the characteristic polynomial of the matrix (stack level 1)

Notice that the equation $(x \cdot I-A) \cdot p(x)=m(x) \cdot I$ is similar, in form, to the eigenvalue equation $A \cdot x = \lambda x$.

As an example, try:

[[4 1 -2] [1 2 -1][-2 -1 0]] [ENTER] [←][MATRICES][OPER][NXT][MAD].

The result is:

```
4:    -8.
3:    [[ 0.13 -0.25 -0.38][-0.25 0.50 -0.25][-0.38 -0.25 -0.88]]
2: {[[1 0 0][0 1 0][0 0 1]] [[ -2 1 -2][1 -4 -1][-2 -1 -6] [[-1 2 3][2 -4 2][3 2 7]]}
1:    'X^3+-6*x^2+2*X+8'
```

Matrix factorization

Matrix factorization or decomposition, introduced earlier, consists of obtaining matrices that when multiplied result in a given matrix. We presented earlier on, the basic idea of LU decomposition, as well as of SVD decomposition. We present these and other types of matrix decomposition through the use of the functions contained in the matrix FACT menu. This menu is accessed through [←][MATRICES][FACT]. The functions contained in this menu are:

[LQ][LU][QR][SCHUR][SVD][SVL].

The function LU

The function LU takes as input a square matrix A, and returns a lower-triangular matrix L, an upper triangular matrix U, and a permutation matrix P, in stack levels 3, 2, and 1, respectively. The results L, U, and P, satisfy the equation P·A = L·U. When you call the LU function, the calculator performs a Crout LU decomposition of A using partial pivoting.
For example:

[[-1 2 5][3 1 -2][7 6 5]] [ENTER][↤][MATRICES][FACT][LU]

produces:

3: [[7 0 0][-1 2.86 0][3 -1.57 -1]]
2: [[1 0.86 0.71][0 1 2][0 0 1]]
1: [[0 0 1][1 0 0][0 1 0]]

Orthogonal matrices and singular value decomposition (SVD)

A square matrix is said to be orthogonal if its columns represent unit vectors that are mutually orthogonal. Thus, if we let matrix $U = [v_1\ v_2\ ...\ v_n]$ where the v_i, i = 1, 2, ..., n, are column vectors, and if $v_i \cdot v_j = \delta_{ij}$, where δ_{ij} is the Kronecker's delta function, then U will be an orthogonal matrix. This conditions also imply that $U \cdot U^T = I$.

The Singular Value Decomposition (SVD) of a rectangular matrix $A_{m \times n}$ consists in determining the matrices U, S, and V, such that

$$A_{m \times n} = U_{m \times m} \cdot S_{m \times n} \cdot V^T_{n \times n},$$

Where U and V are orthogonal matrices, and S is a diagonal matrix. The diagonal elements of S are called the <u>singular values</u> of A and are usually ordered so that $s_i \geq s_{i+1}$, for i = 1, 2, ..., n-1. The columns $[u_j]$ of U and $[v_j]$ of V are the corresponding <u>singular vectors</u>.

The function SVD

The function SVD (Singular Value Decomposition) takes as input a matrix $A_{n \times m}$, and returns the matrices $U_{n \times n}$, $V_{m \times m}$, and a vector s in stack levels 3, 2, and 1, respectively. The dimension of vector s is equal to the minimum of the values n and m. The matrices U and V are as defined earlier for singular value decomposition, while the vector s represents the main diagonal of the matrix S used earlier.

For example, try:

[[5 4 -1][2 -3 5][7 2 8]] [ENTER][↤][MATRICES][FACT][LU] produces:

3: [[-0.27 0.81 -0.53][-0.37 -0.59 -0.72][-0.89 3.09E-3 0.46]]
2: [[-0.68 -0.14 -0.72][0.42 0.73 -0.54][-0.60 0.67 0.44]]
1: [12.15 6.88 1.42]

The function SVL

The function SVL (Singular VaLues) returns the singular values of a matrix $A_{n \times m}$ as a vector s whose dimension is equal to the minimum of the values n and m. For example,

[[5 4 -1][2 -3 5][7 2 8]] [ENTER][←][MATRICES][FACT][SVL] produces [12.15 6.88 1.42].

The function SCHUR

The function SCHUR produces the Schur decomposition of a square matrix A returning matrices Q and T, in stack levels 2 and 1, respectively, such that $A = Q \cdot T \cdot Q^T$, where Q is an orthogonal matrix, and T is a triangular matrix. For example,

[[2 3 -1][5 4 -2][7 5 4]][ENTER] [←][MATRICES][FACT][SCHUR] results in:

2: [[0.66 -0.29 -0.70][-0.73 -0.01 -0.68][-0.19 -0.96 0.21]]
1: [[-1.03 1.02 3.86][0 5.52 8.23][0 -1.82 5.52]]

The function LQ

The LQ function produces the LQ factorization of a matrix $A_{n \times m}$ returning a lower $L_{n \times m}$ trapezoidal matrix, a $Q_{m \times m}$ orthogonal matrix, and a $P_{n \times n}$ permutation matrix, in stack levels 3, 2, and 1. The matrices A, L, Q and P are related by $P \cdot A = L \cdot Q$. (A trapezoidal matrix out of an n×m matrix is the equivalent of a triangular matrix out of an n×n matrix).

For example,

[[1 -2 1][2 1 -2][5 -2 1]][ENTER] [←][MATRICES][FACT][LQ] produces

3: [[-5.48 0 0][-1.10 -2.79 0][-1.83 1.43 0.78]]
2: [[-0.27 0.81 -0.18][-0.36 -0.50 -0.79][-0.20 -0.78 -0.59]]
1: [[0 0 1][0 1 0][1 0 0]]

The function QR

The function QR produces the QR factorization of a matrix $A_{n \times m}$ returning a $Q_{n \times n}$ orthogonal matrix, a $R_{n \times m}$ upper trapezoidal matrix, and a $P_{m \times m}$ permutation matrix, in stack levels 3, 2, and 1. The matrices A, P, Q and R are related by $A \cdot P = Q \cdot R$.

For example,

[[1 -2 1][2 1 -2][5 -2 1]][ENTER] [←][MATRICES][FACT][QR] produces

3: [[-0.18 0.39 0.90][-0.37 -0.88 0.30][-0.91 0.28 -0.30]]
2: [[-5.48 -0.37 1.83][0 2.42 -2.20][0 0 -0.90]]
1: [[1 0 0][0 0 1][0 1 0]]

Matrix Quadratic Forms

A <u>quadratic form</u> from a square matrix A is a polynomial expression originated from x·A·x^T. For example, if we use A = [[2 1 -1][5 4 2][3 5 -1]], and x = [X Y Z]^T, the corresponding quadratic form is calculated as

$$\mathbf{x} \cdot \mathbf{A} \cdot \mathbf{x}^T = \begin{bmatrix} X & Y & Z \end{bmatrix} \cdot \begin{bmatrix} 2 & 1 & -1 \\ 5 & 4 & 2 \\ 3 & 5 & -1 \end{bmatrix} \cdot \begin{bmatrix} X \\ Y \\ Z \end{bmatrix} = \begin{bmatrix} X & Y & Z \end{bmatrix} \cdot \begin{bmatrix} 2X + Y - Z \\ 5X + 4Y + 2Z \\ 3X + 5Y - Z \end{bmatrix}$$

Finally,

$$x \cdot A \cdot x^T = 2X^2 + 4Y^2 - Z^2 + 6XY + 2XZ + 7ZY$$

The QUADF menu

The HP 49 G calculator provides the QUADF menu for operations related to QUADratic Forms. The QUADF menu is accessed through [←][MATRICES][QUADF], and includes the functions

[AXQ][GAUSS][QXA][SYLVE].

The function AXQ

The function AXQ produces the quadratic form corresponding to a matrix A$_{n \times n}$ in stack level 2 using the n variables in a vector placed in stack level 1. The function returns the quadratic form in stack level 1 and the vector of variables in stack level 1. For example,

[[2 1 -1][5 4 2][3 5 -1]] [ENTER] ['X' 'Y' 'Z'] [ENTER] [←][MATRICES][QUADF] [AXQ]

returns

```
2: '2*X^2+(6*Y+2*Z)*X+(4*y^2+7*Z*y-Z^2)'
1: ['X' 'Y' 'Z']
```

The function QXA

The function QXA takes as arguments a quadratic form in stack level 2 and a vector of variables in stack level 1, returning the square matrix A from which the quadratic form is derived in stack level 2, and the list of variables in stack level 1. For example,

'X^2+Y^2-Z^2+4*X*Y-16*X*Z' [ENTER] ['X' 'Y' 'Z'] [ENTER] [←][MATRICES][QUADF] [AXQ]

258 © 2000 – Gilberto E. Urroz
All rights reserved

returns

```
2: [[1 2 -8][2 1 0][-8 0 -1]]
1: ['X' 'Y' 'Z']
```

Diagonal representation of a quadratic form

Given a symmetric square matrix A, it is possible to "diagonalize" the matrix A by finding an orthogonal matrix P such that $P^T \cdot A \cdot P = D$, where D is a diagonal matrix. If $Q = x \cdot A \cdot x^T$ is a quadratic form based on A, it is possible to write the quadratic form Q so that it only contains square terms from a variable y, such that $x = P \cdot y$, by using
$$Q = x \cdot A \cdot x^T = (P \cdot y) \cdot A \cdot (P \cdot y)^T = y \cdot (P^T \cdot A \cdot P) \cdot y^T = y \cdot D \cdot y^T.$$

The function SYLVESTER

The function SYLVESTER takes as argument a symmetric square matrix A and returns a vector containing the diagonal terms of a diagonal matrix D, and the orthogonal matrix P, so that $P^T \cdot A \cdot P$ = D. For example,

[[2 1 -1][1 4 2][-1 2 -1]] [ENTER] [↵][MATRICES][QUADF] [SYLVE] produces

```
2: [ '1/2' '2/7' '-23/7']
1: [[2 1 -1][0 '7/2' '5/2'][0 0 1]]
```

The function GAUSS

The function GAUSS returns the diagonal representation of a quadratic form $Q = x \cdot A \cdot x^T$ taking as arguments the quadratic form in stack level 2 and the vector of variables in stack level 1. The result of this function call is the following:

An array of coefficients representing the diagonal terms of D (stack level 4)
A matrix P such that $A = P^T \cdot D \cdot P$ (stack level 3)
The diagonalized quadratic form (stack level 2)
The list of variables (stack level 1)

For example,

X^2+Y^2-Z^2+4*X*Y-16*X*Z' [ENTER] ['X' 'Y' 'Z'] [ENTER] [↵][MATRICES][QUADF] [GAUSS]

returns

```
4: [1 -0.333 20.333]
3: [[1 2 -8][0 -3 16][0 0 1]]
2: '20.333*Z^2+-.333*(16*Z+-3*Y)^2+(-8+8*z+2*Y+X)^2'
1: ['X' 'Y' 'Z']
```

Matrix applications

In this section we explore some applications of matrices in the physical sciences.

Electric circuits

Consider the simple electrical circuit shown in the figure below.

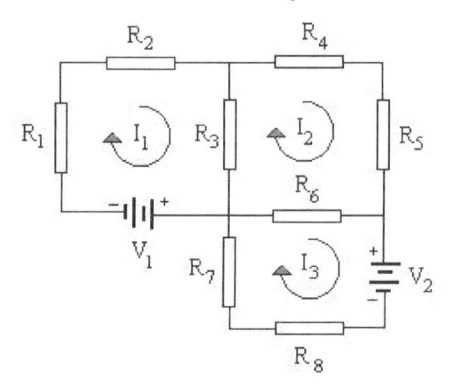

Given the values of the electric resistance, $R_1 = R_3 = R_5 = R_7 = 1.5$ kΩ, $R_2 = R_4 = R_6 = R_8 = 800$ Ω, and the known steady voltages $V_1 = 12$ V, $V_2 = 24$ V. We are asked to determine the electrical currents I_1, I_2, and I_3, associated with the circulation loops shown in the figure.

The circulation loops shown pre-determine for us a preferred direction in each loop to write Kirchoff law of voltage in a closed loop. Basically, we start at a node in the circuit and move around a given loop subtracting voltages $R \cdot I$ if the current is in the same direction as the loop direction, or adding voltages if the current and the loop directions are opposite. When encountering a voltage source, the voltage from the source is added or subtracted according to the orientation of the voltage source with respect to the loop circulation direction. We stop back at the same node were we started to complete the voltage equation for a given loop.

For the case shown in the figure we can write:

I_1: $\qquad -R_1 \cdot I_1 - R_2 \cdot I_1 - R_3 \cdot (I_1 - I_2) - V_1 = 0$
I_2: $\qquad -R_4 \cdot I_2 - R_5 \cdot I_2 - R_6 \cdot (I_2 - I_3) - R_3 \cdot (I_2 - I_1) = 0$
I_3: $\qquad -V_2 - R_8 \cdot I_3 - R_7 \cdot I_3 - R_6 \cdot I_3 = 0$

Replacing the values of the resistances and voltage sources:

I_1: $\qquad -1500 \cdot I_1 - 800 \cdot I_1 - 1500 \cdot (I_1 - I_2) - 12 = 0$
I_2: $\qquad -800 \cdot I_2 - 1500 \cdot I_2 - 800 \cdot (I_2 - I_3) - 1500 \cdot (I_2 - I_1) = 0$
I_3: $\qquad -24 - 800 \cdot I_3 - 1500 \cdot I_3 - 800 \cdot (I_3 - I_2) = 0$

We can solve this system of equations by using the following keystrokes:

`'-1500*I1-800*I1-1500*(I1-I2)-12=0'[ENTER]` Enter equation I1
`'-800*I2-1500*I2-800*(I2-I3)-1500*(I2-I1)=0' [ENTER]` Enter equation I2
`'-24-800*I3-1500*I3-800*(I3-I2)=0'[ENTER]` Enter equation I3

[3][↵][PRG][TYPE][→ARRY]	Create an array to hold the equations
['I1' 'I2' 'I3'][ENTER]	Create an array of variables to solve for
[↵][MATRICES][LIN S][LINSO]	Solves for I1, I2, I3.

To see the solution press [▼], then use [◄] [►] to move about the solution vector. You should get ['I1=-1602/373175' 'I2=-1073/373175' 'I3 = -3166/373175']

Press [ENTER] to return to normal calculator display.

To see the results in decimal form, use:

[↵][PRG][TYPE][OBJ→][⇦]	Separates elements of vector, drop level
1	
[OBJ→][⇦][⇦][↳][→NUM]	-8.4839E-3, or I_3 = - 8.4839 mA.
[⇦][⇦][OBJ→][⇦][⇦][↳][→NUM]	-2.8753E-3, or I_2 = - 2.8753 mA.
[⇦][⇦][OBJ→][⇦][⇦][↳][→NUM]	-4.2928E-3, or I_1 = - 4.2928 mA.

Structural mechanics

Consider the truss structure shown in the figure below. Horizontal and vertical bars are of length 1.0 m, and diagonal bars 1.4142 m. All acute angles in the truss are 45°.

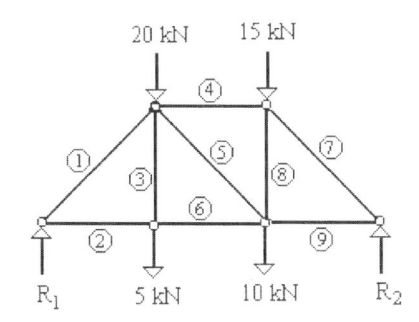

By isolating each node, as shown in the figure below, we can write the following equations for node equilibrium (i.e., $\Sigma F_x = 0$, $\Sigma F_y = 0$):

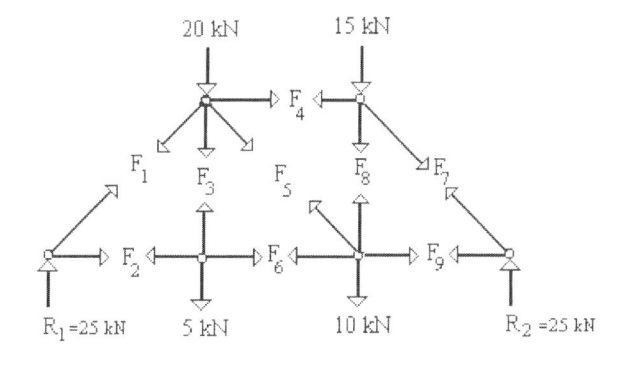

$$F_2 + F_1 \cos 45° = 0,$$
$$25 + F_1 \sin 45° = 0,$$
$$-F_2 + F_6 = 0,$$
$$-5 + F_3 = 0,$$
$$F_4 - F_1 \cos 45° + F_5 \cos 45° = 0,$$
$$-20 - F_3 - F_1 \cos 45° - F_5 \cos 45° = 0,$$
$$-F_4 + F_7 \cos 45° = 0,$$
$$-15 - F_8 - F_7 \cos 45° = 0,$$
$$-F_6 + F_9 - F_5 \cos 45° = 0,$$
$$-10 + F_8 + F_5 \cos 45° = 0,$$
$$-F_9 - F_7 \cos 45° = 0,$$
$$25 + F_7 \cos 45° = 0.$$

With $\sin 45° = \cos 45° = 0.866$, then we have:

```
0.866 F₁   + F₂                                                = 0
0.866 F₁                                                       = -25
           -F₂              + F₆                               = 0
                  F₃                                           = 5
-0.866F₁                +F₄   +0.866F₅                         = 0
-0.866F₁          -F₃         -0.866F₅                         = 20
                        -F₄           + 0.866F₇                = 0
                                      -0.866F₇   - F₈          = 15
                        -0.866F₅ - F₆                   + F₉   = 0
                         0.866F₅                 +F₈           = 10
                                      -0.866F₇   - F₉          = 0
                                       0.866 F₇                = -25
```

We have a total of 12 equations with 9 unknowns. The system is over-determined, so we choose, arbitrarily, the first 9 equations:

```
0.866 F₁   + F₂                                                = 0
0.866 F₁                                                       = -25
           -F₂              + F₆                               = 0
                  F₃                                           = 5
-0.866F₁                +F₄   +0.866F₅                         = 0
-0.866F₁          -F₃         -0.866F₅                         = 20
                        -F₄           + 0.866F₇                = 0
                                      -0.866F₇   - F₈          = 15
                        -0.866F₅ - F₆                   + F9   = 0
```

Writing the system as a matrix equation:

$$\begin{bmatrix} 0.866 & 1 & 0 & 0 & 0 & 0 & 0 & 0 & 0 \\ 0.866 & 0 & 0 & 0 & 0 & 0 & 0 & 0 & 0 \\ 0 & -1 & 0 & 0 & 0 & 1 & 0 & 0 & 0 \\ 0 & 0 & 1 & 0 & 0 & 0 & 0 & 0 & 0 \\ -0.866 & 0 & 0 & 1 & 0.866 & 0 & 0 & 0 & 0 \\ -0.866 & 0 & -1 & 0 & -0.866 & 0 & 0 & 0 & 0 \\ 0 & 0 & 0 & -1 & 0 & 0 & 0.866 & 0 & 0 \\ 0 & 0 & 0 & 0 & 0 & 0 & -0.866 & -1 & 0 \\ 0 & 0 & 0 & 0 & -0.866 & -1 & 0 & 0 & 1 \end{bmatrix} \cdot \begin{bmatrix} F_1 \\ F_2 \\ F_3 \\ F_4 \\ F_5 \\ F_6 \\ F_7 \\ F_8 \\ F_9 \end{bmatrix} = \begin{bmatrix} 0 \\ -25 \\ 0 \\ 5 \\ 0 \\ 20 \\ 0 \\ 15 \\ 0 \end{bmatrix}$$

The coefficient matrix for this problem, where most of the elements are zero, is known as a sparse matrix. To solve this problem we can simply enter the vector:

[0 -25 0 5 0 20 0 15 0][ENTER]

Then the matrix:

[[0.866 1 0 0 0 0 0 0 0][0.866 0 0 0 0 0 0 0 0]...etc.. Make sure that the elements of the matrix are entered properly, then press [÷]. The result is: [-28.87 25. 5. -25. 0. 25. -28.87 -40. 25.], i.e.,

F_1 = -28.87 kN, F_2 = 25 kN, F_3 = 5 kN,
F_4 = -25 kN, F_5 = 0 kN, F_6 = 25 kN,
F_7 = -28.87 kN, F_8 = -40 kN, F_9 = 25 kN.

Dimensionless numbers in fluid mechanics

Dimensional analysis is a technique used in fluid mechanics, and other sciences, to reduce the number of variables involved in an experiment by creating dimensionless numbers that combine the original set of variables. In order to obtain these dimensionless numbers, we make use of the principle of dimensional homogeneity, which basically states that an equation derived from conservation laws and material properties should have the same dimensions on both sides of the equation. For example, the equation for the distance traveled by a projectile dropped from rest at a certain elevation above the ground is given by $d = \frac{1}{2} gt^2$, where g = 9.806 m/s^2, is the acceleration of gravity, and t is the time in seconds. The distance d is given in meters. Instead of dealing with units, we refer to three (sometimes more) fundamental dimensions: length (L), time (T), and mass (M). We use brackets to refer to the dimensions of a quantity, thus, [d] = L, g = [LT^{-2}], and t = [T]. Replacing dimensions in the formula for d we have:

$$[d] = [1/2][g][t]^2 = 1 \cdot LT^{-2} \cdot T^2 = L,$$

as expected. Thus, we say that the equation d = ½ gt^2 is dimensionally homogeneous.

Suppose that we have an experiment that involves the following variables (showed with their dimensions attached):

D = a diameter (L)
V = a flow velocity (LT^{-1})
ν = kinematic viscosity of the fluid (L^{2}T^{-1})
ρ = density of the fluid (ML^{-3})
E = bulk density of the fluid (ML^{-1}T^{-2})
σ = surface tension of the fluid (MT^{-2})
Δp = a characteristic pressure drop in the flow (ML^{-1}T^{-2})
g = acceleration of gravity (LT^{-2})

There are m = 8 variables which need n = 3 dimensions to be expressed (i.e., L, T, and M). Buckingham's Π theorem indicates that you can form r = m - n = 8 - 3 = 5 dimensionless parameters. The technique consists in selecting one geometric variable, in this case we have no choice but to select D, the only variable that represents geometry alone; a kinematic variable, V (you can also choose ν), i.e., a variable involving length and time; and, finally, a dynamic variable, say ρ, i.e., a variable involving length, time, and mass. These three variables, D, V and ρ, become repeating variables, i.e., variables that will participate in each of the dimensionless parameters to be formed. Each dimensionless parameter ,or Π number, is formed by multiplying the repeating variables raised to a certain unknown power and multiplying one of the remaining variables. For example, we can form for this case the following Π parameters:

$$\Pi_1 = \rho^x \cdot D^y \cdot V^z \cdot \nu,$$
$$\Pi_2 = \rho^x \cdot D^y \cdot V^z \cdot E,$$
$$\Pi_3 = \rho^x \cdot D^y \cdot V^z \cdot \sigma,$$
$$\Pi_4 = \rho^x \cdot D^y \cdot V^z \cdot \Delta p,$$
$$\Pi_5 = \rho^x \cdot D^y \cdot V^z \cdot g.$$

Since the Π numbers are dimensionless, we can write [Π_i] = L$^0 \cdot$T$^0 \cdot$M^0, for i = 1,2, 3, 4, 5. Replacing the dimensions of the variables involved in each dimensionless parameters we can write, for example, for Π_1:

$$L^0 \cdot T^0 \cdot M^0 = (ML^{-3})^x \cdot (L)^y \cdot (LT^{-1})^z \cdot (L^2 T^{-1}) = (L)^{-3x+y+z+2} \cdot (T)^{-z-1} \cdot (M)^x,$$

From which we get the following equations:

$$-3x+y+z+2 = 0$$
$$-z - 1 = 0$$
$$x = 0$$

Or,

$$\begin{bmatrix} -3 & 1 & 1 \\ 0 & 0 & -1 \\ 1 & 0 & 0 \end{bmatrix} \cdot \begin{bmatrix} x \\ y \\ z \end{bmatrix} = \begin{bmatrix} -2 \\ 1 \\ 0 \end{bmatrix}$$

If we replace the dimensions of the non-repeating variables in the remaining Π parameters, we can expand the matrix equation shown above to read:

$$\begin{bmatrix} -3 & 1 & 1 \\ 0 & 0 & -1 \\ 1 & 0 & 0 \end{bmatrix} \cdot \begin{bmatrix} x \\ y \\ z \end{bmatrix} = \begin{bmatrix} -2 & 1 & 0 & 1 & -1 \\ 1 & 2 & 2 & 2 & 2 \\ 0 & -1 & -1 & -1 & 0 \end{bmatrix}$$

So, the independent vector b has become a matrix B, and we can write the matrix equation A·X = B. The columns of B are the negatives of the exponents of the dimensions, L, T, and M, in that order, of each of the non-repeating variables as shown in the Π parameters that we set up.

To solve for the variables x,y,z for each parameter use:

[[-2 1 0 1 -1][1 2 2 2 2][0 -1 -1 -1 0]][ENTER]   Enter matrix B
[[-3 1 1][0 0 -1][1 0 0]][ENTER]   Enter matrix A
[÷]   Solve for X

The result is the matrix

$$\mathbf{X} = \begin{pmatrix} 0 & -1 & -1 & -1 & 0 \\ -1 & 0 & -1 & 0 & 1 \\ -1 & -2 & -2 & -2 & -2 \end{pmatrix},$$

each column representing the values of x,y,z, for the repeating variables in each of the dimensionless parameters, thus we have:

$$\Pi_1 = \rho^0 \cdot D^{-1} \cdot V^{-1} \cdot \nu = \nu/DV,$$
$$\Pi_2 = \rho^{-1} \cdot D^0 \cdot V^{-2} \cdot E = E/\rho V^2,$$
$$\Pi_3 = \rho^{-1} \cdot D^{-1} \cdot V^{-2} \cdot \sigma = \sigma/\rho DV^2,$$
$$\Pi_4 = \rho^{-1} \cdot D^0 \cdot V^{-2} \cdot \Delta p = \Delta p/\rho V^2,$$
$$\Pi_5 = \rho^0 \cdot D^1 \cdot V^{-2} \cdot g = gD/V^2.$$

Stress at a point in a solid in equilibrium

Consider a solid body in equilibrium under the action of a system of forces and moments, as illustrated in the figure below. If we were to make an imaginary cut through the solid body, so that we can separate it into two parts at section S.

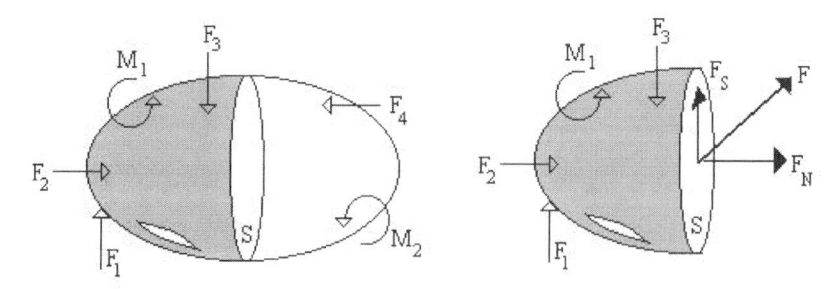

The effect of the part that we remove to the right of the cut surface S is replaced by the force F, which in turn can be decomposed into a normal component, F_N, and a shear or tangential component F_S.

Suppose now that we isolate a small particle off this solid body, and we do it by cutting the body with four planes so that we can draw the particle as shown in the left-hand side of the figure below.

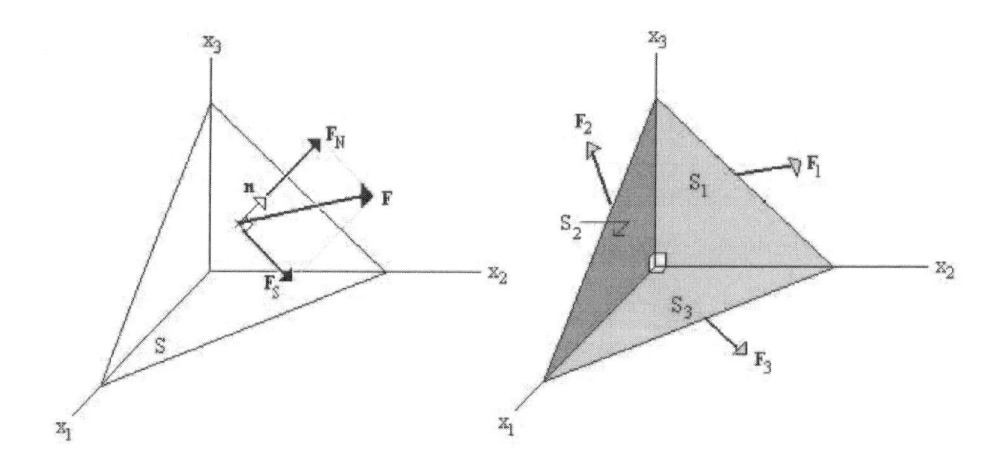

Three of the planes are chose to be perpendicular to each other so that they help us identify a Cartesian coordinate system (x_1, x_2, x_3) as shown above. The surface S, limiting the particle from above, has a normal unit vector $n = [\cos \alpha_1, \cos \alpha_2, \cos \alpha_3]$, where $\cos \alpha_1$, $\cos \alpha_2$, and $\cos \alpha_3$ are the direction cosines of n. The other three surfaces limiting the particle are S_1, S_2, and S_3, where the sub-index indicates the axis that is normal to the surface. The effect of the solid body on this particle is represented by the forces F, F_1, F_2, and F_3, acting, respectively, upon surfaces

S, S_1, S_2, and S_3. Let the areas corresponding to each surface S, S_1, S_2, and S_3 be given by A, A_1, A_2, and A_3. It is possible to show, from the geometry of the figure, that

$$A_1 = A \cdot \cos \alpha_1, \quad A_2 = A \cdot \cos \alpha_2, \quad \text{and} \quad A_3 = A \cdot \cos \alpha_3.$$

The force F on surface S can be decomposed into a normal component,

$$F_N = F_N \cdot n = F_N \cdot [\cos \alpha_1, \cos \alpha_2, \cos \alpha_3] = F_N \cdot (\cos \alpha_1 \cdot e_1 + \cos \alpha_2 \cdot e_2 + \cos \alpha_3 \cdot e_3) = F_N \cdot \cos \alpha_j \cdot e_j, \quad (*)$$

(*) using Einstein's repeated index convention.

and a shear component,

$$F_S = F - F_N,$$

as shown in the figure above. The vectors are the unit vectors corresponding to the three coordinate directions.

The forces on surfaces S_1, S_2, and S_3 can be written in terms of the stress components, σ_{ij}, shown in the figure below, as

$$F_i = [-\sigma_{i1}, -\sigma_{i2}, -\sigma_{i3}] \cdot A_i = (-\sigma_{i1} \cdot e_1 - \sigma_{i2} \cdot e_2 - \sigma_{i3} \cdot e_3) \cdot A_i = (-\sigma_{i1} \cdot e_1 - \sigma_{i2} \cdot e_2 - \sigma_{i3} \cdot e_3) \cdot A \cdot \cos \alpha_i \quad (i = 1,2,3).$$

Using Einstein's repeated index convention we can write

$$F_i = [-\sigma_{i1}, -\sigma_{i2}, -\sigma_{i3}] \cdot A_i = -\sigma_{ij} \cdot e_j \cdot A_i = -\sigma_{ij} \cdot e_j \cdot A \cdot \cos \alpha_i \quad (i = 1,2,3).$$

The sub-indices identifying each stress components are chosen so that the first sub-index represents the sub-index of the axis normal to the surface of interest, and the second represents the direction along which the stress acts.

Stresses with the same sub-index, σ_{ii} ($i = 1,2,3$) act normal to the appropriate surface and are known as normal stresses. The other two components on each of the surfaces S_1, S_2, and S_3, are known as shear stresses, i.e., σ_{ij}, $i \neq j$. The direction of action as shown in the figure below is the conventional way to represent the stresses, namely, the stresses are positive when acting in the negative coordinate directions, so that the resulting forces have a negative sign, as shown in the equation above.

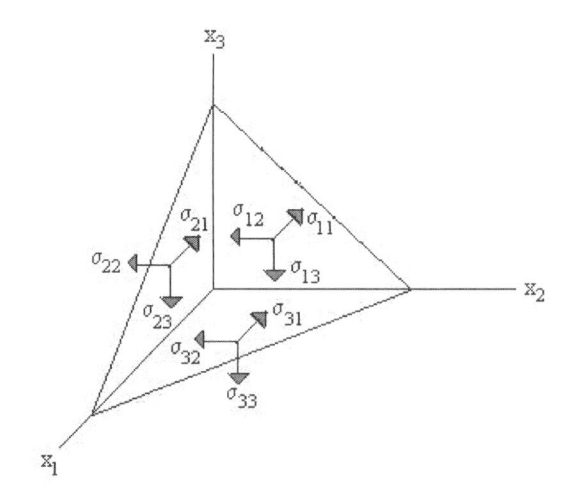

The stress components illustrated in the figure above can be written as a matrix known as the stress tensor,

$$T = \begin{bmatrix} \sigma_{11} & \sigma_{12} & \sigma_{13} \\ \sigma_{21} & \sigma_{22} & \sigma_{23} \\ \sigma_{31} & \sigma_{32} & \sigma_{33} \end{bmatrix}.$$

The set up of the Cartesian coordinate system and the stresses in the particle under consideration can be used to define the stress condition at a point in the limit when the dimensions of the particle tend to zero. Under such conditions you can prove that the stress tensor is symmetric, i.e., $\sigma_{ij} = \sigma_{ji}$. Therefore, to define completely the state of stress at a point we need only to know the three normal stresses and three of the shear stresses.

For the equilibrium of force on the particle we can write

$$F + \Sigma F_i = F + \Sigma(-\sigma_{ij} \cdot e_j \cdot A \cdot \cos \alpha_i) = 0, \text{ or } F = \sigma_{ij} \cdot e_j \cdot A \cdot \cos \alpha_i$$

[using Einstein's convention, with both i and j repeated]

If we let

$$F = \sigma \cdot A,$$

where s is the stress vector on surface S, and replace this value in the previous equation we get

$$\sigma = \sigma_{ij} \cdot \cos \alpha_i \cdot e_j = \cos \alpha_i \cdot e_j \cdot \sigma_{ij} = n \cdot T$$

To find the magnitude of the normal component of the stress vector, i.e., the projection of the stress σ along the unit normal vector n, we use

$$\sigma_n = \boldsymbol{\sigma} \bullet n / |n| = \boldsymbol{\sigma} \bullet n = (\sigma_{ij} \cos \alpha_i \cdot e_j) \bullet (\cos \alpha_k \cdot e_k) = \sigma_{ij} \cos \alpha_i \cdot \cos \alpha_k \cdot (e_j \bullet e_k).$$

We can prove that for the unit vectors in the Cartesian coordinate system,

$$e_j \bullet e_k = \delta_{jk},$$

where δ_{jk} is Dirac's delta function. Thus, the normal component of the stress on surface S is

$$\sigma_n = \sigma_{ij} \cos \alpha_i \cdot \cos \alpha_k \cdot \delta_{jk}$$

Since the product indicated in this expression is zero if $j \neq k$, then the only terms surviving are those for which $j = k$, i.e.,

$$\sigma_n = \sigma_{ij} \cos \alpha_i \cdot \cos \alpha_j = \cos \alpha_j \cdot \sigma_{ij} \cdot \cos \alpha_i.$$

You can prove that this latter result can be written in vector and matrix notation as

$$\sigma_n = n \cdot T \cdot n^T,$$

where

$$n = \cos \alpha_j \cdot e_j.$$

Thus, the normal stress magnitude can be written as a quadratic form for any normal unit vector $n = n_j \cdot e_j$, written as a row vectors, with $n_j = \cos \alpha_j$, $j = 1, 2, 3$. Also, the normal stress as a vector will be written as

$$\boldsymbol{\sigma}_n = \sigma_n \cdot n = (n \cdot T \cdot n^T) \cdot n$$

The normal force is given by

$$F_N = \boldsymbol{\sigma}_n \cdot A = (\sigma_n \cdot A) \cdot n.$$

The shear force can be written in terms of shear stress on surface S, $F_S = F - F_N = \boldsymbol{\sigma}_S \cdot A$, so that

$$\boldsymbol{\sigma}_S = \boldsymbol{\sigma} - \boldsymbol{\sigma}_n.$$

Example:

Let the stress at a point be given by

$$T = \begin{bmatrix} 25 & -10 & 20 \\ -10 & 30 & 15 \\ 20 & 15 & -40 \end{bmatrix} \cdot Pa$$

Determine the total stress $\boldsymbol{\sigma}$, the normal stress $\boldsymbol{\sigma}_n$, and shear stress $\boldsymbol{\sigma}_S$, if the surface S has a normal unit vector n = [0.5 0.25 0.8292]. What are the total force F, the normal force F_n, and the shear force F_S, if the surface S has an area of 0.00001 m²?

Solution:

To calculate the total stress we use

$$\overset{\rho}{\sigma} = n \cdot T \cdot \overset{\rho}{e} = \begin{bmatrix} 25 & -10 & 20 \\ -10 & 30 & 15 \\ 20 & 15 & -40 \end{bmatrix} \cdot \begin{bmatrix} 0.5 \\ 0.25 \\ 0.8292 \end{bmatrix}$$

This result can be obtained from the calculator by using:

25 -10 20][-10 30 15][20 15 -40]] [ENTER] [0.5 0.25 0.8292][ENTER] [×]

The result is   σ = [26.548 14.938 -19.418] Pa.

To calculate the normal stress, use:

$$\sigma_n = \mathbf{n} \cdot \mathbf{T} \cdot \mathbf{n}^T = \begin{bmatrix} 0.5 & 0.25 & 0.8292 \end{bmatrix} \cdot \begin{bmatrix} 25 & -10 & 20 \\ -10 & 30 & 15 \\ 20 & 15 & -40 \end{bmatrix} \cdot \begin{bmatrix} 0.5 \\ 0.25 \\ 0.8292 \end{bmatrix}$$

In the calculator, use

[0.5 0.25 0.8292][ENTER][ENTER] [[25 -10 20][-10 30 15][20 15 -40]] [ENTER] [▶] [×]
[↵][MTH][VECTR][DOT] [↪][ALG][EXPAN]

The result is   σ_n = 0.9251 Pa.

The shear stress is given by

σ_s = σ - σ_n = [26.548 14.938 -19.418] - 0.9251·[0.5 0.25 0.8292] = [26.085 14.707 -20.185] Pa.

The forces can be calculated by multiplying the stresses times the area of the surface S, i.e., F = σ·A, F_n = σ_n·A, and F = σ·A.

Principal stresses at a point

Given the stress tensor T representing the state of stress at a point P in a Cartesian coordinate system (x_1, x_2, x_3), suppose that you want to find the normal vector, or vectors, n for which the stress is only in the normal direction. In other words, we are trying to find n and σ_n such that

T·n = σ_n·n.

This equation is the eigenvalue equation for the matrix T with eigenvalues σ_n and eigenvectors n.

Recall that this equation can be written also as

(T - σ_n·I) ·n = 0,

which has non-trivial solution if

$$\det(T - \sigma_n \cdot I) = 0.$$

For the previous example, we can write

$$\det(T - \sigma_n \cdot I) = \begin{vmatrix} 25-\sigma_n & -10 & 20 \\ -10 & 30-\sigma_n & 15 \\ 20 & 15 & -40-\sigma_n \end{vmatrix} = 0.$$

To obtain the characteristic polynomial, eigenvalues and eigenvectors of T with the HP 49 G calculator use:

[[25 -10 20][-10 30 15][20 15 -40]] [ENTER] [ENTER] [ENTER]

[←][MATRICES][EIGEN][PCAR] provides 'X^3+-15*X^2+-2175*X+49625', or

$$\sigma_n^3 - 15\sigma_n^2 - 2175\sigma_n + 49625 = 0.$$

[⇔][EGV] results in

2: [[-0.30 1. -0.80][-0.23 0.81 1.][1. 0.43 -1.31E-2]]
1: [-49.39 26.57 37.81]
i.e.,

$$n_1 = [-0.30\ -0.23\ 1],\ (\sigma_n)_1 = -49.39,$$
$$n_2 = [1.\ 0.81\ 0.43],\ (\sigma_n)_2 = 26.57,$$
$$n_3 = [-0.80\ 1.\ -1.31E-2],\ (\sigma_n)_3 = 37.81.$$

The three normal stresses found are known as the principal stresses at the point. The eigenvalues represent the normal vectors to the surfaces where those principal stresses act. These directions are known as the principal axes.

Multiple linear fitting

Consider a data set of the form

x_1	x_2	x_3	...	x_n	y
x_{11}	x_{21}	x_{31}	...	x_{n1}	y_1
x_{12}	x_{22}	x_{32}	...	x_{n2}	y_2
x_{13}	x_{32}	x_{33}	...	x_{n3}	y_3
.	.	.		.	.
.	.	.		.	.
$x_{1,m-1}$	$x_{2,m-1}$	$x_{3,m-1}$	...	$x_{n,m-1}$	y_{m-1}
$x_{1,m}$	$x_{2,m}$	$x_{3,m}$	...	$x_{n,m}$	y_m

Suppose that we search for a data fitting of the form

$$y = b_0 + b_1 \cdot x_1 + b_2 \cdot x_2 + b_3 \cdot x_3 + \ldots + b_n \cdot x_n.$$

You can obtain the least-square approximation to the values of the coefficients

$$b = [b_0 \quad b_1 \quad b_2 \quad b_3 \ldots b_n],$$

by putting together the matrix X

$$\begin{bmatrix} 1 & x_{11} & x_{21} & x_{31} & \ldots & x_{n1} \\ 1 & x_{12} & x_{22} & x_{32} & \ldots & x_{n2} \\ 1 & x_{13} & x_{32} & x_{33} & \ldots & x_{n3} \\ \cdot & \cdot & \cdot & \cdot & & \cdot \\ \cdot & \cdot & \cdot & \cdot & & \cdot \\ 1 & x_{1,m} & x_{2,m} & x_{3,m} & \ldots & x_{n,m} \end{bmatrix}$$

Then, the vector of coefficients is obtained from

$$b = (X^T \cdot X)^{-1} \cdot X^T \cdot y,$$

where y is the vector

$$y = [y_1 \; y_2 \; \ldots \; y_m]^T.$$

For <u>example</u>, use the following data to obtain the multiple linear fitting

$$y = b_0 + b_1 \cdot x_1 + b_2 \cdot x_2 + b_3 \cdot x_3.$$

x_1	x_2	x_3	y
1.20	3.10	2.00	5.70
2.50	3.10	2.50	8.20
3.50	4.50	2.50	5.00
4.00	4.50	3.00	8.20
6.00	5.00	3.50	9.50

With the calculator you can proceed as follows:

First, within your HOME directory, create a sub-directory to be called MPFIT (Multiple linear and Polynomial data FITting) by using:

[↵][`][ALPHA][ALPHA][M][P][F][I][T][ALPHA] [↵][PRG][MEM][DIR][CRDIR]

Then, press [VAR][MPFIT] to get into the MPFIT sub-directory.

Within the sub-directory, type this program:

`<< → X y << X TRAN X * INV X TRAN * y * >> >>`

by using:

[→][<<>>] [→][→] [ALPHA][X] [SPC] [ALPHA][←][Y] [→][<<>>] [ALPHA][X]
[←][MATRICES][OPER][NXT][NXT][TRAN] [ALPHA][X] [×] [1/x] [ALPHA][X]
[←][MATRICES][OPER][NXT][NXT][TRAN] [×][ALPHA][←][Y] [×] [ENTER]

and store it in a variable called MTREG (MulTiple REGression)

[→][`][ALPHA][ALPHA][M][T][R][E][G][ALPHA][STO▶]

Next, enter the matrices X and b into the stack:

[[1 1.2 3.1 2][1 2.5 3.1 2.5][1 3.5 4.5 2.5][1 4 4.5 3][1 6 5 3.5]][ENTER][ENTER] (keep extra copy)
[5.7 8.2 5.0 8.2 9.5][ENTER]

Press [VAR][MTREG]. The result is: [-4.1721 -1.09 -3.37E-2 5.7976], i.e.,

$$y = -4.1721 - 1.09 \cdot x_1 - 3.37 \times 10^{-2} \cdot x_2 + 5.7976 \cdot x_3 .$$

You should have in your calculator's stack the value of the matrix X and the vector b, the fitted values of y are obtained from y = X·b, thus, just press [×] to obtain:

[6.01 7.49 4.99 8.71 9.41].

Compare these fitted values with the original data as shown in the table below:

x_1	x_2	x_3	y	y-fitted
1.20	3.10	2.00	5.70	6.01
2.50	3.10	2.50	8.20	7.49
3.50	4.50	2.50	5.00	4.99
4.00	4.50	3.00	8.20	8.71
6.00	5.00	3.50	9.50	9.41

Polynomial fitting

Consider the x-y data set

x	y
x_1	y_1
x_2	y_2
x_3	y_3
.	.
.	.
x_{n-1}	y_{n-1}
x_n	y_n

Suppose that we want to fit a polynomial or order p to this data set. In other words, we seek a fitting of the form

$$y = b_0 + b_1 \cdot x + b_2 \cdot x^2 + b_3 \cdot x^3 + \ldots + b_p \cdot x^p.$$

You can obtain the least-square approximation to the values of the coefficients

$$b = [b_0 \quad b_1 \quad b_2 \quad b_3 \ldots b_p],$$

by putting together the matrix X

$$\begin{bmatrix} 1 & x_1 & x_1^2 & x_1^3 & \ldots & x_1^{p-1} & y_1^p \\ 1 & x_2 & x_2^2 & x_2^3 & \ldots & x_2^{p-1} & y_2^p \\ 1 & x_3 & x_3^2 & x_3^3 & \ldots & x_3^{p-1} & y_3^p \\ \cdot & \cdot & \cdot & \cdot & & \cdot & \cdot \\ \cdot & \cdot & \cdot & \cdot & & \cdot & \cdot \\ 1 & x_n & x_n^2 & x_n^3 & \ldots & x_n^{p-1} & y_n^p \end{bmatrix}$$

Then, the vector of coefficients is obtained from $b = (X^T \cdot X)^{-1} \cdot X^T \cdot y$, where y is the vector $y = [y_1 \, y_2 \ldots y_n]^T$.

Earlier on, in this chapter, we defined the Vandermonde matrix corresponding to a vector $x = [x_1 \, x_2 \ldots x_m]$ as

$$\begin{bmatrix} 1 & x_1 & x_1^2 & x_1^3 & \ldots & x_1^{n-1} \\ 1 & x_2 & x_2^2 & x_2^3 & \ldots & x_2^{n-1} \\ 1 & x_3 & x_3^2 & x_3^3 & \ldots & x_3^{n-1} \\ \cdot & \cdot & \cdot & \cdot & & \cdot \\ \cdot & \cdot & \cdot & \cdot & & \cdot \\ 1 & x_n & x_n^2 & x_n^3 & \ldots & x_n^{n-1} \end{bmatrix}$$

is similar to the matrix X of interest to the polynomial fitting, but having only n, rather than (p+1) columns.

We can take advantage of the VANDERMONDE function to create the matrix X if we observe the following rules:

If p = n-1, X = V_n.
If p < n-1, then we need to remove columns p+2, ..., n-1, n from matrix V_n to form matrix X.
If p > n-1, then we need to add columns n+1, ..., p-1, p+1, to matrix V_n to form matrix X.

In step 3 from this list, we have to be aware that column i (i= n+1, n+2, ..., p+1) is the vector $[x_1^i \, x_2^i \ldots x_n^i]$. If we were to use a list of data values for x rather than a vector, i.e., x = { $x_1 \, x_2 \ldots x_n$ }

we can easily calculate the sequence { x_1^i x_2^i ... x_n^i }. Then, we can transform this list into a vector and use the COL menu to add those columns to the matrix V_n until X is completed.

After X is ready, and having the vector y available, the calculation of the coefficient vector b is the same as in multiple linear fitting (the previous matrix application). Thus, we can write a program to calculate the polynomial fitting that can take advantage of the program already developed for multiple linear fitting. We need to add to this program the steps 1 through 3 listed above.

The <u>algorithm</u> for the program, therefore, can be written as follows:

Enter vectors x and y, of the same dimension, as lists. (Note: since the function VANDERMONDE uses a list as input, it is more convenient to enter the (x,y) data as a list.) Also, enter the value of p.

Determine n = size of vector x.

Use the function VANDERMONDE to generate the Vandermonde matrix V_n for the list x entered.

If p = n-1, then

       X = V_n,
Else If p < n-1

       Remove columns p+2, ..., n from V_n to form X (Use a FOR loop and COL-)
Else

       Add columns n+1, ..., p+1 to V_n to form X (FOR loop, calculate x^i, convert to vector, use COL+)

Convert y to vector

Calculate b using program MTREG (see example on multiple linear fitting)

Here is the <u>translation of the algorithm</u> to a program in User RPL language. (The commands can be entered using the keystroke sequences listed in the last page of this chapter):

```
<<                          Open program
  → x y p                   Enter lists x and y, and number p (levels 3,2,1,
respectively)
  <<                        Open subprogram
  x SIZE → n
    <<
      x VANDERMONDE         Place x in stack, obtain $V_n$
      IF 'p<n-1' THEN       This IF implements step 3 in the algorithm
        n                   Place n in stack
        p 2 +               Calculate p+1
        FOR j               Start a FOR-STEP loop with j = n-1, n-2, ..., p+1, step = -1
          j COL– DROP       Remove column and drop it from stack
        -1 STEP             Close FOR-STEP loop
      ELSE
        IF 'p>n-1' THEN
          n 1 +             Calculate n+1
          p 1 +             Calculate p+1
```

```
        FOR j                   Start a FOR-NEXT loop with j = n, n+1, ..., p+1.
          x j ^                 Calculate $x^j$, as a list
          OBJ→ →ARRY            Convert list to array
          j COL+                Add column to matrix
        NEXT                    Close FOR-NEXT loop
      END                       Ends second IF clause.
    END                         Ends first IF clause.  Its result is matrix X
    y OBJ→ →ARRY                Convert list y to an array
    MTREG                       X and y used by program MTREG (from previous example)
    →NUM                        Convert to decimal format
    >>                          Close sub-program
    >>                          Close sub-program
  >>                            Close main program
```

Save it into a variable called POLY (POLYnomial fitting).

As an example, use the following data to obtain a polynomial fitting with p = 2, 3, 4, 5, 6.

x	y
2.30	179.72
3.20	562.30
4.50	1969.11
1.65	65.87
9.32	31220.89
1.18	32.81
6.24	6731.48
3.45	737.41
9.89	39248.46
1.22	33.45

Because we will be using the same x-y data for fitting polynomials of different orders, it is advisable to save the lists of data values x and y into variables xx and yy, respectively. This way, we will not have to type them all over again in each application of the program POLY. Thus, proceed as follows:

{ 2.3 3.2 4.5 1.65 9.32 1.18 6.24 3.45 9.89 1.22 } [ENTER] 'xx' [STO▶]
{179.72 562.30 1969.11 65.87 31220.89 32.81 6731.48 737.41 39248.46 33.45} [ENTER] 'yy' [STO▶]

To fit the data to polynomials use the following:

[xx][yy][2][POLY], Result: [4527.73 -3958.52 742.23]

i.e., $y = 4527.73 - 39.58x + 742.23x^2$

[xx][yy][3][POLY], Result: [-998.05 1303.21 -505.27 79.23]

i.e., $y = -998.05 + 1303.21x - 505.27x^2 + 79.23x^3$

[xx][yy][4][POLY], Result: [20.92 -2.61 -1.52 6.05 3.51]

i.e., $y = 20.97 - 2.61x - 1.52x^2 + 6.05x^3 + 3.51x^4$

[xx][yy][5][POLY], Result: [19.08 0.18 -2.94 6.36 3.48 0.00]

i.e., $$y = 19.08+0.18x-2.94x^2+6.36x^3+3.48x^4+0.00x^5$$

[xx][yy][6][POLY], Result: [-16.73 67.17 -48.69 21.11 1.07 0.19 0.00]

i.e., $$y = -16.73+67.17x-48.69x^2+21.11x^3+1.07x^4+0.19x^5+0.00x^6$$

Selecting the best fitting

As you can see from the results above, you can fit any polynomial to a set of data. The question arises, which is the best fitting for the data? To help one decide on the best fitting we can use several criteria:

The correlation coefficient, r. This value is constrained to the range $-1 < r < 1$. The closer r is to +1 or -1, the better the data fitting.
The sum of squared errors, SSE. This is the quantity that is to be minimized by least-square approach.
A plot of residuals. This is a plot of the error corresponding to each of the original data points. If these errors are completely random, the residuals plot should show no particular trend.
Before attempting to program these criteria, we present some definitions:

Given the vectors x and y of data to be fit to the polynomial equation, we form the matrix X and use it to calculate a vector of polynomial coefficients b. We can calculate a vector of fitted data, y', by using
$$y' = X \cdot b.$$
An error vector is calculated by
$$e = y - y'.$$
The sum of square errors is equal to the square of the magnitude of the error vector, i.e.,
$$SSE = |e|^2 = e \bullet e = \Sigma e_i^2 = \Sigma (y_i - y'_i)^2.$$
To calculate the correlation coefficient we need to calculate first what is known as the sum of squared totals, SST, defined as
$$SST = \Sigma (y_i - \bar{y})^2,$$
where $\bar{y}$ is the mean value of the original y values, i.e.,
$$\bar{y} = (\Sigma y_i)/n.$$
In terms of SSE and SST, the correlation coefficient is defined by
$$r = [1-(SSE/SST)]^{1/2}.$$

Here is the new program including calculation of SSE and r (Once more, consult the last page of this chapter to see how to produce the variable and command names in the program):

<<	Open program
→ x y p	Enter lists x and y, and number p (levels 3,2,1, respectively)
<<	Open subprogram
x SIZE → n	
<<	
x VANDERMONDE	Place x in stack, obtain V_n
IF 'p<n-1' THEN	This IF implements step 3 in the algorithm
n	Place n in stack
p 2 +	Calculate p+1
FOR j	Start a FOR-STEP loop with j = n-1, n-2, ..., p+1, step = -1
j COL– DROP	Remove column and drop it from stack
-1 STEP	Close FOR-STEP loop
ELSE	
IF 'p>n-1' THEN	
n 1 +	Calculate n+1
p 1 +	Calculate p+1
FOR j	Start a FOR-NEXT loop with j = n, n+1, ..., p+1.
x j ^	Calculate x^j, as a list
OBJ→ →ARRY	Convert list to array
j COL+	Add column to matrix
NEXT	Close FOR-NEXT loop
END	Ends second IF clause.
END	Ends first IF clause. Its result is matrix X
y OBJ→ →ARRY	Convert list y to an array
→ X yv	Enter matrix and array as X and y
<<	Open sub-program
X yv MTREG	X and y used by program MTREG
→NUM	If needed, converts result to decimal format
→ b	Resulting vector is passed as b into sub-program
<<	Open sub-program
b yv	Place b and yv (y as a vector) in stack
X b *	Calculate X·b
-	Calculate e = y - X·b
ABS SQ DUP	Calculate SSE, place two copies in stack
y ΣLIST n /	Calculate $\bar{y}$
n 1 →LIST SWAP CON	Create vector of n values of $\bar{y}$
yv – ABS SQ	Calculate SST
/	Calculate SSE/SST
NEG 1 + √	Calculate $r = [1-SSE/SST]^{1/2}$
"r" →TAG	Tag result as "r"
SWAP	Exchange stack levels 1 and 2
"SSE" →TAG	Tag result as SSE
>>	Close sub-program
>>	Close sub-program
>>	Close sub-program
>>	Close sub-program
>>	Close main program

Save this program under the name POLYR, to emphasize calculation of the correlation coefficient r.

Using the POLYR program for values of p between 2 and 6 produce the following table of values of the correlation coefficient, r, and the sum of square errors, SSE:

p	r	SSE
2	0.9971908	10731140.01
3	0.9999768	88619.36
4	0.9999999	7.48
5	0.9999999	8.92
6	0.9999998	432.61

While the correlation coefficient is very close to 1.0 for all values of p, the values of SSE vary widely. The smallest value of SSE corresponds to p = 4. Thus, the preferred data fitting for the original x-y data would be:

$$y = 20.97 - 2.61x - 1.52x^2 + 6.05x^3 + 3.51x^4.$$

Keystroke combinations for writing the programs POLY and POLYR

The following list of keystroke combinations will let you access all the variables and functions used in the programs that follow. The keystroke combinations are grouped by type:

Letters and words:

X	[ALPHA][↵][X]	y	[ALPHA][↵][Y]
p	[ALPHA][↵][P]	n	[ALPHA][↵][N]
X	[ALPHA][X]	j	[ALPHA][↵][J]
yv	[ALPHA][↵][Y][ALPHA][↵][V]	"r"	[→]["][ALPHA][↵][R][▶]
"SSE"	[→]["][ALPHA][S][S][E][▶]		

Characters and functions directly accessible in keyboard:

<<	[→][<< >>]	→	[→][→]
ABS	[↵][ABS]	SQ	[↵][x²]
NEG	[+/-]	+	[+]
–	[–]	*	[×]
/	[÷]	→NUM	[→][→NUM]

Matrix related commands:

SIZE	[↵][MTH][MATRX][MAKE][SIZE]
COL–	[↵][MTH][MATRX][COL][COL-]
COL+	[↵][MTH][MATRX][COL][COL+]
CON	[↵][MTH][MATRX][MAKE][CON]
VANDERMONDE	[↵][MATRICES][CREAT][NXT][NXT][VANDE]

Program branching commands:

IF	[←][PRG][BRCH][IF][IF]	THEN	[←][PRG][BRCH][IF][THEN]
ELSE	[←][PRG][BRCH][IF][ELSE]	END	[←][PRG][BRCH][IF][END]
FOR	[←][PRG][BRCH][FOR][FOR]	STEP	[←][PRG][BRCH][FOR][STEP]
NEXT	[←][PRG][BRCH][FOR][NEXT]		

Stack programming commands:

DROP	[←][PRG][STACK][DROP]	DUP	[←][PRG][STACK][DUP]
SWAP	[←][PRG][STACK][SWAP]		

Type programming commands:

OBJ→	[←][PRG][TYPE][OBJ→]	→ARRY	[←][PRG][TYPE][→ARRY]
→LIST	[←][PRG][TYPE][→LIST]	→TAG	[←][PRG][TYPE][→TAG]

User-defined commands:

MTREG   [MTREG] (user defined soft key)

11 Graphics and character strings

In this chapter we introduce some of the graphics capabilities of the HP 49 G calculator particularly regarding graphics in the x-y plane. We will present graphics of functions in Cartesian coordinates and polar coordinates, parametric plots, graphics of conics, bar plots, scatterplots, and a variety of three-dimensional graphs. We also discuss programming using graphics, and applications of character strings.

Graphs options in the HP 49 G

To provide an overall view of all the graphic capabilities of the HP 49 G, we will first let you see a list of all the options for graphics that the calculator provides. To access this list press in RPN mode, press, simultaneously, the keys [←] and [2D/3D] (the F4 soft menu key). The calculator will produce the PLOT SETUP window, which includes the TYPE field. Right in front of the TYPE field you will, most likely, see the option Function highlighted. This is the default type of graph for the calculator. To see the list of graph types, press the soft menu key labeled [CHOOS]. This will produce a drop down menu with the following options (use the up- and down-arrow keys to see all the options):

Function:	for plotting equations of the form y = f(x) in Cartesian coordinates in the plane
Polar:	for plotting equations of the from r = f(θ) in polar coordinates in the plane
Parametric:	for plotting equations of the form x = x(t), y = y(t) in the plane
Diff Eq:	for plotting the numerical solution of a linear differential equation
Conic:	for plotting conic equations (circles, ellipses, hyperbolas, and parabolas)
Truth:	for plotting inequalities in the plane
Histogram:	for plotting frequency histograms (associated with statistical applications)
Bar:	for plotting simple bar charts
Scatter:	for plotting scatter plots of discrete data sets (also associated with statistics)
Slopefield:	for plotting traces of the slopes of a function of the form f(x,y) = 0.
Fast3D:	for plotting curved surfaces and solid bodies in space
Wireframe:	for plotting curved surfaces in space showing wireframe grids
Ps-Contour:	for plotting contour plots of surfaces
Y- Slice:	for plotting a slicing view of a function f(x,y).
Gridmap:	for plotting real and imaginary part traces of a function of a complex variable
Pr-Surface:	for plotting parametric surfaces given by x = x(u,v), y = y(u,v), z = z(u,v).

Plotting an expression of the form y = f(x)

In this section we present an example of a plot of a function of the form y = f(x). In order to proceed with the plot, first, purge the variable x, if it is defined in the current directory (x will be the independent variable in the calculator's PLOT feature, therefore, you don't want to have it pre-defined). Create a sub-directory called 'TPLOT' (for test plot), or other meaningful name, to perform the following exercise.

Let's plot the function,

$$f(x) = \frac{1}{\sqrt{2\pi}} \exp(-\frac{x^2}{2})$$

⬇ First, enter the PLOT SETUP environment by pressing, simultaneously, [←][2D/3D]. Make sure that the option Function is selected as the TYPE, and that 'X' is selected as the independent variable (INDEP). Press [NXT][OK] to return to normal calculator display. The PLOT SET UP window should look similar to this:

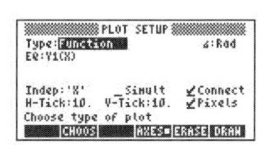

> Note: You will notice that a new variable, called PPAR, shows up in your soft menu key labels. This stands for Plot PARameters. To see its contents, press [→][PPAR]. A detailed explanation of the contents of PPAR is provided later in this chapter. Press [⇦] to drop the contents of the stack.

⬇ Enter the PLOT environment by pressing, simultaneously, [←][Y =]. Press [ADD] to get you into the equation writer. You will be prompted to fill the right-hand side of an equation Y1(x) = ▪. Type the function using the following keystrokes:

[←][e^x][ALPHA][X][y^x][2][▶][+/-][÷][2][▶][▶] [÷][√][2][×] [←][π][ENTER]

The expression 'Y1(X) = EXP(-X^2/2)/)/√ (2*π)' will be highlighted. Press [NXT][OK] to return to normal calculator display. The PLOT screen for a FUNCTION type of graph looks like this:

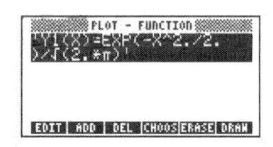

282 © 2000 – Gilberto E. Urroz
All rights reserved

> Note: Two new variables show up in your soft menu key labels, namely EQ and Y1. To see the contents of EQ, use [→][EQ]. The content of EQ is simply the function name 'Y1(X)'. The variable EQ is used by the calculator to store the equation, or equations, to plot. (EQ is also to store the equation to solve, as we will see in a subsequent chapter).
>
> To see the contents of Y1 press [→][Y1]. You will get the function Y1(X) defined as the program:
>
> $$<< \rightarrow X \; 'EXP(-X^2/2)/\sqrt{(2*\pi)}' >>.$$
>
> Press [⇐], twice, to drop the contents of the stack.

↯ Enter the PLOT WINDOW environment by pressing, simultaneously, [←][WIN]. Use a range of -4 to 4 for H-VIEW by using [4][+/-][OK][4][OK], then press [AUTO] to generate the V-VIEW automatically. The PLOT WINDOW screen looks as follows:

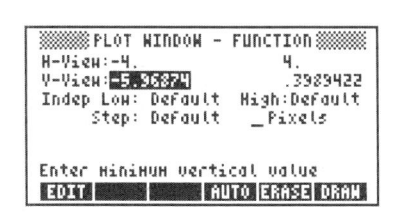

↯ Plot the graph: [ERASE][DRAW].

↯ To see labels: [EDIT][NXT][LABEL][MENU]

↯ To recover the first graphics menu: [NXT][NXT][PICT]

↯ To trace the curve: [TRACE][(X,Y)]

Here is the way this graph looks without soft menu keys:

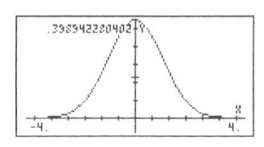

Use [▶] and [◀] to move along the curve. Check that for x = 1.05 , y = 0.231. Also, check that for x = -1.48 , y = 0.134.

⬇ To recover the menu, and return to the PLOT WINDOW environment, press [NXT][CANCL], then [NXT][OK].

Some useful PLOT operations for FUNCTION plots

In order to discuss these PLOT options, we'll modify the function to force it to have some real roots (Since the current curve is totally contained above the x axis, it has no real roots.)   Press [↱][Y1] to list the contents of the function Y1 on the stack: << →X 'EXP(-X^2/2)/√(2*π) ' >>. To edit this expression use:

[▼]	Launches the line editor
[↱][▼]	Moves cursor to the end of the line
[◄][◄][◄][-][.][1]	Modifies the expression
[ENTER]	Returns to normal calculator display

Next, store the modified expression into variable y by using [↰][Y1].

The function to be plotted is now,

$$f(x) = \frac{1}{\sqrt{2\pi}} \exp(-\frac{x^2}{2}) - 0.1$$

Enter the PLOT WINDOW environment by pressing, simultaneously, [↰][WIN].   Keep the range of -4 to 4   for H-VIEW, press [▼][AUTO] to generate the V-VIEW. To plot the graph, press [ERASE][DRAW].

⬇ Once the graph is plotted, press [FCN] to access the function menu. With this menu you can obtain additional information about the plot such as intersects with the x-axis, roots, slopes of the tangent line, area under the curve, etc.

⬇ For example, to find the root on the left side of the curve, move the cursor near that point, and press [ROOT]. You will get the result: ROOT: -1.6635.... Press [NXT] to recover the menu.   Here is the result of ROOT in the current plot:

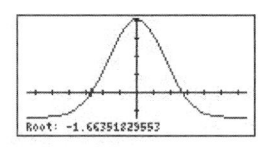

⬇ If you move the cursor towards the right-hand side of the curve and press [ROOT], the result now is ROOT: 1.6635... The calculator indicated, before showing the root, that it was found through SIGN REVERSAL. Press [NXT] to recover the menu.

⬇ Pressing [ISECT] will give you the intersection of the curve with the x-axis, which is essentially the root.   Press [ISECT].   You will get the same message as before, namely SIGN REVERSAL, before getting the result I-SECT: 1.6635.... The [ISECT] function is intended to determine the intersection of any two curves closest to the location of the cursor. In this case, where only one curve, namely, Y1(X) as defined above, is involved, the intersection sought is that of f(x) with the x-axis. Press [NXT] to recover the menu.

⁌ Place the cursor on the curve at any point and press [SLOPE] to get the value of the slope at that point. For example, at the negative root, SLOPE: 0.16670.... Press [NXT] to recover the menu.

⁌ To determine the highest point in the curve, place the cursor near the vertex and press [EXTR]. The result is EXTRM: 0.. Press [NXT] to recover the menu.

⁌ Other buttons available in the first menu are [AREA] to calculate the area under the curve, and [SHADE] to shade an area under the curve. Press [NXT] to see more options. The second menu includes one button called [VIEW] that flashes for a few seconds the equation plotted. Press [VIEW]. Alternatively, you can press the button [NXEQ] to see the name of the function Y1(x). Press [NXT] to recover the menu.

⁌ The button [F(X)] gives the value of f(x) corresponding to the cursor position. Place the cursor anywhere in the curve and press [F(X)]. The value will be shown in the lower left corner of the display. Press [NXT] to recover the menu.

⁌ Place the cursor in any given point of the trajectory and press [TANL] to obtain the equation of the tangent line to the curve at that point. The equation will be displayed on the lower left corner of the display. Press [NXT] to recover the menu.

⁌ If you press [F'] the calculator will plot the derivative function, $f'(x) = df/dx$, as well as the original function, f(x). Notice that the two curves intercept at two points. Move the cursor near the left intercept point and press [FCN][ISECT], to get I-SECT: (-0.6834...,0.21585). Press [NXT] to recover the menu.

⁌ To leave the FCN environment, press [PICT] (or [NXT][PICT]).

⁌ Press [CANCL] to return to the PLOT WINDOW environment. Then, press [NXT][OK] to return to normal calculator display.

> Note: the stack will show all the graph operations performed properly identified.

⁌ Enter the PLOT environment by pressing, simultaneously, [↵][Y =]. Notice that the highlighted field in the PLOT environment now contains the derivative of Y1(X). Press [NXT][OK] to return to return to normal calculator display.

⁌ Press [↵][EQ] to check the contents of EQ. You will notice that it contains a list instead of a single expression. The list has as elements an expression for the derivative of Y1(X) and Y1(X) itself. Originally, EQ contained only Y1(x). After we pressed the button [F'] in the [FCN] environment, the calculator automatically added the derivative of Y1(x) to the list of equations in EQ.

Saving a graph for future use

If you want to save your graph to a variable, get into the PICTURE environment by pressing [◄]. Then, press [EDIT][NXT][NXT][PICT→]. This captures the current picture into a graphics object. To return to the stack, press [PICT][CANCL].

In level 1 of the stack you will see a graphics object described as Graphic 131 × 64. To store it into a variable, say FIG, type ['][ALPHA][ALPHA][F][I][G][ALPHA][STO]. Your figure is now stored in variable FIG.

To display your figure again, recall it to level 1 of the stack, by pressing [FIG]. Level 1 now reads Graphic 131 × 64. Enter the PICTURE environment, press [◄].

Clear the current picture, [EDIT][NXT][ERASE].

Move the cursor to the upper left corner of the display, by using the [◄] and [▲] keys.

To display the figure currently in level 1 of the stack press [NXT][REPL].

To return to normal calculator function, press [PICT][CANCL].

Graphics of transcendental functions

In this section we use some of the graphics features of the calculator to show the typical behavior of the natural log, exponential, trigonometric and hyperbolic functions. You will not see any graphs in this chapter, for I want you to see them in your calculator.

Graph of ln(X)

Using the calculator in RPN mode, press, simultaneously, the left-shift key [↰] and the [2D/3D] (F4) key to produce the PLOT SETUP window. The field labeled **Type** will be highlighted. If the option Function is not already selected press the soft key labeled [CHOOS], use the up and down keys ([▲] [▼]) to select Function, and press [OK] to complete the selection. Check that the field labeled Indep: contains the variable 'X'. If that is not so, press the down arrow key [▼] twice until the Indep field is highlighted, press the soft key labeled [EDIT] and modify the value of the independent variable to read 'X'. Press [OK] when done. Press [NXT][OK] to return to normal calculator display.

Next, press, simultaneously, the left-shift key [↰] and the [Y=] (F1) key to produce the PLOT-FUNCTION window. If there is any equation highlighted in this window, press [DEL] as needed to clear the window completely. When the PLOT-FUNCTION window is empty you will get a prompt message that reads: **No Equ., Press ADD**. Press the soft key labeled [ADD]. This will trigger the equation writer with the expression Y1(X)=◄ . Type LN(X) by using [↪][LN][X]. Press [ENTER] to return to the PLOT-FUNCTION window. Press [NXT][OK] to return to normal calculator display.

The next step is to press, simultaneously, the left-shift key [↰] and the [WIN] (F2) key to produce the PLOT WINDOW - FUNCTION window. Most likely, the display will show the horizontal (H-View) and vertical (V-View) ranges as:

```
                    H-View: -6.5        6.5
                    V-View: -3.1        3.2
```

These are the default values for the x- and y-range, respectively, of the current graphics display window. Change the H-View values to read:

```
                    H-View: -1          10
```

by using [1][+/-][OK][1][0][OK]. Next, press the soft key labeled [AUTO] to let the calculator determine the corresponding vertical range. After a couple of seconds this range will be shown in the PLOT WINDOW-FUNCTION window. At this point we are ready to produce the graph of ln(X). Press [ERASE][DRAW] to plot the natural logarithm function.

To add labels to the graph press [EDIT][NXT][LABEL]. Press [MENU] to remove the menu labels, and get a full view of the graph. Press [NXT] to recover the current graphic menu. Press [NXT][PICT] to recover the original graphical menu.

To determine the coordinates of points on the curve press [TRACE] (the cursor moves on top of the curve at a point located near the center of the horizontal range). Next, press [(X,Y)] to see the coordinates of the current cursor location. These coordinates will be shown at the bottom of the screen. Use the right- and left-arrow keys to move the cursor along the curve. As you move the cursor along the curve the coordinates of the curve are displayed at the bottom of the screen. Check that when Y:1.00E0, X:2.72E0. This is the point (e,1), since ln(e) = 1. Press [NXT] to recover the graphics menu.

Next, we will find the intersection of the curve with the x-axis by pressing [FCN][ROOT]. The calculator returns the value `Root: 1`, confirming that ln(1) = 0. Press [NXT][NXT][PICT][CANCL] to return to the PLOT WINDOW - FUNCTION. Press [ENTER] to return to normal calculator display. You will notice that the root found in the graphics environment was copied to the calculator stack.

Note: When you press [VAR], your variables list will show new variables called [X] and [Y1]. Press [→][Y1] to see the contents of this variable. You will get the program

$$<< \to X \; 'LN(X)' >>,$$

which you will recognize as the program that may result from defining the function

$$'Y1(X) = LN(X)'$$

by using [←][DEF]. This is basically what happens when you [ADD] a function in the PLOT - FUNCTION window (the window that results from pressing [←][Y=], simultaneously), i.e., the function gets defined and added to your variable list.

Next, press [→][X] to see the contents of this variable. A value of 10.275 is placed in the stack. This value is determined by our selection for the horizontal display range. We selected a range between -1 and 10 for X. To produce the graph, the calculator generates values between the range limits using a constant increment, and storing the values generated, one at a time, in the variable [X] as the graph is drawn. For the horizontal range (-1,10), the increment used seems to be 0.275. When the value of X becomes larger than the maximum value in the range (in this case, when X = 10.275), the drawing of the graph stops. The last value of X for the graphic under consideration is kept in variable X.

> If you want to purge X and Y1 when finished working with the graphics, create the list { X Y1} in stack level 1 by using [←][{}][→][X][→][Y1][ENTER], and then use [TOOL][PURGE].

Graph of the exponential function

First, load the function exp(X), by pressing, simultaneously, the left-shift key [←] and the [Y=] (F1) key to access the PLOT-FUNCTION window. Press [DEL] to remove the function LN(X), if you didn't delete Y1 as suggested in the previous note. Press [ADD] and type [←][e^x][X][ENTER] to enter EXP(X) and return to the PLOT-FUNCTION window. Press [NXT][OK] to return to normal calculator display.

Next, press, simultaneously, the left-shift key [←] and the [WIN] (F2) key to produce the PLOT WINDOW - FUNCTION window. Change the H-View values to read:

$$H\text{-View}: \quad -8 \qquad 2$$

by using [8][+/-][OK][2][OK]. Next, press [AUTO]. After the vertical range is calculated, press [ERASE][DRAW] to plot the exponential function.

To add labels to the graph press [EDIT][NXT][LABEL]. Press [MENU] to remove the menu labels, and get a full view of the graph. Press [NXT] to recover the current graphic menu. Press [NXT] to recover the graphics menu. Press [CANCL] to return to the PLOT WINDOW - FUNCTION. Press [ENTER] to return to normal calculator display.

The PPAR variable

Press [VAR] to recover your variables menu, if needed. In your variables menu you should have a variable labeled [PPAR]. Press [→][PPAR] to get the contents of this variable in the stack. Press the down-arrow key, , to launch the stack editor, and use the up- and down-arrow keys to view the full contents of PPAR. The screen will show the following values:

```
{
(-8.,-1.10797263281)
(2.,7.38905609893) X
0. { (0.,0.) {# 10d
# 10d }} FUNCTION Y
}
```

PPAR stands for Plot PARameters, and its contents include two ordered pairs of real numbers,

(-8.,-1.10797263281) and (2.,7.38905609893),

which represent the coordinates of the lower left corner and the upper right corner of the plot, respectively. Next, PPAR lists the name of the independent variable, X, followed by a number that specifies the increment of the independent variable in the generation of the plot. The value shown here is the default value, zero (0.), which specifies increments in X corresponding to 1 pixel in the graphics display. The next element in PPAR is a list containing first the coordinates of the point of intersection of the plot axes, i.e., (0.,0.), followed by a list that specifies the tick mark annotation on the x- and y-axes, respectively {# 10d # 10d}. Next, PPAR lists the type of plot that is to be generated, i.e., FUNCTION, and, finally, the y-axis label, i.e., Y.

The variable PPAR, if non-existent, is generated every time you create a plot. The contents of the function will change depending on the type of plot and on the options that you select in the PLOT window (the window generated by the simultaneous activation of the [↰] and [WIN](F2) keys.

Inverse functions and their graphs

Let $y = f(x)$, if we can find a function $y = g(x)$, such that, $g(f(x)) = x$, then we say that $g(x)$ is the inverse function of $f(x)$. Typically, the notation $g(x) = f^{-1}(x)$ is used to denote an inverse function. Using this notation we can write: if $y = f(x)$, then $x = f^{-1}(y)$. Also, $f(f^{-1}(x)) = x$, and $f^{-1}(f(x)) = x$.

As indicated earlier, the ln(x) and exp(x) functions are inverse of each other, i.e., $\ln(\exp(x)) = x$, and $\exp(\ln(x)) = x$. This can be verified in the calculator by using:

[EQW][↪][LN][↰][e^x][ALPHA][↰][X][ENTER]   Result: 'LN(EXP(x))'
[↪][EVAL]   Result: 'x', i.e., $\ln(\exp(x)) = x$.

[EQW][↰][e^x] [↪][LN][ALPHA][↰][X][ENTER]   Result: 'EXP(LN(x))'
[↪][EVAL]   Result: 'x', i.e., $\exp(\ln(x)) = x$.

When a function $f(x)$ and its inverse $f^{-1}(x)$ are plotted simultaneously in the same set of axes, their graphs are reflections of each other about the line $y = x$. Let's check this fact with the calculator for the functions LN(X) and EXP(X) by following this procedure:

Press, simultaneously, [↰][Y=]. The function Y2(X) = EXP(X) should be available in the PLOT - FUNCTION window from the previous exercise. Press [ADD], and type [↪][LN][X][ENTER], to load the function Y3(X) = LN(X). Also, press [ADD][X][ENTER], to load the function Y4(X) = X. Press [NXT][OK] to return to normal calculator display.

Press, simultaneously, [↰][WIN], and change the H-View range to read:

                    H-View: -8   8

Press [AUTO] to generate the vertical range. Press [ERASE][DRAW] to produce the graph of y = ln(x), y = exp(x), and y = x, simultaneously.

You will notice that only the graph of y = exp(x) is clearly visible. Something went wrong with the [AUTO] selection of the vertical range. What happens is that, when you press [AUTO] in the PLOT FUNCTION - WINDOW screen, the calculator produces the vertical range corresponding to the first function in the list of functions to be plotted. Which, in this case, happens to be Y2(X) = EXP(X). We will have to enter the vertical range ourselves in order to display the other two functions in the same plot.

Press [CANCL] to return to the PLOT FUNCTION - WINDOW screen. Modify the vertical and horizontal ranges to read:

```
              H-View: -8      8
              V-View: -4      4
```

By selecting these ranges we ensure that the scale of the graph is kept 1 vertical to 1 horizontal. Press [ERASE][DRAW] and you will get the plots of the natural logarithm, exponential, and y = x functions. It will be evident from the graph that LN(X) and EXP(X) are reflections of each other about the line y = X. Press [CANCL] to return to the PLOT WINDOW - FUNCTION. Press [ENTER] to return to normal calculator display.

Summary of FUNCTION plot operation

In this section we present information regarding the PLOT SETUP, PLOT-FUNCTION, and PLOT WINDOW screens accessible through the left-shift key combined with the soft-menu keys F1 through F4. Based on the graphing examples presented above, the procedure to follow to produce a FUNCTION plot (i.e., one that plots one or more functions of the form Y = F(X)), is the following:

- [↵][2D/3D], simultaneously: Access to the PLOT SETUP window. If needed, change TYPE to FUNCTION, and enter the name of the independent variable.

Settings:

- A check on _Simult means that if you have two or more plots in the same graph, they will be plotted simultaneously when producing the graph.
- A check on _Connect means that the curve will be a continuous curve rather than a set of individual points.
- A check on _Pixels means that the marks indicated by H-Tick and V-Tick will be separated by that many pixels.
- The default value for both by H-Tick and V-Tick is 10.

Soft key menu options:

- Use [EDIT] to edit functions of values in the selected field.

- Use [CHOOS] to select the type of plot to use when the Type: field is highlighted. For the current exercises, we want this field set to FUNCTION.

> Note: the soft menu keys [EDIT] and [CHOOS] are not available at the same time. One or the other will be selected depending on which input field is highlighted.

- Press the [AXES] soft menu key to select or deselect the plotting of axes in the graph. If the option 'plot axes' is selected, a square dot will appear in the key label: [AXES■]. Absence of the square dot indicates that axes will not be plotted in the graph.

- Use [ERASE] to erase any graph currently existing in the graphics display window.

- Use [DRAW] to produce the graph according to the current contents of PPAR for the equations listed in the PLOT-FUNCTION window.

- Press [NXT] to access the second set of soft menu keys in this screen.

- Use [RESET] to reset any selected field to its default value.

- Use [CANCL] to cancel any changes to the PLOT SETUP window and return to normal calculator display.

- Press [OK] to save changes to the options in the PLOT SETUP window and return to normal calculator display.

- [←][Y=], simultaneously: . Access to the PLOT window (in this case it will be called PLOT -FUNCTION window).

Soft menu key options:

- Use [EDIT] to edit the highlighted equation.

- Use [ADD] to add new equations to the plot.

Note: [ADD] and [EDIT] will trigger the equation writer EQW that you can use to write new equations or edit old equations.

- Use [DEL] to remove the highlighted equation.

- Use [CHOOS] to add an equation that is already defined in your variables menu, but not listed in the PLOT - FUNCTION window.

- Use [ERASE] to erase any graph currently existing in the graphics display window.

- Use [DRAW] to produce the graph according to the current contents of PPAR for the equations listed in the PLOT-FUNCTION window.

- Press [NXT] to activate the second menu list.

- Use [MOVE] and [MOVE] to move the selected equation one location up or down, respectively.

- Use [CLEAR] if you want to clear all the equations currently active in the PLOT - FUNCTION window. The calculator will verify whether or not you want to clear all the functions before erasing all of them. Select YES, and press [OK] to proceed with clearing all functions. Select NO, and press [OK] to de-activate the option CLEAR.

- Press [OK] when done to return to normal calculator display.

- [←][WIN], simultaneously: Access to the PLOT WINDOW screen.

Settings:

⬇ Enter lower and upper limits for horizontal view (H-View) and vertical view (V-View) ranges in the plot window. Or,

⬇ Enter lower and upper limits for horizontal view (H-View), and press [AUTO], while the cursor is in one of the V-View fields, to generate the vertical view (V-View) range automatically. Or,

⬇ Enter lower and upper limits for vertical view (V-View), and press [AUTO]], while the cursor is in one of the H-View fields, to generate the horizontal view (H-View) range automatically.

⬇ The calculator will use the horizontal view (H-View) range to generate data values for the graph, unless you change the options Indep Low, (Indep) High, and (Indep) Step. These values determine, respectively, the minimum, maximum, and increment values of the independent variable to be used in the plot. If the option Default is listed in the fields Indep Low, (Indep) High, and (Indep) Step, the calculator will use the minimum and maximum values determined by H-View.

⬇ A check on _Pixels means that the value of the independent variable increment (Step:) are given in pixels rather than in plot coordinates.

Soft menu key options:

⬇ Use [EDIT] to edit any entry in the window.

⬇ Use [AUTO] as explained in Settings, above.

⬇ Use [ERASE] to erase any graph currently existing in the graphics display window.

⬇ Use [DRAW] to produce the graph according to the current contents of PPAR for the equations listed in the PLOT-FUNCTION window.

⬇ Press [NXT] to activate the second menu list.

⬇ Use [RESET] to reset the field selected (i.e., where the cursor is positioned) to its default value.

⬇ Use [CALC] to access calculator stack to perform calculations that may be necessary to obtain a value for one of the options in this window. When the calculator stack is made available to you, you will also have the soft menu key options [CANCL] and [OK] .

⬇ Use [CANCL] in case you want to cancel the current calculation and return to the PLOT WINDOW screen. Or,

⬇ Use [OK] to accept the results of your calculation and return to the PLOT WINDOW screen.

⬇ Use TYPES to get information on the type of objects that can be used in the selected option field.

✦ Use [CANCL] to cancel any changes to the PLOT WINDOW screen and return to normal calculator display.

✦ Press [OK] to accept changes to the PLOT WINDOW screen and return to normal calculator display.

✦ [↵][GRAPH], simultaneously: Plots the graph based on the settings stored in variable PPAR and the current functions defined in the PLOT - FUNCTION screen. If a graph, different from the one you are plotting, already exists in the graphic display screen, the new plot will be superimposed on the existing plot. This may not be the result you desire, therefore, I recommend to use the [ERASE][DRAW] soft menu keys available in the PLOT SETUP, PLOT-FUNCTION or PLOT WINDOW screens.

Plots of trigonometric and hyperbolic functions and their inverses

The procedures used above to plot LN(X) and EXP(X), separately or simultaneously, can be used to plot any function of the form y = f(x). It is left as an exercise to the reader to produce the plots of trigonometric and hyperbolic functions and their inverses. The table below suggests the values to use for the vertical and horizontal ranges in each case. You can include the function Y=X when plotting simultaneously a function and its inverse to verify their 'reflection' about the line Y = X.

Function to plot	H-View range Minimum	H-View range Maximum	V-View range Minimum	V-View range Maximum
SIN(X)	-3.15	3.15	AUTO	
ASIN(X)	-1.2	1.2	AUTO	
SIN & ASIN	-3.2	3.2	-1.6	1.6
COS(X)	-3.15	3.15	AUTO	
ACOS(X)	-1.2	1.2	AUTO	
COS & ACOS	-3.2	3.2	-1.6	1.6
TAN(X)	-3.15	3.15	-10	10
ATAN(X)	-10	10	-1.8	1.8
TAN & ATAN	-2	-2	-2	-2
SINH(X)	-2	2	AUTO	
ASINH(X)	-5	5	AUTO	
SINH & ASINH	-5	5	-5	5
COSH(X)	-2	2	AUTO	
ACOSH(X)	-1	5	AUTO	
COS & ACOS	-5	5	-1	5
TANH(X)	-5	5	AUTO	
ATANH(X)	-1.2	1.2	AUTO	
TAN & ATAN	-5	5	-2.5	2.5

Plots in polar coordinates

First of all, let's delete the following variables in the sub-directory where we are practicing these graphics exercises: X, EQ, Y1, PPAR, by using:

[↵][{}][X][EQ][Y1][PPAR][ENTER] [TOOL][PURGE].

By doing this, all parameters related to graphics will be cleared. Press [VAR] to check that the variables were indeed purged.

We will try to plot the function f(θ) = 2(1-sin(θ)), as follows:

⬥ First, make sure that your calculator's angle measure is set to radians, by using [MODE][▼][▼], and toggling [+/-] until the option Radians is highlighted in front of the Angle Measure... field. Press [OK] when done.

⬥ Press [↵][2D/3D], simultaneously to access to the PLOT SETUP window.

⬥ Change TYPE to POLAR, by pressing [CHOOS][▼][OK].

⬥ Press [▼] and type [↦]['][2][×][↵][( )][1][-][SIN][ALPHA][↦][T][OK].

⬥ The cursor is now in the Indep field. Press [↦]['][ALPHA][↦][T][OK] to change the independent variable to θ.

⬥ Press [NXT][OK] to return to normal calculator display.

⬥ Press [↵][WIN], simultaneously, to access the PLOT window (in this case it will be called PLOT -POLAR window).

⬥ Change the H-VIEW range to -8 to 8, by using [8][+/-][OK][8][OK], and the V-VIEW range to -6 to 2 by using [6][+/-][OK][2][OK].

> Note: the H-VIEW and V-VIEW determine the scales of the display window only, and their ranges are not related to the range of values of the independent variable in this case.

⬥ Change the Indep Low value to 0, and the High value to 6.28 (≈ 2π), by using:
[0][OK]   [6][.][2][8][OK].

⬥ Press [ERASE][DRAW] to plot the function in polar coordinates. The result is a curve shaped like a hearth. This curve is known as a cardiod (cardios, Greek for heart).

⬥ Press [EDIT][NXT][LABEL][MENU] to see the graph with labels. Press [NXT] to recover the menu. Press [NXT][PICT] to recover the original graphics menu.

⬥ Press [TRACE][(x,y)] to trace the curve. The data shown at the bottom of the display is the angle θ and the radius r, although the latter is labeled Y (default name of dependent variable).

⬥ Press [NXT][CANCL] to return to the PLOT WINDOW screen. Press [NXT][OK] to return to normal calculator display.

In this exercise we entered the equation to be plotted directly in the PLOT SETUP window. We can also enter equations for plotting using the PLOT window, i.e., simultaneously pressing [↰][Y =]. For example, when you press [↰][Y =] after finishing the previous exercise, you will get the equation '2*(1-SIN(θ))' highlighted. Let's say, we want to plot also the function '2*(1-COS(θ))' along with the previous equation.

⬇ Press [ADD], and type [2][×][↰][( )][1][-][COS][ALPHA][↱][T][ENTER], to enter the new equation.

⬇ Press [ERASE][DRAW] to see the two equations plotted in the same figure. The result is two intersecting cardioids. Press [CANCL][ON] to return to normal calculator display.

Plotting conic curves

The most general form of a conic curve in the x-y plane is:

$$Ax^2+By^2+Cxy+Dx+Ey+F = 0$$

We also recognize as conic equations those given in the canonical form for the following figures:

- circle: $(x-x_o)^2+(y-y_o)^2 = r^2$
- ellipse: $(x-x_o)^2/a^2 + (y-y_o)^2/b^2 = 1$
- parabola: $(y-b)^2 = K(x-a)$ or $(x-a)^2 = K(y-b)$
- hyperbola: $(x-x_o)^2/a^2 + (y-y_o)^2/b^2 = 1$ or $xy = K$,

where x_o, y_o, a, b, and K are constant.

The name conic curves follows because these figures (circles, ellipses, parabolas or hyperbolas) result from the intersection of a plane with a cone. For example, a circle is the intersection of a cone with a plane perpendicular to the cone's main axis.

The HP49 G calculator has the ability of plotting one or more conic curves by selecting Conic as the function TYPE in the PLOT environment. Make sure to delete the variables PPAR and EQ before continuing. For example, let's store the list of equations

{'(X-1)^2+(Y-2)^2=3' 'X^2/4+Y^2/3=1'},

into the variable EQ, by using:

[EQW]	Launch equation writer
[↰][( )][ALPHA][X][-][1][▶][▶][▶][y^x][2][▶][+]	Type (X-1)^2+
[↰][( )][ALPHA][Y][-][2][▶][▶][▶][y^x][2][▶][▶]	Type (Y-2)^2
[↱][=][3][ENTER]	Type =3, and enter the equation to stack.
[EQW]	Launch equation writer once more
[ALPHA][X][y^x][2][▶][÷][4][▶][+]	Type X^2/4 +
[ALPHA][Y][y^x][2][▶][÷][3][▶][▶]	Type Y^2/3
[↱][=][1][ENTER]	Type =1, and enter the equation to stack.
[2][↰][PRG][LIST][→LIST]	Create the list.

[VAR] [→]['][ALPHA][ALPHA][E][Q][STO▶]   Store list of equations into variable EQ.

These equations we recognize as those of a circle centered at (1,2) with radius √3, and of an ellipse centered at (0,0) with semi-axis lengths a = 2 and b = √3.

- Enter the PLOT environment, by pressing, simultaneously, [←][2D/3D], and select Conic as the TYPE ([CHOOS][▼][▼][▼][▼][OK]). The list of equations will be listed in the EQ field.

- Make sure that the independent variable (Indep) is set to 'X' and the dependent variable (Depnd) to 'Y'.

- Press [NXT][OK] to return to normal calculator display.

- Enter the PLOT WINDOW environment, by pressing, simultaneously, [←][WIN].

- Change the range for H-VIEW to -3 to 3, by using [3][+/-][OK][3][OK]. Also, change the V-VIEW range to -1.5 to 2 by using [1][.][5][+/-][OK][2][OK].

- Plot the graph (be patient here):   [ERASE][DRAW].

Note: I selected the H-View and V-View ranges to show the intersection of the two curves. There is no general rule to select those ranges, except based on what we know about the curves. For example, for the equations shown above, we know that the circle will extend from -3+1 = -2 to 3+1 = 4 in x, and from
-3+2=-1 to 3+2=5 in y. In addition, the ellipse, which is centered at the origin (0,0), will extend from -2 to 2 in x, and from -√3 to √3 in y.

Notice that for the circle and the ellipse the region corresponding to the left and right extremes of the curves are not plotted. This is the case with all circles or ellipses plotted using Conic as the TYPE.

- To see labels: [EDIT][NXT][LABEL][MENU]

- To recover the menu: [NXT][NXT][PICT]

- To estimate the coordinates of the point of intersection, press the [(X,Y)] button and move the cursor as close as possible to those points using the arrow keys. The coordinates of the cursor are shown in the display. For example, the left point of intersection is close to (-0.692, 1.67), while the right intersection is near (1.89,0.5).

- To recover the menu and return to the PLOT environment, press [NXT][CANCL].

- To return to normal calculator display, press [NXT][OK].

Parametric plots

Parametric plots in the plane are those plots whose coordinates are generated through the system of equations x = x(t) and y = y(t), where t is know as the parameter. An example of such graph is the trajectory of a projectile, $x(t) = x_0 + v_0 \cos\theta_0 \cdot t$, $y(t) = y_0 + v_0 \sin\theta_0 \cdot t - \frac{1}{2}g \cdot t^2$. To plot equations like these, which involve constant values x_0, y_0, v_0, and θ_0, we need to store the values of those parameters in variables. To develop this example, create a sub-directory called 'PROJM' for PROJectile Motion, and within that sub-directory enter the following variables:

[0][▶']['][ALPHA][X][0][STO▶]   Stores X0 = 0
[1][0][▶']['][ALPHA][Y][0][STO▶]   Stores Y0 = 10
[1][0][▶']['][ALPHA][V][0][STO▶]   Stores V0 = 10
[2][0][▶']['][ALPHA][▶][T][0][STO▶]   Stores θ0 = 30
[9][.][8][0][6][▶']['][ALPHA][◀][G][STO▶]   Stores g = 9.806

To ensure that we work in degrees for the angle, use [◀][PRG][NXT][MODES][ANGLE][DEG]. Press [VAR] to recover the variable list. You should have in your soft menu keys the following labels:

[g][θ0][V0][Y0][X0].

Next, define the functions:

'X(t) = X0 + V0*COS(θ0)*t'   [ENTER][◀][DEF]
'Y(t) = Y0 + V0*SIN(θ0)*t - 0.5*g*t^2' [ENTER][◀][DEF]

which will add the variables [Y] and [X] to the soft menu key labels.

To produce the graph itself, follow these steps:

▪ Press [◀][2D/3D], simultaneously to access to the PLOT SETUP window.

▪ Change TYPE to Parametric, by pressing [CHOOS][▼][▼][OK].

▪ Press [▼] and type 'X(t)+i*Y(t)'[OK] to define the parametric plot as that of a complex variable. (The real and imaginary parts of the complex variable correspond to the x- and y-coordinates of the curve.)

▪ The cursor is now in the Indep field. Press [▶]['][ALPHA][◀][T][OK] to change the independent variable to t.

▪ Press [NXT][OK] to return to normal calculator display.

▪ Press [◀][WIN], simultaneously, to access the PLOT window (in this case it will be called PLOT -PARAMETRIC window). Instead of modifying the horizontal and vertical views first, as done for other types of plot, we will set the lower and upper values of the independent variable first as follows:

▪ Select the Indep Low field by pressing [▼][▼]. Change this value to [0][OK]. Then, change the value of High to [2][OK]. Enter [0][.][1][OK] for the Step value.

> Note: Through these settings we are indicating that the parameter t will take values of t = 0, 0.1, 0.2, ..., etc., until reaching the value of 2.0.

▪ Press [▲][▲] to highlight one of the fields corresponding to V-View. Then, press [AUTO]. This will generate automatic values of the H-View and V-View ranges based on the values of the independent variable t and the definitions of X(t) and Y(t) used. The result will be:

    H-View: 0   1.732050

V-View: -1.24493   11.27425

⯈ Press [ERASE][DRAW] to draw the parametric plot.

⯈ Press [EDIT][NXT][LABEL][MENU] to see the graph with labels. The window parameters are such that you only see half of the labels in the x-axis. Press [NXT] to recover the menu. Press [NXT][PICT] to recover the original graphics menu.

⯈ Press [TRACE][(X,Y)] to determine coordinates of any point on the graph. Use [▶] and [◀] to move the cursor about the curve. At the bottom of the screen you will see the value of the parameter t and coordinates of the cursor as (X,Y).

⯈ Press [NXT][CANCL] to return to the PLOT WINDOW environment. Then, press [ON], or [NXT][OK], to return to normal calculator display.

A review of your soft menu key labels shows that you now have the following variables:

[t][EQ][PPAR][Y][X][g]

Pressing [NXT] shows

[θ0][V0][Y0][X0]

The variables t, EQ, and PPAR are generated by the calculator to store the current values of the parameter, t, of the equation to be plotted EQ (which contains 'X(t) + I*Y(t)'), and the plot parameters. The other variables contain the values of constants used in the definitions of X(t) and Y(t). Because these value may change from plot to plot for projectile trajectories, I suggest we reorder the variables by using:

[↰][{}][X0][Y0][V0][θ0][NXT][g][ENTER]   Creates list {X0 Y0 V0 θ0 g}
[↰][PRG][MEM][DIR][NXT][ORDER]   Re-orders variable list
[VAR]   Recovers soft key menu

The ordering of the variables is now:

[X0][Y0][V0][θ0][g][t].

The variable t is under calculator control when plotting. To define a new projectile trajectory, we need to manipulate the values X0, Y0, V0, and θ0, only.

For example, if we were to change the angle of the trajectory to, say, $\theta_0 = 20°$, we can use:

[2][0][↰][θ0]

We can generate the new plot, together with the existing plot, by pressing, simultaneously, [↰][GRAPH]. What this command does is to plot the graph with the existing plot parameters, but without first erasing the contents of the current graphics window. Thus, for the current example, you will see two projectile trajectories plotted, the one generated earlier, and a lower trajectory corresponding to the new value of the initial angle.

If you want to erase the current picture contents before producing a new plot, you need to access either the PLOT, PLOT WINDOW, or PLOT SETUP screens, by pressing, [↰][Y =], [↰][WIN], or [↰][2D/3D] (the two keys must be pressed simultaneously). Then, press

[ERASE][DRAW]. Press [CANCL] to return to the PLOT, PLOT WINDOW, or PLOT SETUP screen. Press [ON], or [NXT][OK], to return to normal calculator display.

Creating a table of results

In chapter 6 we presented an example in which we generated a table of values (X,Y) for an expression of the form Y=f(X), i.e., a Function type of graph. In this section, we present the procedure for generating a table corresponding to a parametric plot.

- First, let's access the TABLE SETUP window by pressing, simultaneously, [←][TBLSET]. This window shows a Starting value of 0.0, and a Step value of 0.1 for the independent variable. Let's keep those values active, and leave the TABLE SETUP environment by pressing [OK].

- Generate the table by pressing, simultaneously, [←][TABLE]. The resulting table has three columns representing the parameter t, and the coordinates of the corresponding points. For this table the coordinates are labeled X1 and Y1.

- Use the arrow keys, [▲][▼][◄][►], to move about the table.

- Press [ON] to return to normal calculator display.

This procedure for creating a table corresponding to the current type of plot can be applied to other plot types.

Plotting the solution to simple differential equations

The plot of a simple differential equation can be obtained by selecting Diff Eq in the TYPE field of the PLOT SETUP environment as follows: suppose that we want to plot x(t) from the differential equation

$$dx/dt = \exp(-t^2), \text{ with initial conditions: } x = 0 \text{ at } t = 0.$$

The HP 49 G series calculator allows for the plotting of the solution of differential equations of the form

$$Y'(T) = F(T,Y).$$

For our case, we let Y→x and T→t, therefore, F(T,Y)→f(t,x) = $\exp(-t^2)$.

Before plotting the solution, x(t), for t = 0 to 5, delete the variables EQ and PPAR.

- Press [←][2D/3D], simultaneously to access to the PLOT SETUP window.

- Change TYPE to Diff Eq, by pressing [CHOOS][▼][▼][▼][OK].

- Press [▼] and type [→]['][←][e^x][-][ALPHA][←][T][y^x][2][OK].

- The cursor is now in the H-Var field. It should show H-Var:0 and also V-Var:1. This is the code used by the HP 49 G to identify the variables to be plotted. H-Var:0 means the independent variable (to be selected later) will be plotted in the horizontal axis. Also, V-Var:1 means the dependent variable (default name 'Y') will be plotted in the vertical axis.

⁃ Press [▼]. The cursor is now in the Indep field. Press [→]['][ALPHA][←][T][OK] to change the independent variable to t.

⁃ Press [NXT][OK] to return to normal calculator display.

⁃ Press [←][WIN], simultaneously, to access the PLOT window (in this case it will be called PLOT -POLAR window).

⁃ Change the H-VIEW and V-VIEW parameters to read:

```
H-VIEW:  -1    5
V-VIEW:  -1    1.5
```

⁃ Change the Init value to 0, and the Final value to 5 by using: [0][OK] [5][OK].

⁃ The values Step and Tol represent the step in the independent variable and the tolerance for convergence to be used by the numerical solution. Let's leave those values with their current settings. Press [▼].

⁃ The Init-Soln value represents the initial value of the solution to start the numerical result. For the present case, we have for initial conditions x(0) – 0, thus, we need to change this value to 0.0, by using [0][OK].

⁃ Press [ERASE][DRAW] to plot the solution to the differential equation.

⁃ Press [EDIT][NXT][LABEL][MENU] to see the graph with labels. Press [NXT] to recover the menu. Press [NXT][PICT] to recover the original graphics menu.

⁃ When we observed the graph being plotted, you'll notice that the graph is not very smooth. That is because the plotter is using a time step that is too large. To refine the graph and make it smoother, use a step of 0.1. Try the following keystrokes:

[CANCL] [▼][▼][▼][.][1][OK][ERASE][DRAW]

The plot will take longer to be completed, but the shape is definitely smoother than before.

⁃ Press [EDIT][NXT][LABEL][MENU], to see axes labels and range. Notice that the labels for the axes are shown as 0 (horizontal) and 1 (vertical). These are the definitions for the axes as given in the PLOT WINDOW screen (see above), i.e., H-VAR(t): 0, and V-VAR(x): 1.

⁃ Press [NXT][NXT][PICT] to recover menu and return to PICT environment.

⁃ Press [(X,Y)] to determine coordinates of any point on the graph. Use [▶] and [◀] to move the cursor in the plot area. At the bottom of the screen you will see the coordinates of the cursor as (X,Y). The HP48G is using X and Y as the default names for the horizontal and vertical axes, respectively.

⁜ Press [NXT][CANCL] to return to the PLOT WINDOW environment. Then, press [ON] to return to normal calculator display.

More details on using graphical solutions of differential equations are presented in Chapter ....

Truth plots

Truth plots are used to produce two-dimensional plots of regions that satisfy a certain mathematical condition that can be either true or false. For example, suppose that you want to plot the region for which
$X^2/36 + Y^2/9 < 1$, proceed as follows:

⁜ Press [↵][2D/3D], simultaneously to access to the PLOT SETUP window.

⁜ Change TYPE to Truth, by pressing [CHOOS][▼][▼][▼][▼][▼] [OK].

⁜ Press [▼] and type 'X^2/36+Y^2/9 < 1'[OK] to define the conditions to be plotted.

⁜ The cursor is now in the Indep field. Leave that as 'X' if already set to that variable, or change it to 'X' if needed.

⁜ Press [NXT][OK] to return to normal calculator display.

⁜ Press [↵][WIN], simultaneously, to access the PLOT window (in this case it will be called PLOT -PARAMETRIC window). Let's keep the default value for the window's ranges:

```
H-View: -6.5  6.5
V-View: -3.1  3.1
```

⁜ Press [ERASE][DRAW] to draw the truth plot. Because the calculator samples the entire plotting domain, point by point, plot, it takes a few minutes to produce a truth plot. The present plot should produce a shaded ellipse of semi-axes 6 and 3 (in x and y, respectively), centered at the origin.

⁜ Press [EDIT][NXT][LABEL][MENU] to see the graph with labels. The window parameters are such that you only see half of the labels in the x-axis. Press [NXT] to recover the menu. Press [NXT][PICT] to recover the original graphics menu.

⁜ Press [(X,Y)] to determine coordinates of any point on the graph. Use the arrow keys to move the cursor about the region plotted. At the bottom of the screen you will see the value of the coordinates of the cursor as (X,Y).

⁜ Press [NXT][CANCL] to return to the PLOT WINDOW environment. Then, press [ON], or [NXT][OK], to return to normal calculator display.

You can have more than one condition plotted at the same time if you multiply the conditions. For example, to plot the graph of the points for which $X^2/36 + Y^2/9 < 1$, and $X^2/16 + Y^2/9 > 1$, use the following:

⁜ Press [↵][2D/3D], simultaneously to access to the PLOT SETUP window.

⬇ Press [▼] and type '(X^2/36+Y^2/9 < 1)· (X^2/16+Y^2/9 > 1)'[OK] to define the conditions to be plotted.

⬇ Press [ERASE][DRAW] to draw the truth plot. Again, you have to be patient while the calculator produces the graph. If you want to interrupt the plot, press [ON], once. Then press [CANCEL].

Plotting histograms, bar plots, and scatter plots

Histograms, bar plots and scatter plots are used to plot discrete data stored in the reserved variable ΣDAT. This variable is used not only for these types of plots, but also for all kind of statistical applications as shown in Chapter .... As a matter of fact, the use of histogram plots is postponed until we get to that chapter, for the plotting of a histogram requires to perform a grouping of data and a frequency analysis before the actual plot.   In this section we will show how to load data in the variable ΣDAT and how to plot bar plots and scatter plots.

We will use the following data for plotting bar plots and scatter plots:

x	y	z
3.1	2.1	1.1
3.6	3.2	2.2
4.2	4.5	3.3
4.5	5.6	4.4
4.9	3.8	5.5
5.2	2.2	6.6

Bar plots

First, make sure your calculator's CAS is in Exact mode:

[MODES][CAS] [▼][▼]

Uncheck Approx option if checked, then press [OK][OK].

Next, enter the data shown above as a matrix, i.e.,

[[3.1 2.1 1.1][3.6 3.2 2.2][4.2 4.5 3.3][4.5 5.6 4.4][4.9 3.8 5.5][5.2 2.2 6.6]] [ENTER]

to store it in ΣDAT, use:

[↰]['] [↰][Σ] [▶][⇦][⇦] [ALPHA][ALPHA][D][A][T] [ALPHA] [STO▶]

Press [VAR] to recover your variables menu. A soft menu key labeled [ΣDAT] should be available in the stack.

To produce the graph:

- Press [←][2D/3D], simultaneously to access to the PLOT SETUP window.

- Change TYPE to Bar, by pressing [CHOOS][▼][▼][▼][▼][▼] [▼][▼][OK].

- A matrix will be shown at the ΣDAT field. This is the matrix we stored earlier into ΣDAT.

- Press [▼][▼] to highlight the Col: field. This field lets you choose the column of ΣDAT that is to be plotted. The default value is 1. Keep it to plot column 1 in ΣDAT.

- Press [NXT][OK] to return to normal calculator display.

- Press [←][WIN], simultaneously, to access the PLOT WINDOW screen.

- Change the V-View to read, V-View: 0 5.

- Press [ERASE][DRAW] to draw the bar plot.

- Press [NXT][CANCL] to return to the PLOT WINDOW environment. Then, press [ON], or [NXT][OK], to return to normal calculator display.

The number of bars to be plotted determines the width of the bar. The H- and V-VIEW are set to 10, by default. We changed the V-VIEW to better accommodate the maximum value in column 1 of ΣDAT. Bar plots are useful when plotting categorical (i.e., non-numerical) data.

Suppose that you want to plot the data in column 2 of the ΣDAT matrix:

- Press [←][2D/3D], simultaneously to access to the PLOT SETUP window.

- Press [▼][▼] to highlight the Col: field and type 2 [OK].

- Press [←][WIN], simultaneously to access to the PLOT SETUP window.

- Change V-View to read V-View: 0 6

- Press [ERASE][DRAW].

- Press [CANCL] to return to the PLOT WINDOW screen, then [ON] to return to normal calculator display.

Scatter plots

We will use the same SDAT matrix to produce scatter plots. First, we will plot the values of y vs x, then those of y vs z, as follows:

- Press [←][2D/3D], simultaneously to access to the PLOT SETUP window.

- Change TYPE to Scatter, by pressing [CHOOS][▼][OK].

⬇ Press [▼][▼] to highlight the Cols: field. Enter [1][OK] [2][OK] to select column 1 as X and column 2 as Y in the Y-vs-X scatter plot.

⬇ Press [NXT][OK] to return to normal calculator display.

⬇ Press [↵][WIN], simultaneously, to access the PLOT WINDOW screen.

⬇ Change the plot window ranges to read:

```
H-View: 0  6.
V-View: 0  6.
```

⬇ Press [ERASE][DRAW] to draw the bar plot. Press [EDIT][NXT][LABEL][MENU] to see the plot unencumbered by the menu and with identifying labels.

⬇ Press [NXT][NXT][PICT] to leave the EDIT environment.

⬇ Press [NXT][CANCL] to return to the PLOT WINDOW environment. Then, press [ON], or [NXT][OK], to return to normal calculator display.

To plot y vs z, use:

⬇ Press [↵][2D/3D], simultaneously to access to the PLOT SETUP window.

⬇ Press [▼][▼] to highlight the Cols: field. Enter [3][OK] [2][OK] to select column 3 as X and column 2 as Y in the Y-vs-X scatter plot.

⬇ Press [NXT][OK] to return to normal calculator display.

⬇ Press [↵][WIN], simultaneously, to access the PLOT WINDOW screen.

⬇ Change the plot window ranges to read:

```
H-View: 0  7.
V-View: 0  7.
```

⬇ Press [ERASE][DRAW] to draw the bar plot. Press [EDIT][NXT][LABEL][MENU] to see the plot unencumbered by the menu and with identifying labels.

⬇ Press [NXT][NXT][PICT] to leave the EDIT environment.

Slope fields

Slope fields are used to visualize the solutions to a differential equation of the form $y' = f(x,y)$. Basically, what is presented in the plot are segments tangential to the solution curves, since $y' = dy/dx$, evaluated at any point (x,y), represents the slope of the tangent line at point (x,y).

For example, to visualize the solution to the differential equation $y' = f(x,y) = x+y$, use the following:

- Press [↵][2D/3D], simultaneously to access to the PLOT SETUP window.

- Change TYPE to Slopefield.

- Press [▼] and type 'X+Y' [OK].

- Make sure that 'X' is selected as the Indep: and 'Y' as the Depnd: variables.

- Press [NXT][OK] to return to normal calculator display.

- Press [↵][WIN], simultaneously, to access the PLOT WINDOW screen.

- Change the plot window ranges to read:

```
X-Left:-5  X-Right:5
Y-Near:-5  Y-Far: 5
```

- Press [ERASE][DRAW] to draw the slope field plot. Press [EDIT][NXT][LABEL][MENU] to see the plot unencumbered by the menu and with identifying labels.

- Press [NXT][NXT][PICT] to leave the EDIT environment.

- Press [NXT][CANCL] to return to the PLOT WINDOW environment. Then, press [ON], or [NXT][OK], to return to normal calculator display.

If you could reproduce the slope field plot in paper, you can trace by hand lines that are tangent to the line segments shown in the plot. This lines constitute lines of $y(x,y)$ = constant, for the solution of $y' = f(x,y)$. Thus, slope fields are useful tools for visualizing particularly difficult equations to solve.

Try also a slope field plot for the function $y' = f(x,y) = -(y/x)^2$, by using:

- Press [↵][2D/3D], simultaneously to access to the PLOT SETUP window.

- Change TYPE to Slopefield.

- Press [▼] and type '– (Y/X)^2' [OK].

- Press [ERASE][DRAW] to draw the slope field plot. Press [EDIT][NXT][LABEL][MENU] to see the plot unencumbered by the menu and with identifying labels.

- Press [NXT][NXT][PICT] to leave the EDIT environment.

⬇ Press [NXT][CANCL] to return to the PLOT WINDOW environment. Then, press [ON], or [NXT][OK], to return to normal calculator display.

Fast 3D plots

Fast 3D plots are used to visualize three-dimensional surfaces represented by equations of the form z = f(x,y). For example, if you want to visualize z = f(x,y) = x^2+y^2, we can use the following:

⬇ Press [↵][2D/3D], simultaneously to access to the PLOT SETUP window.

⬇ Change TYPE to Fast3D.

⬇ Press [▼] and type 'X^2+Y^2' [OK].

⬇ Make sure that 'X' is selected as the Indep: and 'Y' as the Depnd: variables.

⬇ Press [NXT][OK] to return to normal calculator display.

⬇ Press [↵][WIN], simultaneously, to access the PLOT WINDOW screen.

⬇ Keep the default plot window ranges to read:

```
X-Left:-1  X-Right:1
Y-Near:-1  Y-Far: 1
Z-Low: -1  Z-High: 1

Step Indep: 10   Depnd: 8
```

> Note: The Step Indep: and Depnd: values represent the number of gridlines to be used in the plot. The larger these number, the slower it is to produce the graph, although, the times utilized for graphic generation are relatively fast. For the time being we'll keep the default values of 10 and 8 for the Step data.

⬇ Press [ERASE][DRAW] to draw the three-dimensional surface. The result is a wireframe picture of the surface with the reference coordinate system shown at the lower left corner of the screen. By using the arrow keys ([◄][►][▲][▼]) you can move the surface around. The orientation of the reference coordinate system will change accordingly. Go ahead and try using the arrow keys to move the surface around. When done, press [EXIT].

This is one view of the FAST3D plot for this example:

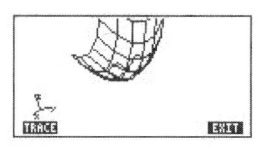

⁜ Press [CANCL] to return to the PLOT WINDOW environment.

Change the Step data to read: Step Indep: 20 Depnd: 16

⁜ Press [ERASE][DRAW] to see the surface plot. When done, press [EXIT].

⁜ Press [CANCL] to return to PLOT WINDOW.

⁜ Press [ON], or [NXT][OK], to return to normal calculator display.

Try also a Fast 3D plot for the surface $z = f(x,y) = \sin(x^2+y^2)$

⁜ Press [↵][2D/3D], simultaneously to access the PLOT SETUP window.

⁜ Press [▼] and type 'SIN(X^2+Y^2)' [OK].

⁜ Press [ERASE][DRAW] to draw the slope field plot. Press [EDIT][NXT][LABEL][MENU] to see the plot unencumbered by the menu and with identifying labels.

⁜ Press [NXT][NXT][PICT] to leave the EDIT environment.

⁜ Press [CANCL] to return to the PLOT WINDOW environment. Then, press [ON], or [NXT][OK], to return to normal calculator display.

Wireframe plots

Wireframe plots are plots of three-dimensional surfaces described by $z = f(x,y)$. Unlike Fast 3D plots, wireframe plots are static plots. The user can choose the viewpoint for the plot, i.e., the point from which the surface is seen.

For example, to produce a wireframe plot for the surface $z = x + 2y - 3$, use the following:

⁜ Press [↵][2D/3D], simultaneously to access to the PLOT SETUP window.

⁜ Change TYPE to Wireframe.

⁜ Press [▼] and type 'X+2*Y-3' [OK].

⁜ Make sure that 'X' is selected as the Indep: and 'Y' as the Depnd: variables.

⁜ Press [NXT][OK] to return to normal calculator display.

⁜ Press [↵][WIN], simultaneously, to access the PLOT WINDOW screen.

⁜ Keep the default plot window ranges to read:

```
            X-Left:-1   X-Right:1
            Y-Near:-1   Y-Far: 1
            Z-Low: -1   Z-High: 1
            XE:0    YE:-3    ZE:0
            Step Indep: 10  Depnd: 8
```

The coordinates XE, YE, ZE, stand for "eye coordinates," i.e., the coordinates from which an observer sees the plot. The values shown are the default values. The Step Indep: and Depnd: values represent the number of gridlines to be used in the plot. The larger these number, the slower it is to produce the graph. For the time being we'll keep the default values of 10 and 8 for the Step data.

⁜ Press [ERASE][DRAW] to draw the three-dimensional surface. The result is a wireframe picture of the surface.

⁜ Press [EDIT][NXT][LABEL][MENU] to see the graph with labels and ranges. This particular version of the graph is limited to the lower part of the display. We can change the viewpoint to see a different version of the graph.

⁜ Press [NXT][NXT][PICT][CANCL] to return to the PLOT WINDOW environment.

⁜ Change the eye coordinate data to read :   XE:0   YE:-3   ZE:3

⁜ Press [ERASE][DRAW] to see the surface plot.

⁜ Press [EDIT][NXT][LABEL][MENU] to see the graph with labels and ranges. This version of the graph occupies more area in the display than the previous one. We can change the viewpoint, once more, to see another version of the graph.

⁜ Press [NXT][NXT][PICT][CANCL] to return to the PLOT WINDOW environment.

⁜ Change the eye coordinate data to read :   XE:3   YE:3   ZE:3

⁜ Press [ERASE][DRAW] to see the surface plot. This time the bulk of the plot is located towards the right -hand side of the display.

⁜ Press [NXT][NXT][PICT][CANCL] to return to the PLOT WINDOW environment.

⁜ Press [ON], or [NXT][OK], to return to normal calculator display.

Try also a Wireframe plot for the surface $z = f(x,y) = \sin(x^2+y^2)$

⁜ Press [↵][2D/3D], simultaneously to access the PLOT SETUP window.

⁜ Press [▼] and type 'X^2+Y^2' [OK].

⁜ Press [ERASE][DRAW] to draw the slope field plot. Press [EDIT][NXT][LABEL][MENU] to see the plot unencumbered by the menu and with identifying labels.

◆ Press [NXT][NXT][PICT] to leave the EDIT environment.

◆ Press [CANCL] to return to the PLOT WINDOW environment. Then, press [ON], or [NXT][OK], to return to normal calculator display.

Ps-Contour plots

Ps-Contour plots are contour plots of three-dimensional surfaces described by z = f(x,y). The contours produced are projections of level surfaces z = constant on the x-y plane.

For example, to produce a Ps-Contour plot for the surface $z = x^2+y^2$, use the following:

◆ Press [←][2D/3D], simultaneously to access to the PLOT SETUP window.

◆ Change TYPE to Ps-Contour.

◆ Press [▼] and type 'X^2+Y^2' [OK].

◆ Make sure that 'X' is selected as the Indep: and 'Y' as the Depnd: variables.

◆ Press [NXT][OK] to return to normal calculator display.

◆ Press [←][WIN], simultaneously, to access the PLOT WINDOW screen.

◆ Keep the default plot window ranges to read:

```
X-Left:-2  X-Right:2
Y-Near:-1  Y-Far: 1

Step Indep: 10  Depnd: 8
```

◆ Press [ERASE][DRAW] to draw the contour plot. This operation will take some time, so, be patient. The result is a contour plot of the surface. Notice that the contour are not necessarily continuous, however, they do provide a good picture of the level surfaces of the function.

◆ Press [EDIT][NXT][LABEL][MENU] to see the graph with labels and ranges.

◆ Press [NXT][NXT][PICT][CANCL] to return to the PLOT WINDOW environment.

◆ Press [ON], or [NXT][OK], to return to normal calculator display.

Try also a Ps-Contour plot for the surface z = f(x,y) = sin x cos y.

◆ Press [←][2D/3D], simultaneously to access the PLOT SETUP window.

◆ Press [▼] and type 'SIN(X)*COS(Y)' [OK].

- Press [ERASE][DRAW] to draw the slope field plot. Press [EDIT][NXT][LABEL][MENU] to see the plot unencumbered by the menu and with identifying labels.

- Press [NXT][NXT][PICT] to leave the EDIT environment.

- Press [CANCL] to return to the PLOT WINDOW environment. Then, press [ON], or [NXT][OK], to return to normal calculator display.

Y-Slice plots

Y-Slice plots are animated plots of z-vs-y for different values of x from the function z = f(x,y).

For example, to produce a Y-Slice plot for the surface $z = x^3 - xy^3$, use the following:

- Press [←][2D/3D], simultaneously to access to the PLOT SETUP window.

- Change TYPE to Y-Slice.

- Press [▼] and type 'X^3+X*Y^3' [OK].

- Make sure that 'X' is selected as the Indep: and 'Y' as the Depnd: variables.

- Press [NXT][OK] to return to normal calculator display.

- Press [←][WIN], simultaneously, to access the PLOT WINDOW screen.

- Keep the default plot window ranges to read:

```
X-Left:-1  X-Right:1
Y-Near:-1  Y-Far: 1
Z-Low:-1   Z-High:1
```

Step Indep: 10 Depnd: 8

- Press [ERASE][DRAW] to draw the three-dimensional surface. You will see the calculator produce a series of curves on the screen, that will immediately disappear. When the calculator finishes producing all the y-slice curves, then it will automatically go into animating the different curves.

- Press [ON] to stop the animation. Press [CANCL] to return to the PLOT WINDOW environment.

- Press [ON], or [NXT][OK], to return to normal calculator display.

Try also a Ps-Contour plot for the surface z = f(x,y) = (x+y) sin y.

- Press [←][2D/3D], simultaneously to access the PLOT SETUP window.

- Press [▼] and type '(X+Y)*SIN(Y)' [OK].

- Press [ERASE][DRAW] to produce the Y-Slice animation.

- Press [ON] to stop the animation.

- Press [CANCL] to return to the PLOT WINDOW environment. Then, press [ON], or [NXT][OK], to return to normal calculator display.

Gridmap plots

Gridmap plots produce a grid of orthogonal curves describing a function of a complex variable of the form w =f(z) = f(x+iy), where z = x+iy is a complex variable. The functions plotted correspond to the real and imaginary part of w = Φ(x,y) + iΨ(x,y), i.e., they represent curves Φ(x,y) =constant, and Ψ(x,y) = constant.

For example, to produce a Gridmap plot for the function w = sin(z), use the following:

- Press [←][2D/3D], simultaneously to access to the PLOT SETUP window.

- Change TYPE to Gridmap.

- Press [▼] and type 'SIN(X+i*Y)' [OK].

- Make sure that 'X' is selected as the Indep: and 'Y' as the Depnd: variables.

- Press [NXT][OK] to return to normal calculator display.

- Press [←][WIN], simultaneously, to access the PLOT WINDOW screen.

- Keep the default plot window ranges to read:

```
X-Left:-1  X-Right:1
Y-Near:-1  Y-Far: 1
XXLeft:-1  XXRight:1
YYNear:-1  yyFar: 1
Step Indep: 10  Depnd: 8
```

- Press [ERASE][DRAW] to draw the gridmap plot. The result is a grid of functions corresponding to the real and imaginary parts of the complex function.

- Press [EDIT][NXT][LABEL][MENU] to see the graph with labels and ranges.

- Press [NXT][NXT][PICT][CANCL] to return to the PLOT WINDOW environment.

- Press [ON], or [NXT][OK], to return to normal calculator display.

Other functions of a complex variable worth trying for Gridmap plots are:

(1) SIN((X,Y))   i.e., F(z) = sin(z)   (2) (X,Y)^2   i.e., F(z) = z^2
(3) EXP((X,Y))   i.e., F(z) = e^z (4) SINH((X,Y))   i.e., F(z) = sinh(z)

(5) TAN((X,Y))   i.e., F(z) = tan(z)   (6) ATAN((X,Y)) i.e., F(z) = $\tan^{-1}(z)$
(7) (X,Y)^3   i.e., F(z) = z^3 (8) 1/(X,Y)   i.e., F(z) = 1/z
(9) √ (X,Y)   i.e., F(z) = $z^{1/2}$

Pr-Surface plots

Pr-Surface (parametric surface) plots are used to plot a three-dimensional surface whose coordinates (x,y,z) are described by x = x(X,Y), y = y(X,Y), z=z(X,Y), where X and Y are independent parameters.

> Note: The equations x = x(X,Y), y = y(X,Y), z=z(X,Y) represent a parametric description of a surface. X and Y are the independent parameters. Most textbooks will use (u,v) as the parameters, rather than (X,Y). Thus, the parametric description of a surface is given as x = x(u,v), y = y(u,v), z=z(u,v).

For example, to produce a Pr-Surface plot for the surface x = x(X,Y) = X sin Y, y = y(X,Y) = x cos Y, z=z(X,Y)=X, use the following:

- Press [←][2D/3D], simultaneously to access to the PLOT SETUP window.

- Change TYPE to Pr-Surface.

- Press [▼] and type '{X*SIN(Y) X*COS(Y) X}' [OK].

- Make sure that 'X' is selected as the Indep: and 'Y' as the Depnd: variables.

- Press [NXT][OK] to return to normal calculator display.

- Press [←][WIN], simultaneously, to access the PLOT WINDOW screen.

- Keep the default plot window ranges to read:

```
X-Left:-1  X-Right:1
Y-Near:-1  Y-Far: 1
Z-Low: -1  Z-High:1
XE: 0   YE:-3   zE:0
Step Indep: 10 Depnd: 8
```

- Press [ERASE][DRAW] to draw the three-dimensional surface. The result is a contour plot of the surface. Notice that the contour are not necessarily continuous, however, they do provide a good picture of the level surfaces of the function.

- Press [EDIT][NXT][LABEL][MENU] to see the graph with labels and ranges.

- Press [NXT][NXT][PICT][CANCL] to return to the PLOT WINDOW environment.

- Press [ON], or [NXT][OK], to return to normal calculator display.

The VPAR variable

> The VPAR (Volume Parameter) variable contains information regarding the "volume" used to produce a three dimensional graph. Therefore, you will see it produced whenever you create a three dimensional plot such as Fast3D, Wireframe, or Pr-Surface.

Programming with graphics

While the examples presented above have been operated interactively, it is possible to use graphs in your programs to complement numerical results with graphics. For example, in Chapter 10, in the section entitled "Matrix applications," we indicated that it is possible to check how well a polynomial expression fits a set of data by looking at the plot of residuals. This plot is simply the difference between the fitted and original values of y plotted against the values of x as a scatter plot. In that section we created a program called POLYR that obtains the coefficients of the polynomial, the correlation coefficient, and the sum of squared errors. In this section we present a program, that follows the general lines of POLYR, but also produces a plot of residuals.

Example: Residual plots for polynomial fitting

The program to use is written as:

<<	Open program
→ x y p	Enter lists x and y, and number p (levels 3,2,1, respectively)
<<	Open subprogram
x SIZE → n	
<<	
x VANDERMONDE	Place x in stack, obtain V_n
IF 'p<n-1' THEN	This IF implements step 3 in the algorithm
n	Place n in stack
p 2 +	Calculate p+1
FOR j	Start a FOR-STEP loop with j = n-1, n-2, ..., p+1, step = -1
j COL− DROP	Remove column and drop it from stack
-1 STEP	Close FOR-STEP loop
ELSE	
IF 'p>n-1' THEN	
n 1 +	Calculate n+1
p 1 +	Calculate p+1
FOR j	Start a FOR-NEXT loop with j = n, n+1, ..., p+1.
x j ^	Calculate x^j, as a list
OBJ→ →ARRY	Convert list to array
j COL+	Add column to matrix
NEXT	Close FOR-NEXT loop
END	Ends second IF clause.
END	Ends first IF clause. Its result is matrix X
y OBJ→ →ARRY	Convert list y to an array
→ X yv	Enter matrix and array as X and y
<<	Open sub-program

```
    X  yv  MTREG              X and y used by program MTREG
    →NUM                      If needed, converts result to decimal format
    → b                       Resulting vector is passed as b into sub-program
    <<                        Open sub-program
      b yv                    Place b and yv (y as a vector) in stack
      X b *                   Calculate X·b
      -                       Calculate e = y - X·b
      x OBJ→ →ARRY            Convert list x into a vector x
      SWAP 2 COL→             Swap vector x with vector e and create a matrix [x e]
      'ΣDAT' STO              Store matrix into statistical matrix ΣDAT
      SCATRPLOT               Produce a scatterplot, plotting col.2 vs. col.1 in ΣDAT
      DRAX LABEL              Draw axes and place labels in the plot
      PICTURE                 Activate picture screen to show graph
    >>                        Close sub-program
    >>                        Close sub-program
  >>                          Close sub-program
  >>                          Close sub-program
>>                            Close
```

Notice that this program is basically the same as POLYR, with modifications in the innermost sub-program. The commands used for the plot, such as SCATRPLOT, DRAX (DRAw aXes), LABEL, and PICTURE, are accessible only through the command catalog. Thus, to enter DRAX, for example, you want to use [CAT][ALPHA][D], then use the down arrow key ([▼]) to find DRAX. When the command is highlighted, press [OK]. To type 'ΣDAT' use: [↱]['][↱][Σ][▶][⇐][⇐][ALPHA][ALPHA][D][A][T][▶].

To run the program with the data stored in variables xx and yy, as indicated in the example of Chapter 10, use:

[xx][yy][2][POLYG]

The calculator will produce the scatterplot of the residuals. To remove the menu labels press [NXT][MENU].

Note: For an acceptable residual plot the points should be distributed randomly about the graph. If a trend (e.g., linear trend, quadratic trend) is appreciable, the fitting may not be a good one. The graph for p = 2, shows a clear "sinusoidal" distribution of points, therefore, a second-order polynomial is not a good fitting to the data. (This confirms the observations presented in the example in Chapter 10 based on the values of the sum of squared errors).

To recover the menu labels press [NXT]. Press [NXT][PICT][CANCEL] to return to normal calculator display. The vector with values of the coefficients, b, from the polynomial fitting will be available in the stack.

To try other values of p, use:

[xx][yy][3][POLYG]
[xx][yy][4][POLYG]
[xx][yy][5][POLYG]
[xx][yy][5][POLYG]

The corresponding graphs are shown below. Notice that....

Plotting commands for programming

Commands for setting up and producing plots are available through the PLOTm menu adapted from the HP 48 G/G+/GX series calculator. If you did not program this menu in your calculator, you can access it also by using:

[8][1][.][0][1] [SPC] [ALPHA][ALPHA]
[←][PRG][NXT][MODES][MENU][MENU]

The menu thus produced provides the user access to a variety of graphics functions. Since it is very useful, I suggest we user-define the F3 (GRAPH) key to provide access to this menu. This way, regardless of which is your working sub-directory, you will have access to the PLOT menu.

User-defined key for the PLOT menu

Enter the following keystrokes to determine whether you have any user-defined keys already stored in your calculator:

[←][PRG][NXT][MODES][KEYS][RCLK]

Unless you have user-defined some keys, you should get in return a list containing an S, i.e., {S}. This indicates that the Standard keyboard is the only key definition stored in your calculator.

To user-define a key you need to add to this list a command or program followed by a reference to the key. The keys in the HP 49 G calculator are referred two by a number consisting of two digits, e.g., 23, 31, 52, etc., where the first digit represents the number of the row and the second the number of the column where the key is located. Thus, the key for SIN will be referred to as key 53. Row 1 is the soft menu key row, while column 1 is the leftmost column.

Since there could be up to 6 different functions associated with a key, through combinations with [←],[→], and [ALPHA], the function that we want to user-define is referred to by an additional number:

        0 or 1, If the key is unshifted
        2, If the key is combined with [←]
        3, If the key is combined with [→]
        4, If the key is combined with [ALPHA]
        5, If the key is combined with [ALPHA][←]
        6, If the key is combined with [ALPHA][→]

This last number is attached as the decimal part of the two-number reference mentioned before. Thus, if we want to user-define the F3 key (key number 13), so that the unshifted key (function 13.0) produces the PLOT menu (menu 81.01), we need to type the following:

[←][{}][ALPHA][S][SPC][→] [<<>>] [8][1][.][0][1]   Type in { S << 81.01
[←][PRG][NXT][MODES][MENU][MENU] [▶][▶]   Type in MENU >>_
[1][3][.][0] [ENTER]   Type in 13.0 }, enter result in stack
[←][PRG][NXT][MODES][KEYS][STOKE]   Stores key definition by user

To verify that the list { S << 81.01 MENU >> 13.0 } was stored in the user-defined keys, press [RCLKE]. Press [VAR] to return to your list of variables.

The PLOT menu

> Note: We will not work any exercise while presenting the PLOT menu, its functions or sub-menus. It will be more like a tour of the contents of PLOT as they relate to the different type of graphs available in the HP 49 G calculator.

To activate a user defined key you need to press [←][USER] (same as the ALPHA key) before pressing the key or keystroke combination of interest. To activate the PLOT menu, with the key definition used above, press:

[←][USER][F3 C].

You will get the following menu:

[PTYPE][PPAR][EQ][ERASE][DRAX][DRAW]

Pressing [NXT] you obtain:

[3D][STAT][FLAG][LABEL][AUTO][INFO]

The following diagram shows the menus in PLOT. The number accompanying the different menus and functions in the diagram are used as reference in the subsequent description of those objects.

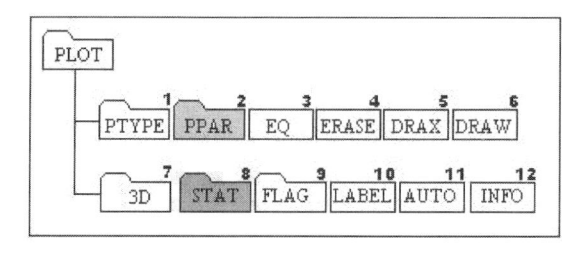

The soft menu key labeled 3D, STAT, FLAG, PTYPE, and PPAR, produce additional menus, which will be presented in more detail later. At this point we describe the functions directly accessible through soft menu keys for menu number 81.02. These are:

The function LABEL (10)

The function LABEL is used to label the axes in a plot including the variable names and minimum and maximum values of the axes. The variable names are selected from information contained in the variable PPAR.

The function AUTO (11)

The function AUTO (AUTOscale) calculates a display range for the y-axis or for both the x- and y-axes in two-dimensional plots according to the type of plot defined in PPAR. For any of the three-dimensional graphs the function AUTO produces no action. For two-dimensional plots, the following actions are performed by AUTO:

- FUNCTION: based on the plotting range of x, it samples the function in EQ and determines the minimum and maximum values of y.
- CONIC: sets the y-axis scale equal to the x-axis scale
- POLAR: based on the values of the independent variable (typically θ), it samples the function in EQ and determines minimum and maximum values of both x and y.
- PARAMETRIC: produces a similar result as POLAR based on the values of the parameter defining the equations for x and y.
- TRUTH: produces no action.
- BAR: the x-axis range is set from 0 to n+1 where n is the number of elements in ΣDAT. The range of values of y is based on the contents of ΣDAT. The minimum and maximum values of y are determined so that the x-axis is always included in the graph.
- HISTOGRAM: similar to BAR.
- SCATTER: sets x- and y-axis range based on the contents of the independent and dependent variables from ΣDAT.

The function INFO (12)

The function INFO is interactive only (i.e., it cannot be programmed). When the corresponding soft menu key is pressed it provides information about the current plot parameters.

The variable EQ (3)

The variable name EQ is reserved by the calculator to store the current equation in plots or solution to equations (see chapter ...). The soft menu key labeled EQ in this menu can be used as it would be if you have your variable menu available, e.g., if you press [EQ] it will list the current contents of that variable.

The function ERASE (4)

The function ERASE erases the current contents of the graphics window. In programming, it can be used to ensure that the graphics window is cleared before plotting a new graph.

The function DRAX (5)

The function DRAX draws the axes in the current plot, if any is visible.

The function DRAW (6)

The function DRAW draws the plot defined in PPAR.

The PTYPE menu under PLOT (1)

The PTYPE menu lists the name of all two-dimensional plot types pre-programmed in the calculator. The menu contains the following menu keys:

[FUNCT][CONIC][POLAR][PARAM][TRUTH][DIFFE]

These keys correspond to the plot types Function, Conic, Polar, Parametric, Truth, and Diff Eq, presented earlier. Pressing one of these soft menu keys, while typing a program, will place the corresponding function call in the program.

Press [NXT] [PLOT] to get back to the main PLOT menu.

The PPAR menu (2)

The PPAR menu lists the different options for the PPAR variable as given by the following soft menu key labels:

[INDEP][DEPND][XRNG][YRNG][RES][RESET]

Pressing [NXT] you get:

[CENTR][SCALE][SCALE][SCALE][AXES][ATICK]

> Note: the SCALE commands shown here actually represent SCALE, SCALEW, SCALEH, in that order.

Pressing [NXT] you get:

[PPAR][INFO][   ][   ][   ][PLOT]

The following diagram illustrates the functions available in the PPAR menu. The letters attached to each function in the diagram are used for reference purposes in the description of the functions shown below.

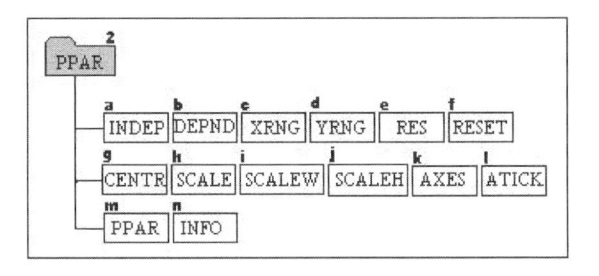

```
         2
    ┌─PPAR
    │   a      b      c     d      e     f
    ├──[INDEP][DEPND][XRNG][YRNG][ RES ][RESET]
    │   g      h      i      j      k      l
    ├──[CENTR][SCALE][SCALEW][SCALEH][AXES][ATICK]
    │   m      n
    └──[PPAR][INFO]
```

INFO (n) and PPAR (m)

If you press [INFO], or enter [→][PPAR], while in this menu, you will get a listing of the current PPAR settings. For example, having just completed the example for the program POLYG for p = 2, this is the display I got in my calculator when pressing [INFO]:

```
Indep: X
Depnd: Y
Xrng:           1.18           9.89
Yrng: -2485.89     1271.181
Res: 0.
```

This information indicates that X is the independent variable (Indep), Y is the dependent variable (Depnd), the x-axis range goes from 1.18 to 9.89 (Xrng), the y-axis range goes from -2485.89 to 1271.181 (Yrng). The last piece of information in the screen, the value of Res (resolution) determines the interval of the independent variable used for generating the plot.

The soft menu key labels included in the PPAR menu represent commands that can be used in programs. These commands include:

INDEP (a)

The command INDEP specifies the independent variable and its plotting range. These specifications are stored as the third parameter in the variable PPAR. The default value is 'X'. The values that can be assigned to the independent variable specification are:

- A variable name, e.g., 'Vel'
- A variable name in a list, e.g., { Vel }
- A variable name and a range in a list, e.g., { Vel 0 20 }
- A range without a variable name, e.g., { 0 20 }
- Two values representing a range, e.g., 0 20

In a program, any of these specifications will be followed by the command INDEP.

DEPND (b)

The command DEPND specifies the name of the dependent variable. For the case of TRUTH plots it also specifies the plotting range. The default value is Y. The type of specifications for the DEPND variable are the same as those for the INDEP variable.

XRNG (c) and YRNG (d)

The command XRNG specifies the plotting range for the x-axis, while the command YRNG specifies the plotting range for the y-axis. The input for any of these commands is two numbers representing the minimum and maximum values of x or y. The values of the x- and y-axis ranges are stored as the ordered pairs (x_{min}, y_{min}) and (x_{max}, y_{max}) in the two first elements of the variable PPAR. Default values for x_{min} and x_{max} are -6.5 and 6.5, respectively. Default values for x_{min} and x_{max} are -3.1 and 3.2, respectively.

RES (e)

The RES (RESolution) command specifies the interval between values of the independent variable when producing a specific plot. The resolution can be expressed in terms of user units as a real number, or in terms of pixels as a binary integer (numbers starting with #, e.g., #10). The resolution is stored as the fourth item in the PPAR variable.

CENTR (g)

The command CENTR takes as argument an ordered pair (x,y) or a value x, and adjusts the first two elements in the variable PPAR, i.e., (x_{min}, y_{min}) and (x_{max}, y_{max}), so that the center of the plot is (x,y) or (x,0), respectively.

SCALE (h)

The SCALE command determines the plotting scale represented by the number of user units per tick mark. The default scale is 1 user-unit per tick mark. When the command SCALE is used, it takes as arguments two numbers, x_{scale} and y_{scale}, representing the new horizontal and vertical scales. The effect of the SCALE command is to adjust the parameters (x_{min}, y_{min}) and (x_{max}, y_{max}) in PPAR to accommodate the desired scale. The center of the plot is preserved.

SCALEW (i)

Given a factor x_{factor}, the command SCALEW multiplies the horizontal scale by that factor. The W in SCALEW stands for 'width.' The execution of SCALEW changes the values of x_{min} and x_{max} in PPAR.

SCALEH (j)

Given a factor y_{factor}, the command SCALEH multiplies the vertical scale by that factor. The H in SCALEH stands for 'height.' The execution of SCALEW changes the values of y_{min} and y_{max} in PPAR.

> Note: Changes introduced by using SCALE, SCALEW, or SCALEH, can be used to zoom in or zoom out in a plot.

ATICK (l)

The command ATICK (Axes TICK mark) is used to set the tick-mark annotations for the axes. The input value for the ATICK command can be one of the following:

⬥ A real value x : sets both the x- and y-axis tick annotations to x units
⬥ A list of two real values { x y }: sets the tick annotations in the x- and y-axes to x and y units, respectively.
⬥ A binary integer #n: sets both the x- and y-axis tick annotations to #n pixels
⬥ A list of two binary integers {#n #m}: sets the tick annotations in the x- and y-axes to #n and #m pixels, respectively.

AXES (k)

The input value for the axes command consists of either an ordered pair (x,y) or a list {(x,y) atick "x-axis label" "y-axis label"}. The parameter atick stands for the specification of the tick marking annotations as described above for the command ATICK. The ordered pair represents the center of the plot. If only an ordered pair is given as input to AXES, only the axes origin is altered. The argument to the command AXES, whether an ordered pair or a list of values, is stored as the fifth parameter in PPAR.

To return to the PLOT menu, press [PLOT].

Press [NXT] to reach the second menu of the PLOT menu set.

RESET (f)

This button will reset the plot parameters to default values.

The 3D menu within PLOT (7)

The 3D menu contains two sub-menus, PTYPE and VPAR, and one variable, EQ. We are familiar already with the meaning of EQ, therefore, we will concentrate on the contents of the PTYPE and VPAR menus. The diagram below shows the branching of the 3D menu.

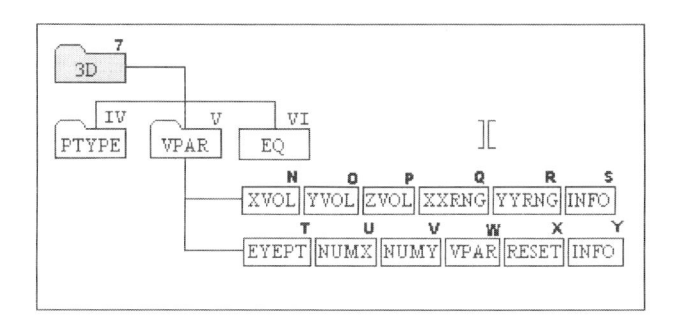

The PTYPE menu within 3D (IV)

The PTYPE menu under 3D contains the following functions:

[SLOPE][WIREF][YSLIC][PCONT][GRIDM][PARSU]

These functions correspond to the graphics options Slopefield, Wireframe, Y-Slice, Ps-Contour, Gridmap and Pr-Surface presented earlier in this chapter. Pressing one of these soft menu keys, while typing a program, will place the corresponding function call in the program.

Press [NXT] [3D] to get back to the main 3D menu.

The VPAR menu within 3D (V)

The variable VPAR stands for Volume PARameters, referring to a parallelepiped in space within which the three-dimensional graph of interest is constructed. When press [VPAR] in the 3D menu, you will get the following functions:

[XVOL][YVOL][ZVOL][XXRNG][YYRNG][INFO]

Press [NXT] to get:

[EYEPⅠ][NUMX][NUMY][VPAR][RESET][INFO]

Here we describe the meaning of these functions:

INFO (S) and VPAR (W)

When you press [INFO] (S) you get the following information in your screen:

```
Xvol:      -1.        1.
Yvol:      -1.        1.
Xvol:      -1.        1.
Xrng:      -1.        1.
Yrng:      -1.        1.
```

The ranges in Xvol, Yvol, and Zvol describe the extent of the parallelepiped in space where the graph will be generated. Xrng and Yrng describe the range of values of x and y, respectively, as independent variables in the x-y plane that will be used to generate functions of the form z = f(x,y).

Press [NXT] and [INFO] (Y) (a second INFO key) to obtain this additional information:

```
Xeye:   0.
Yeye:  -3.
Zeye:   0.
Xstep: 10.
Ystep:  8.
```

These are the value of the location of the viewpoint for the three-dimensional graph (Xeye, Yeye, Zeye), and of the number of steps in x and y to generate a grid for surface plots.

XVOL (N), YVOL (O), and ZVOL (P)

These functions take as input a minimum and maximum value and are used to specify the extent of the parallelepiped where the graph will be generated (the viewing parallelepiped). These values are stored in the variable VPAR. The default values for the ranges XVOL, YVOL, and ZVOL are -1 to 1.

XXRNG (Q) and YYRNG (R)

These functions take as input a minimum and maximum value and are used to specify the ranges of the variables x and y to generate functions $z = f(x,y)$. The default value of the ranges XXRNG and YYRNG will be the same as those of XVOL and YVOL.

EYEPT (T)

The function EYEPT takes as input real values x, y, and z representing the location of the viewpoint for a three-dimensional graph. The viewpoint is a point in space from which the three-dimensional graph is observed. Changing the viewpoint will produce different views of the graph. The figure below illustrates the idea of the viewpoint with respect to the actual graphic space and its projection in the plane of the screen.

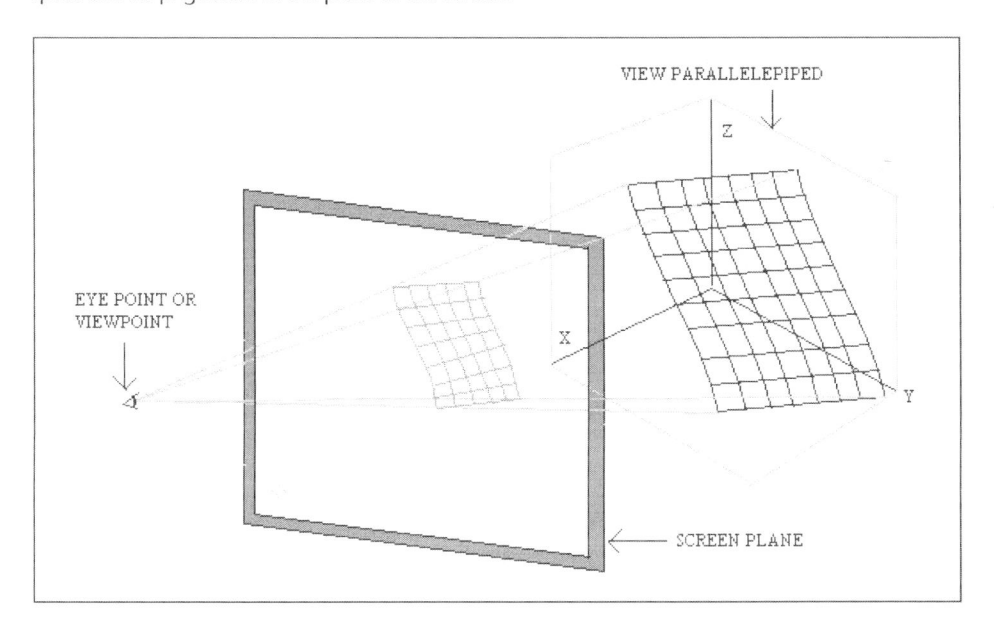

NUMX(U) and NUMY (V)

The functions NUMX and NUMY are used to specify the number of points or steps along each direction to be used in the generation of the base grid from which to obtain values of $z = f(x,y)$.

VPAR (W)

This is just a reference to the variable VPAR.

RESET (X)

Resets parameters in screen to their default values.

Press [NXT][3D] to return to the 3D menu.

Press [PLOT] to return to the PLOT menu.

The STAT menu within PLOT

The STAT menu provides access to plots related to statistical analysis. Within this menu we find the following menus:

[PTYPE][DATA][ΣPAR][   ][   ][PLOT]

The diagram below shows the branching of the STAT menu within PLOT. The numbers and letters accompanying each function or menu are used for reference in the descriptions that follow the figure.

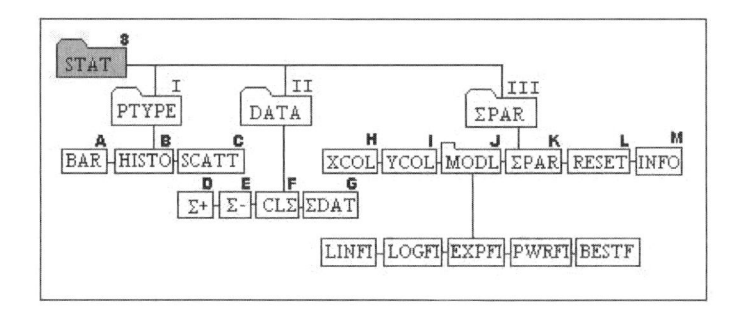

The PTYPE menu within STAT (I)

The PTYPE menu provides the following functions:

[BAR][HISTO][SCATT][   ][   ][STAT]

These keys correspond to the plot types Bar (A), Histogram (B), and Scatter(C), presented earlier. Pressing one of these soft menu keys, while typing a program, will place the corresponding function call in the program.

Press [STAT] to get back to the STAT menu.

The DATA menu within STAT (II)

The DATA menu provides the following functions:

[Σ+][Σ-][CLΣ][ΣDAT][   ][STAT]

The functions listed in this menu are used to manipulate the ΣDAT statistical matrix. The functions Σ+ (D) and Σ- (E), add or remove data rows from the matrix ΣDAT. CLΣ (F) clears the ΣDAT (G) matrix, and the soft menu key labeled ΣDAT is just used as a reference for interactive applications. More details on the use of these functions are presented in a later chapter on statistical applications.

Press [STAT] to return to the STAT menu.

The ΣPAR menu within STAT (III)

The ΣPAR menu provides the following functions:

[XCOL][YCOL][MODL][SPAR][RESET][INFO]

INFO (M) and ΣPAR (K)

The key INFO in ΣPAR provides the following information:

```
Xcol: 1.
Ycol: 2.
Intercept: 0.
Slope: 0.
Model: LINFIT
```

The information listed in the screen is contained in the variable ΣPAR. The values shown are the default values for the x-column, y-column, intercept and slope of a data fitting model, and the type of model to be fit to the data in ΣDAT.

XCOL (H)

The command XCOL is used to indicate which of the columns of ΣDAT, if more than one, will be the x- column or independent variable column.

YCOL (I)

The command YCOL is used to indicate which of the columns of ΣDAT, if more than one, will be the y- column or dependent variable column.

MODL (J)

The command MODL refers to the model to be selected to fit the data in SDAT, if a data fitting is implemented. To see which options are available, press [MODL]. You will get the following menu:

[LINFI][LOGFI][EXPFI][PWRFI][BESTFI][ΣPAR]

These functions correspond to Linear Fit, Logarithmic Fit, Exponential Fit, Power Fit, or Best Fit. Data fitting is described in more detail in a later chapter.

Press [ΣPAR] to return to the ΣPAR menu.

ΣPAR (K)

ΣPAR is just a reference to the variable SPAR for interactive use.

RESET (L)

This function resets the contents of ΣPAR to its default values.

Press [NXT][STAT] to return to the STAT menu. Press [PLOT] to return to the main PLOT menu.

The FLAG menu within PLOT

The FLAG menu is actually interactive, so that you can select any of the following options:

- AXES: when selected, axes are shown if visible within the plot area or volume.

- **CNCT:** when selected the plot is produced so that individual points are connected.
- **SIMU:** when selected, and if more than one graph is to be plotted in the same set of axes, plots all the graphs simultaneously.

Press [PLOT] to return to the PLOT menu.

So, how are plots generated in a program?

Depending on whether we are dealing with a two-dimensional graph defined by a function, by data from ΣDAT, or by a three-dimensional function, you need to set up the variables PPAR, ΣPAR, and /or VPAR before generating a plot in a program. The commands shown in the previous section help you in setting up such variables.

Following we describe the general format for the variables necessary to produce the different types of plots available in the HP 49 G calculator.

Two-dimensional graphics

The two-dimensional graphics generated by functions, namely, Function, Conic, Parametric, Polar, Truth and Differential Equation, use PPAR with the format:

$$\{ (x_{min}, y_{min})\ (x_{max}, y_{max})\ \text{indep res axes ptype depend} \}$$

The two-dimensional graphics generated from data in the statistical matrix ΣDAT, namely, Bar, Histogram, and Scatter, use the ΣPAR variable with the following format:

$$\{ \text{x-column y-column slope intercept model} \}$$

while at the same time using PPAR with the format shown above.

The meaning of the different parameters in PPAR and ΣPAR were presented in the previous section.

Three-dimensional graphics

The three-dimensional graphics available, namely, options Slopefield, Wireframe, Y-Slice, Ps-Contour, Gridmap and Pr-Surface, use the VPAR variable with the following format:

$$\{x_{left},\ x_{right},\ y_{near},\ y_{far},\ z_{low},\ z_{high},\ x_{min},\ x_{max},\ y_{min},\ y_{max},\ x_{eye},\ y_{eye},\ z_{eye},\ x_{step},\ y_{step}\}$$

These pairs of values of x, y, and z, represent the following:

- Dimensions of the view parallelepiped (x_{left}, x_{right}, y_{near}, y_{far}, z_{low}, z_{high})

- Range of x and y independent variables (x_{min}, x_{max}, y_{min}, y_{max})

- Location of viewpoint (x_{eye}, y_{eye}, z_{eye})

- Number of steps in the x- and y-directions (x_{step}, y_{step})

Three-dimensional graphics also require the PPAR variable with the parameters shown above.

The variable EQ

All plots, except those based on ΣDAT, also require that you define the function or functions to be plotted by storing the expressions or references to those functions in the variable EQ.

In summary, to produce a plot in a program you need to load EQ, if required. Then load PPAR, PPAR and SPAR, or PPAR and VPAR. Finally, use the name of the proper plot type: FUNCTION, CONIC, POLAR, PARAMETRIC, TRUTH, DIFFEQ, BAR, HISTOGRAM, SCATTER, SLOPE, WIREFRAME, YSLICE, PCONTOUR, GRIDMAP, or PARSURFACE, to produce your plot.

Examples of interactive plots using the PLOT menu

To better understand the way a program works with the PLOT commands and variables, try the following examples of interactive plots using the PLOT menu.

Example 1 - A function plot:

[↵][USER][F3 C]	Get PLOT menu
[PTYPE][FUNCT]	Select FUNCTION as the plot type
'√r' [ENTER][↵][EQ]	Store function '√r' into EQ
[PPAR]	Show plot parameters
[ALPHA][↵][R][ENTER][INDEP]	Define 'r' as the independent variable
[ALPHA][↵][S][DEPND]	Define 's' as the dependent variable
[1][+/-][SPC][1][0][XRNG]	Define (-1, 10) as the x-range
[1][+/-][SPC][5][YRNG]	Define (-1, 5) as the y-range
{ (0,0) {.4 .2} "Rs" "Sr"}[NXT][AXES]	Define axes center, tick marks, labels
[NXT][PLOT]	Return to PLOT menu
[ERASE][DRAX][NXT][LABEL]	Erase picture screen, draw axes, place labels
[NXT][DRAW]	Draw function and show picture
[EDIT][NXT][MENU]	Removes menu labels
[NXT][NXT][PICT][CANCL]	Returns to normal calculator display

Example 2 - A parametric plot:

[↵][USER][F3 C]	Get PLOT menu
[PTYPE][PARAM]	Select PARAMETRIC as the plot type
{'SIN(t)+i*SIN(2*t)'}[ENTER][↵][EQ]	Store complex function X+iY into EQ
[PPAR]	Show plot parameters
{t 0 6.29}[ENTER][INDEP]	Define 't' as the independent variable

[ALPHA][Y][DEPND]	Define 'Y' as the dependent variable
[2][.][2][+/-][SPC][2][.][2][XRNG]	Define (-2.2,2.2) as the x-range
[1][.][1][+/-][SPC][1][.][1][YRNG]	Define (-1.1,1.1) as the y-range
{ (0,0) {.4 .2} "X(t)" "Y(t)"}[NXT][AXES]	Define axes center, tick marks, labels
[NXT][PLOT]	Return to PLOT menu
[ERASE][DRAX][NXT][LABEL]	Erase picture screen, draw axes, place labels
[NXT][DRAW]	Draw function and show picture
[EDIT][NXT][MENU]	Removes menu labels
[NXT][NXT][PICT][CANCL]	Returns to normal calculator display

Example 3 - A polar plot:

[↰][USER][F3 C]	Get PLOT menu
[PTYPE][POLAR]	Select POLAR as the plot type
'1+SIN(θ)'[ENTER][↰][EQ]	Store complex function r = f(θ) into EQ
[PPAR]	Show plot parameters
{θ 0 6.29}[ENTER][INDEP]	Define 'θ' as the independent variable
[ALPHA][Y][DEPND]	Define 'Y' as the dependent variable
[3][+/-][SPC][3][XRNG]	Define (-3,3) as the x-range
[0][.][5][+/-][SPC][2][.][5][YRNG]	Define (-0.5,2.5) as the y-range
{ (0,0) {.5 .5} "x" "y"}[NXT][AXES]	Define axes center, tick marks, labels
[NXT][PLOT]	Return to PLOT menu
[ERASE][DRAX][NXT][LABEL]	Erase picture screen, draw axes, place labels
[NXT][DRAW]	Draw function and show picture
[EDIT][NXT][MENU]	Removes menu labels
[NXT][NXT][PICT][CANCL]	Returns to normal calculator display

From these examples we see a pattern for the interactive generation of a two-dimensional graph through the PLOT menu:

1 - Select PTYPE.
2 - Store function to plot in variable EQ (using the proper format, e.g., 'X(t)+iY(t)' for PARAMETRIC).
3 - Enter name (and range, if necessary) of independent and dependent variables
4 - Enter axes specifications as a list { center atick x-label y-label }
5 - Use ERASE, DRAX, LABEL, DRAW to produce a fully labeled graph with axes

This same approach can be used to produce plots with a program, except that in a program you need to add the command PICTURE after the DRAW function is called to recall the graphics screen to the stack.

Examples of program-generated plots

In this section we show how to implement with programs the generation of the last three examples.

Example 1 - A function plot

Enter the following program:

<<	Start program
{PPAR EQ} PURGE	Purge current values of PPAR and EQ

`'√r' 'EQ' STO`	Store '√r' into EQ
`'r' INDEP`	Set independent variable to 'r'
`'s' DEPND`	Set dependent variable to 's'
`FUNCTION`	Select FUNCTION as the plot type
`{ (0.,0.) {.4 .2} "Rs" "Sr" } AXES`	Set axes information
`-1. 5. XRNG`	Set x range
`-1. 5. YRNG`	Set y range
`ERASE DRAW DRAX LABEL`	Erase and draw plot, axes, and labels
`PICTURE`	Recall graphics screen to stack
`>>`	End program

Store the program in variable PLOT1. To run it, press [VAR], if needed, then press [PLOT1].

Example 2 - A parametric plot

Enter the following program:

`<<`	Start program
`{PPAR EQ} PURGE`	Purge current values of PPAR and EQ
`'SIN(t)+i*SIN(2*t)' 'EQ' STO`	Store 'X(t)+iY(t)' into EQ
`{ t 0. 6.29} INDEP`	Set independent variable to 'r', range is included
`'Y' DEPND`	Set dependent variable to 'Y'
`PARAMETRIC`	Select PARAMETRIC as the plot type
`{ (0.,0.) {.5 .5} "X(t)" "Y(t)" } AXES`	Set axes information
`-2.2 2.2 XRNG`	Set x range
`-1.1 1.1 YRNG`	Set y range
`ERASE DRAW DRAX LABEL`	Erase and draw plot, axes, and labels
`PICTURE`	Recall graphics screen to stack
`>>`	End program

Store the program in variable PLOT2. To run it, press [VAR], if needed, then press [PLOT2].

Example 3 - A polar plot

Enter the following program:

`<<`	Start program
`{PPAR EQ} PURGE`	Purge current values of PPAR and EQ
`'1+SIN(θ)' 'EQ' STO`	Store 'f(θ)' into EQ
`{ θ 0. 6.29} INDEP`	Set independent variable to 'θ', range is included
`'Y' DEPND`	Set dependent variable to 'Y'
`POLAR`	Select POLAR as the plot type
`{ (0.,0.) {.5 .5} "x" "y"} AXES`	Set axes information
`-3. 3. XRNG`	Set x range
`-.5 2.5 YRNG`	Set y range
`ERASE DRAW DRAX LABEL`	Erase and draw plot, axes, and labels
`PICTURE`	Recall graphics screen to stack
`>>`	End program

Store the program in variable PLOT3. To run it, press [VAR], if needed, then press [PLOT3].

These exercises illustrate the use of PLOT commands in programs. They just scratch the surface of programming applications of plots. I invite the reader to try their own exercises on programming plots.

Interactive drawing with the HP 49 G calculator

Whenever we produce a two-dimensional graph, we find in the graphics screen a soft menu key labeled EDIT. Pressing [EDIT] produces a menu that include the following options:

[DOT+][DOT-][LINE][TLINE][BOX][CIRCL]

Pressing [NXT] results in:

[MARK][+/-][LABEL][DEL][ERASE][MENU]

Pressing [NXT] once more you get:

[SUB][REPL][PICT→][X,Y→][   ][PICT].

Through the examples in this, and previous, chapters we have had the opportunity to use the functions LABEL, MENU, PICT→, and REPL. The function ERASE is the same as provided in the PLOT menu, presented earlier. Many of the remaining functions, such as DOT+, DOT-, LINE, BOX, CIRCL, MARK, DEL, etc., can be used to draw points, lines, circles, etc. on the graphics screen, as described below.

To see how to use these functions we will try the following exercise:

First, we get the graphics screen corresponding to the Function plot in the previous examples, with a few modifications:

[←][USER][F3 C]	Get PLOT menu
[PTYPE][FUNCT]	Select FUNCTION as the plot type
'X' [ENTER][←][EQ]	Store function 'r' into EQ
[PPAR]	Show plot parameters
[RESET]	Reset parameters to their default values
[1][0][+/-][SPC][1][0][XRNG]	Define (-10, 10) as the x-range
[5][+/-][SPC][5][YRNG]	Define (-5, 5) as the y-range
{ (0,0) {.5 .5} "X" "Y"}[NXT][AXES]	Define axes center, tick marks, labels
[NXT][PLOT]	Return to PLOT menu
[ERASE][DRAX][NXT][LABEL]	Erase picture screen, draw axes, place labels
[NXT][DRAW]	Draw function and show picture
[EDIT]	Activate the EDIT menu

Next, we illustrate the use of the different drawing functions on the resulting graphics screen. They require use of the cursor and the arrow keys ([◄][►][▲][▼]) to move the cursor about the graphics screen.

DOT+ and DOT-

When DOT+ is selected, pixels will be activated wherever the cursor moves leaving behind a trace of the cursor position. When DOT- is selected, the opposite effect occurs, i.e., as you move the cursor, pixels will be deleted.

For example, use the [▶][▲] keys to move the cursor somewhere in the middle of the first quadrant of the x-y plane, then press [DOT+]. The label will be selected ([DOT+■]). Press and hold the [▶] key to see a horizontal line being traced. Now, press [DOT-], to select this option ([DOT-■]). Press and hold the [◀] key to see the line you just traced being erased. Press [DOT-] when done to deselect this option.

MARK

This command allows the user to set a mark point which can be used for a number of purposes, such as:
- Start of line with the LINE or TLINE command
- Corner for a BOX command
- Center for a CIRCLE command

Using the MARK command by itself simply leaves an x in the location of the mark.

LINE

This command is used to draw a line between two points in the graph. To see it in action, position the cursor somewhere in the first quadrant, and press [LINE]. A MARK is placed over the cursor indicating the origin of the line. Use the [▶] key to move the cursor to the right of the current position, say about 1 cm to the right, and press [LINE]. A line is draw between the first and the last points.
Notice that the cursor at the end of this line is still activated indicating that the calculator is ready to plot a line starting at that point. Press [▼] to move the cursor downwards, say about ½ inch, and press [LINE] again. Now you should have a straight angle traced by a horizontal and a vertical segments. The cursor is still active. To deactivate it, without moving it at all, press [LINE]. The cursor returns to its normal shape (a cross) and the LINE function is no longer active.

TLINE

(Toggle LINE) Move the cursor to the second quadrant to see this function in action. Press [TLINE]. A MARK is placed at the start of the toggle line. Move the cursor with the arrow keys away from this point, and press [TLINE]. A line is drawn from the current cursor position to the reference point selected earlier. Pixels that are on in the line path will be turned off, and vice versa. To remove the most recent line traced, press [TLINE] again. To deactivate TLINE, move the cursor to the original MARK, and press [LINE][LINE].

BOX

This command is used to draw a box in the graph. Move the cursor to a clear area of the graph, and press [BOX]. This highlights the cursor. Move the cursor with the arrow keys to a point away, and in a diagonal direction, from the current cursor position. Press [BOX] again. A rectangle is drawn whose diagonal joins the initial and ending cursor positions. The initial position of the box is still marked with and x. Moving the cursor to another position and pressing [BOX] will generate a new box containing the initial point. To deselect BOX, move the cursor to the original point where BOX was activated, then press [LINE][LINE].

CIRCL

This command produces a circle. Mark the center of the circle with a MARK command, then move the cursor to a point that will be part of the periphery of the circle, and press [CIRCL]. To deactivate CIRCL, return the cursor to the MARK position and press [LINE].

Try this command by moving the cursor to a clear part of the graph, press [MARK]. Move the cursor to another point, then press [CIRCL]. A circle centered at the MARK, and passing through the last point will be drawn.

LABEL

Pressing [LABEL] places the labels in the x- and y-axes of the current plot. This feature has been used extensively through this chapter.

DEL

This command is used to remove parts of the graph between two MARK positions. Move the cursor to a point in the graph, and press [MARK]. Move the cursor to a different point, press [MARK] again. Then, press [DEL]. The section of the graph boxed between the two marks will be deleted.

ERASE

The function ERASE clears the entire graphics window. This command is available in the PLOT menu, as well as in the plotting windows accessible through the soft menu keys.

MENU

Pressing [MENU] will remove the soft key menu labels to show the graphic unencumbered by those labels. To recover the labels, press [NXT].

SUB

Use this command to extract a subset of a graphics object. The extracted object is automatically placed in the stack. Select the subset you want to extract by placing a MARK at a point in the graph, moving the cursor to the diagonal corner of the rectangle enclosing the graphics subset, and press [SUB]. This feature can be used to move parts of a graphics object around the graph.

REPL

This command places the contents of a graphic object currently in stack level 1 at the cursor location in the graphics window. The upper left corner of the graphic object being inserted in the graph will be placed at the cursor position. Thus, if you want a graph from the stack to completely fill the graphic window, make sure that the cursor is placed at the upper left corner of the display.

PICT→

This command places a copy of the graph currently in the graphics window on to the stack as a graphic object. The graphic object placed in the stack can be saved into a variable name for storage or other type of manipulation.

X,Y→

This command copies the coordinates of the current cursor position, in user coordinates, in the stack.

Drawing commands for use in programming

You can draw figures in the graphics window directly from a program by using commands such as those contained in the PICT menu, accessible by [←][PRG][PICT]. The functions available in this menu are the following:

[PICT][PDIM][LINE][TLINE][BOX][ARC]

Pressing [NXT]:

[PIXON][PIXOF][PIX?][PVIEW][PX→C][C→PX]

Obviously, the commands LINE, TLINE, and BOX, perform the same operation as their interactive counterpart, given the appropriate input. These and the other functions in the PICT menu refer to the graphics windows whose x- and y-ranges are determined in the variable PPAR, as demonstrated above for different graph types. The functions in the PICT command are described next:

PICT

This soft key refers to a variable called PICT that stores the current contents of the graphics window. This variable name, however, cannot be placed within quotes, and can only store graphics objects. In that sense, PICT is not like any other HP 49 G variables.

PDIM

The function PDIM takes as input either two ordered pairs (x_{min}, y_{min}) (x_{max} y_{max}) or two binary integers #w and #h. The effect of PDIM is to replace the current contents of PICT with an empty screen. When the argument is (x_{min}, y_{min}) (x_{max} y_{max}), these values become the range of the user-defined coordinates in PPAR. When the argument is #w and #h, the ranges of the user-defined coordinates in PPAR remain unchanged, but the size of the graph changes to #h × #v pixels.

PICT and the graphics screen

PICT, the storage area for the current graph, can be thought of as a two dimensional graph with a minimum size of 131 pixels wide by 64 pixels high. The maximum width of PICT is 2048 pixels, with no restriction on the maximum height. A pixel is each one of the dots in the calculator's screen that can be turned on (dark) or off (clear) to produce text or graphs. The calculator screen has 131 pixels by 64 pixels, i.e., the minimum size for PICT. If your PICT is larger than the screen, then the PICT graph can be thought of as a two dimensional domain that can be scrolled through the calculator's screen, as illustrated in the diagram shown next.

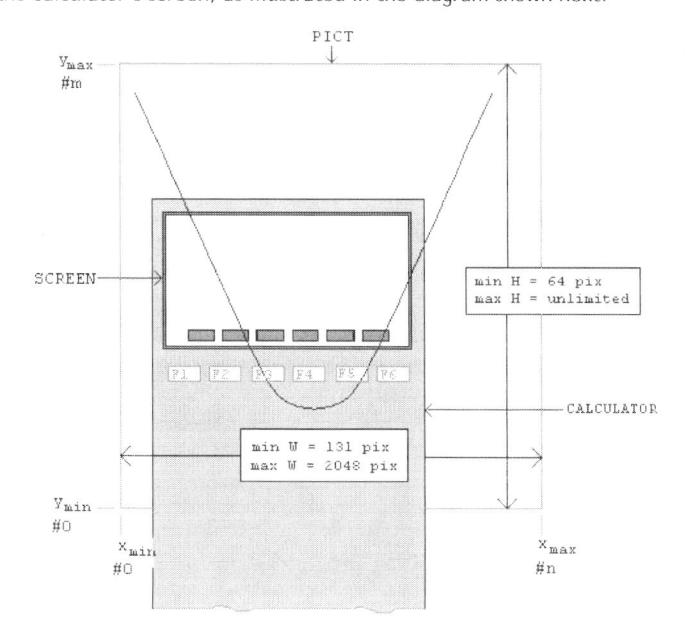

LINE

This command takes as input two ordered pairs (x_1, y_1) (x_2, y_2), or two pairs of pixel coordinates {#n_1 #m_1} {#n_2 #m_2}. It draws the line between those coordinates.

TLINE

This command (Toggle LINE) takes as input two ordered pairs (x_1,y_1) (x_2, y_2), or two pairs of pixel coordinates $\{\#n_1\ \#m_1\}$ $\{\#n_2\ \#m_2\}$. It draws the line between those coordinates, turning off pixels that are on in the line path and vice versa.

BOX

This command takes as input two ordered pairs (x_1,y_1) (x_2, y_2), or two pairs of pixel coordinates $\{\#n_1\ \#m_1\}$ $\{\#n_2\ \#m_2\}$. It draws the box whose diagonals are represented by the two pairs of coordinates in the input.

ARC

This command is used to draw an arc. ARC takes as input the following objects:

- Coordinates of the center of the arc as (x,y) in user coordinates or {#n, #m} in pixels.
- Radius of arc as r (user coordinates) or #k (pixels).
- Initial angle θ_1 and final angle θ_2.

PIX?, PIXON, and PIXOFF

These functions take as input the coordinates of point in user coordinates, (x,y), or in pixels {#n, #m}.

- PIX? Checks if pixel at location (x,y) or {#n, #m} is on.
- PIXOFF turns off pixel at location (x,y) or {#n, #m}.
- PIXON turns on pixel at location (x,y) or {#n, #m}.

PVIEW

This command takes as input the coordinates of a point as user coordinates (x,y) or pixels {#n, #m}, and places the contents of PICT with the upper left corner at the location of the point specified. You can also use an empty list as argument, in which case the picture is centered in the screen. PVIEW does not activate the graphics cursor or the picture menu. To activate any of those features use PICTURE.

PX→C

The function PX→C converts pixel coordinates {#n #m} to user-unit coordinates (x,y).

C→PX

The function C→PX converts user-unit coordinates (x,y) to pixel coordinates {#n #m}.

Programming examples using drawing functions

Example 1 - A program that uses drawing commands

The following program produces a drawing in the graphics screen. (This program has no other purpose than to show how to use HP 49 G calculator commands to produce drawings in the display.)

`<<`	Start program
`DEG`	Select degrees for angular measures
`0. 100. XRNG`	Set x range
`0. 50. YRNG`	Set y range
`ERASE`	Erase picture
`(5., 2.5) (95., 47.5) BOX`	Draw a box between points (5,5) and (95,95)
`(50., 50.) 10. 0. 360. ARC`	Draw a circle center (50,50), radius 10.
`(50., 50.) 12. -180. 180. ARC`	Draw a circle center (50,50), radius 12.
`1 8 FOR j`	Draw 8 lines within the circle
`   (50., 50.) DUP`	Lines are centered as (50,50)
`   '12*COS(45*(j-1))' →NUM`	Calculates x, The other end of line located at 50 + x
`   '12*SIN(45*(j-1))' →NUM`	Calculates y, The other end of line located at 50 + y
`   R → C`	Converts x y to (x,y), a complex number
`   +`	Add (50,50) to (x,y)
`   LINE`	Draws the line
`NEXT`	End of FOR loop
`{ }PVIEW`	Show picture
`>>`	

Example 2 - A program to plot a natural river cross-section

This application may be useful for determining area and wetted perimeters of natural river cross-sections. Typically, a natural river cross section is surveyed and a series of points, representing coordinates x and y with respect to an arbitrary set of coordinates axes. These points can be plotted and a sketch of the cross section produced for a given water surface elevation. The figure below illustrate the terms presented in this paragraph.

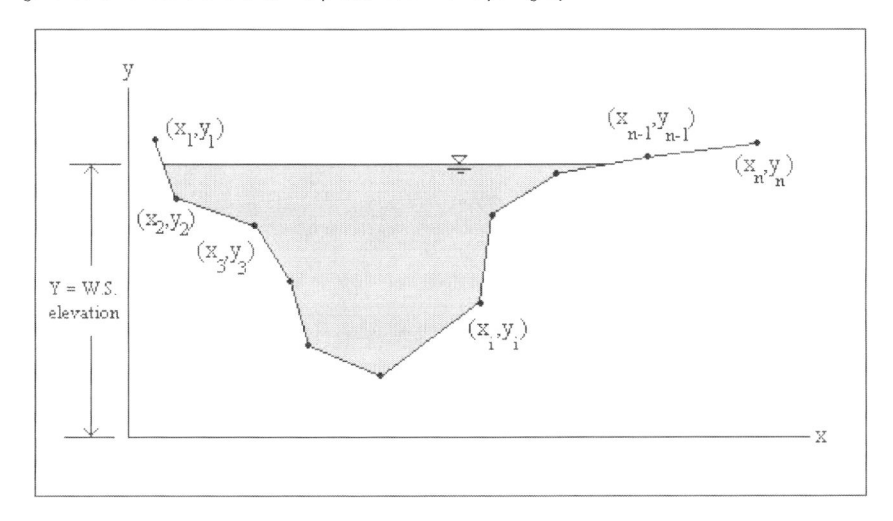

The program, listed below, utilizes four sub-programs FRAME, DXBED, GTIFS, and INTRP. The main program, called XSECT, takes as input a matrix of values of x and y, and the elevation of the water surface Y (see figure above), in that order. The program produces a graph of the cross section indicating the input data with points in the graph, and shows the free surface in the cross-section.

It is suggested that you create a separate sub-directory to store the programs listed below. You could call the sub-directory RIVER, since we are dealing with irregular open channel cross-sections, typical of rivers. Type the following programs and store them in the corresponding variables as listed:

Program XSECT (Meaning of program name: X SECTion, i.e., cross-section)

<<	Open main program
SWAP DUP	Swap levels 1 and 2, yFS goes to level 2
FRAME	Call subprogram FRAME
→COL DROP	Matrix with x-y data split into columns, no. of columns dropped
DUP SIZE EVAL	y-column duplicated to find size (n)
→ yFS x y n	Values in stack include yFS, x-col, y-col, and size of columns
<<	Open sub-program
x y n DXBED	Place x, y, and n in stack as input for program DXBED
x y yFS GTIFS	Place x, y, and yFS in stack as input for program GTIFS
OBJ→ 2. / FLOOR	GTIFS returns list of indices, which is decomposed and truncated
1 SWAP FOR j	Start FOR-NEXT loop to interpolate free surface points
yFS x y INTRP	Interpolate free-surface starting point
SWAP	Exchange levels 1 and 2
yFS x y INTRP	Interpolate free-surface ending point
LINE	Draw a line indicating the free surface
NEXT	End FOR-NEXT loop
{ } PVIEW	Show graphic
>>	Close sub-program
>>	Close program

Program FRAME (Meaning of program name: FRAME, for it establishes the frame where the cross-section will be plotted)

<<	Start program
STOΣ	Store in ΣDAT the matrix with cross-sectional geometry
MINΣ MAXΣ	Obtain minimum and maximum values of x and y
2 COL→ DUP	Form a matrix with max. and min. values, make a copy
→COL DROP -	Decompose the second copy and subtract min from max values
AXL ABS AXL	Convert result to a list, obtain absolute value, convert back to vector
20. /	Divide ranges in x and y by 20
DUP NEG	Copy vector with ranges and change sign
SWAP	Exchange levels 1 and 2
2 COL→	Create matrix with ranges
+	Add and subtract original ranges to ranges/20
→ROW DROP	Separate result in row
OBJ→ DROP YRNG	Define range of x axis
OBJ→ DROP XRNG	Define range of y axis
ERASE	Clear PICT
>>	Close program

Program DXBED (Draw X-section BED, i.e., the program draws the x-section of the channel bed.)

<<	Start program
→ x y n	Accept as input vectors x and y, and size n
<< '(x(1),y(1))' EVAL DUP	Sub-program. Place point (x_1, y_1) and copy it
C→PX DUP	Convert point coordinates to pixel references
{ 1. 1. } -	Subtract one pixel to pixel reference for point (x_1, y_1)
SWAP	Exchange levels 1 and 2
{1. 1. } ADD	Add one pixel to pixel reference for point (x_1, y_1)
BOX	Create a box around point (x_1, y_1)
2 n FOR j	Start FOR-NEXT loop, j = 2,3,...,n
'(x(j),y(j))' EVAL DUP	Place point (x_j, y_j) and copy it
C→PX DUP	Convert point coordinates to pixel references
{ 1. 1. } -	Subtract one pixel to pixel reference for point (x_j, y_j)
SWAP	Exchange levels 1 and 2
{1. 1. } ADD	Add one pixel to pixel reference for point (x_j, y_j)
BOX	Create a box around point (x_j, y_j)
DUP	Copy point (x_j, y_j)
3 ROLLD	Roll the bottom 3 levels of stack
LINE	Create a line between points (x_j, y_j) and (x_{j-1}, y_{j-1})
NEXT	End FOR-NEXT loop
DROP	Delete extra copy of last point
>>	Close sub-program
>>	Close program

Program GTIFS (GeT Indices for Free Surface: the program determines the index of the intervals [y(j),y(j+1)] of the cross-sectional profile where the free surface elevation yFS intersects the profile)

<<	Start program
{ } → x y yF IFS	Accept as input vectors x and y, water surface level yFS, { } for IFS
<<	Start sub-program
1	Place a 1 in the stack (to be used with the FOR-NEXT loop)
x SIZE EVAL 1 -	Obtain n = size of vector x, and calculate n-1
FOR j	Start FOR-NEXT loop for j = 1, 2, ..., n-1
IF	Determine the index for the interval (y(j),y(j+1)) that contains yFS
'yF≤y(j) AND yF>j(j+1)	This is a single logical statement
OR	(Same statement)
yF≥y(j) AND yF<y(j+1)'	(Same statement)
THEN	
IFS j + 'IFS' STO	If the logical statement is true, index j is added to list IFS
END	End of IF statement
NEXT	End of FOR-NEXT loop
IFS REVLIST	Place IFS in stack and revert its order
>>	Close sub-program
>>	Close program

Program INTRP (INTeRPolation: given the index j and water surface elevation yFS, this program
      interpolates the value of x in the interval [x(j), x(j+1)].)

<<	Start program
→ j yF x y	Take an index j, water surface elevation yFS, and vectors x,y as input
<<	Start sub-program
'x(j)+(yF-y(j))* (x(j+1)-x(j))/ (y(j+1)-y(j))' →NUM	Statement between single quotes interpolates value of x in interval (x(j), x(j+1))
yF R→C	Creates (x,yFS) showing intersection of free surface with x-section
>>	Close sub-program
>>	Close program

To see the program XSECT in action, use the following data sets. Enter them as matrices of two columns, the first column being x and the second one y. Store the matrices in variables with names such as XYD1 (X-Y Data set 1) and XYD2 (X-Y Data set 2). To run the program place one of the data sets in the stack, e.g., [VAR][XYD1], then type in a water surface elevation, say 4.0, and press [XSECT]. The calculator will show an sketch of the cross-section with the corresponding water surface. To exit the graph display, press [ON].

Data set 1

x	y
0.4	6.3
1.0	4.9
2.0	4.3
3.4	3.0
4.0	1.2
5.8	2.0
7.2	3.8
7.8	5.3
9.0	7.2

Data set 2

x	y
0.7	4.8
1.0	3.0
1.5	2.0
2.2	0.9
3.5	0.4
4.5	1.0
5.0	2.0
6.0	2.5
7.1	2.0
8.0	0.7
9.0	0.0
10.0	1.5
10.5	3.4
11.0	5.0

Try the following examples:

[XYD1][4][XSECT]
[XYD1][6][XSECT]
[XYD2][4][XSECT]
[XYD2][2][XSECT]

Please be patient when running program XSECT. Due to the relatively large number of graphics functions used, not counting the numerical iterations, it may take about 1 minute to produce a result.

This is the result produced by the input: [XYD2][3][.][5][XSECT]:

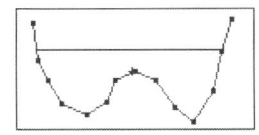

While, the input [XYD2][2][XSECT] produces the following plot:

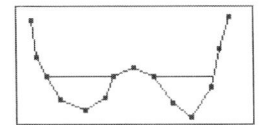

> Note: The program FRAME, as presented above, does not maintain the proper scaling of the graph. If you want to maintain proper scaling, replace FRAME with the following program:
>
> ---
>
> << STOΣ MINΣ MAXΣ 2 COL→ DUP →COL DROP - AXL ABS AXL 20 / DUP NEG SWAP 2 COL→ + →ROW DROP SWAP → yR xR << 131 DUP R→B SWAP yR OBJ→ DROP - xR OBJ→ DROP - / * FLOOR R→B PDIM yR OBJ→ DROP YRNG xR OBJ→ DROP XRNG ERASE >> >>
>
> ---
>
> This program keeps the width of the PICT variable at 131 pixels - the minimum pixel size for the horizontal axis - and adjusts the number of pixels in the vertical axes so that a 1:1 scale is maintained between the vertical and horizontal axes.

Example 3 - A program to visualize a polynomial fitting

In Chapter 10 we developed a program called POLY that generates a vector b with the coefficients
[b_0 b_1 b_2 ... b_p] of a polynomial fitting of the form

$$y = b_0 + b_1 x + b_2 x^2 + ... + b_p x^p.$$

Here we present a program that takes the process a step further showing the polynomial fitting altogether with the original data points. This visual display helps you check how well the polynomial fits the data.

> Note: You should be aware that this is only a gross indication of goodness of fit. Other criteria to evaluate goodness of fit were presented in Chapter 10 and in an earlier example in this Chapter.

The following program assumes that you have program POLY available in your calculator (and in the same sub-directory). The program, to be called PPOLY, is listed following:

```
<< → x y p <<x AXL y AXL 2 COL→ → XX << x y p POLY DUP p PLFIT EQ PLTSU ERASE
DRAX LABEL DRAW 'X' PURGE XX POINTS PICTURE >> >> >>
```

Program PPOLY uses the following sub-programs:

Program PLTSU (PLoT Set-Up):

```
<< MINS OBJ→ DROP DROP MAXS OBJ→ DROP DROP XRNG AUTO >>
```

Program PLFIT (PLot for polynomial FITting):

```
<< → p <<1 1 p FOR j 'X' j ^ NEXT p 1 + →ARRY DOT STEQ FUNCTION >> >>
```

Program POINTS (Plots pOINTS):

```
<< → XX << XX SIZE OBJ→ DROP DROP 1 SWAP FOR j '(XX(j,1),XX(j,2))' EVAL C→PX
DUP { 1 1 } - SWAP { 1 1} ADD BOX NEXT >> >>
```

To tests the program we use as input the same lists xx and yy used in Chapter 10, followed by the degree of the polynomial. Try the following exercises:

[xx][yy][2][PPOLY]
[xx][yy][3][PPOLY]
[xx][yy][4][PPOLY]
[xx][yy][5][PPOLY]
[xx][yy][6][PPOLY]

Because of the number of plotting and drawing functions used, the program may take a minute or more to finish. Be patient and wait for the entire graph to be rendered.

To return to normal calculator display use [CANCL]. The stack will show two additional lines of output, the vector b in level 2, and the polynomial equation fitted in level 1.

Binary, octal, and hexadecimal number systems – a brief introduction

> Note: This section is presented here because graphical operations with pixel coordinates require some knowledge of number systems.

The number system used for everyday arithmetic is known as the decimal system for it uses 10 (Latin, deci) digits, namely 0-9, to write out any real number. Most likely the reason why we use the decimal system is because, in early human development, integer numbers were associated with the ten fingers in our hands. Computers, on the other hand, use a system that is based on two possible states, or binary system. These two states are represented by 0 and 1, ON and

OFF, or high-voltage and low-voltage. Computers also use number systems based on eight digits (0-7) or octal system, and sixteen digits (0-9, A-F) or hexadecimal.

The BASE menu

While the HP 49 G calculator would typically be operated using the decimal system, you can produce calculations using the binary, octal, or hexadecimal system. Many of the functions for manipulating number systems other than the decimal system are available in the BASE menu, accessible through [→][BASE] (the [3] key). The BASE menu includes the functions:

[HEX][DEC][OCT][BIN][R→B][B→R]

Pressing [NXT] you get the following menus (italics) and functions:

[LOGIC][BIT][BYTE][   ][STWS][RCWS].

Writing non-decimal numbers in the calculator

Numbers in non-decimal systems are written preceded by the # symbol in the HP 49 G calculator. The symbol # is readily available as [↲][#] (the [3] KEY). To select which number system will be used for numbers preceded by #, press one of the first four keys in the first BASE menu, i.e., HEX(adecimal), DEC(imal), OCT(al), or BIN(ary). For example, if [HEX■] is selected, any number written in the HP 49 G calculator that starts with # will be a hexadecimal number. Thus, you can write numbers such as #53, #A5B, etc.

Conversion between number systems

Whatever the number system selected, we will refer to it as the binary system for the purpose of using the functions R→B and B→R. For example, if [HEX■] is selected, the function B→R will convert any hexadecimal number (preceded by #) into a decimal number, while the function R→B works in the opposite direction. Try the following exercise:

[→][BASE][HEX]   Select the hexadecimal system
[↲][#][2][3][ENTER]   Enter the hexadecimal number #23h
[B→R]   Convert #23h to floating point (decimal) format

The result is the number 35. Try another exercise:

[↲][#][ALPHA][5][ENTER]   Enter the hexadecimal number #A5h
[B→R]   Convert #A5h to floating point (decimal) format

The result is 165. Now, try converting decimal numbers to hexadecimal. For example,

[1][2][3][4][ENTER][R→B]   Convert 1234 to hexadecimal, result: # 4D2h

Let's try some conversion in the octal system:

[OCT]   Select the octal system
[↲][#][1][7][ENTER][B→R]   Convert # 17o to 15.
[1][2][0][0][ENTER][R→B]   Convert 1200 to octal, result: # 2260o

Or, using the binary system:

[BIN]	Select the octal system
[←][#][1][0][0][1][ENTER][B→R]	Convert # 1001b to 9.
[1][2][1][7][ENTER][R→B]	Convert 1217 to octal, result: # 10011000001b

Notice that every time you enter a number starting with #, you get as the entry the number you entered preceded by # and followed by the letter h, o, or b (hexadecimal, octal, or binary). The type of letter used as suffix depends on which non-decimal number system has been selected, i.e., HEX, OCT, or BIN.

So, what happens if you select the decimal system, by pressing [DEC]? To see what happens in this case, try the following conversions:

[DEC]	Select the octal system
[←][#][1][2][6][ENTER][B→R]	Convert # 126d to 126.
[2][3][7][ENTER][R→B]	Convert 237 to decimal, results in # 237d

The only effect of selecting the DECimal system is that decimal numbers, when started with the symbol #, are written with the suffix d.

The LOGIC menu

The LOGIC menu, obtained through [][BASE][NXT][LOGIC], contains the functions:

[AND][OR][XOR][NOT]

These are exactly the same functions AND, OR, XOR (exclusive OR), and NOT presented in more detail in Chapter 7. The input to these functions are two values or expressions . (one in the case of NOT) that can be expressed as binary logical results, i.e., 0 or 1. The rules for the results of these functions are presented as truth tables in Chapter 7.

The BIT and BYTE menus

The BIT and BYTE menus contain functions used by computer scientists to manipulate bit and byte operations. For more details on these two menus, please refer to the HP 49 G calculator's Advanced User's Manual, available as a pdf file from http://www.hp.com

Hexadecimal numbers for pixel references

As indicated in this chapter, many plot option specifications use pixel references as input, e.g., { #332h #A23h } #Ah 0. 360. ARC, to draw an arc of a circle. Luckily, we can count on the functions C→PX and PX→C to convert quickly between user-unit coordinates and pixel references.

Pixel coordinates

The figure below shows the graphic coordinates for the typical (minimum) screen of 131×64 pixels. Pixels coordinates are measured from the top left corner of the screen {# 0h # 0h}, which corresponds to user-defined coordinates (x_{min}, y_{max}). The maximum coordinates in terms of pixels correspond to the lower right corner of the screen {# 82h #3Fh}, which in user-coordinates is the point (x_{max}, y_{min}). The coordinates of the two other corners both in pixel as well as in user-defined coordinates are shown in the figure.

```
{# 0h #0 h}              {# 82h # 0h}
( x_min, y_max )         ( x_max, y_max )

        131×64 pixels

{# 0h # 3Fh}             {# 82h #3Fh}
( x_min, y_min )         ( x_max, y_min )
```

Zooming in and out in the graphics display

Whenever you produce a two-dimensional FUNCTION graphic interactively, the first soft-menu key, labeled ZOOM, lets you access functions that can be used to zoom in and out in the current graphics display. The ZOOM menu includes the following functions:

[ZFACT][BOXZ][ZIN][ZOUT][ZSQR][ZDFLT]

Press [NXT] to get the following menu:

[HZIN][HZOUT][VZIN][VZOUT][CNTR][ZAUTO]

Pressing [NXT] once more produces this menu:

[ZDECI][ZINTG][ZTRIG][ZLAST][   ][PICT].

We present each of these functions following as applied to one of the graphics resulting from the last exercise on polynomial fitting.

ZFACT, ZIN, ZOUT, and ZLAST

Pressing [ZFACT] produces an input screen that allows you to change the current X and Y-Factors. The X- and Y-Factors relate the horizontal and vertical user-defined unit ranges to their corresponding pixel ranges. Change the H-Factor to read 8., and press [OK], then change the V-Factor to read 2., and press [OK]. Check off the option ✓Recenter on cursor, and press [OK].

Back in the graphics display, press [ZIN]. The graphic is re-drawn with the new vertical and horizontal scale factors, centered at the position where the cursor was located, while maintaining the original PICT size (i.e., the original number of pixels in both directions). Scroll horizontally ([◄] [►]) or vertically ([▲] [▼]) to as much as you can of the zoomed-in graph.

To zoom out, subjected to the H- and V-Factors set with ZFACT, press [ZOOM][ZOUT]. The resulting graph will provide more detail than the zoomed-in graph.

You can always return to the very last zoom window by using ZLAST.

BOXZ

Zooming in and out of a given graph can be performed by using the soft-menu key BOXZ. With BOXZ you select the rectangular sector (the "box") that you want to zoom in into. Move the cursor to one of the corners of the box (using the arrow keys), and press [ZOOM][BOXZ]. Using the arrow keys once more, move the cursor to the opposite corner of the desired zoom box. The cursor will trace the zoom box in the screen. When desired zoom box is selected, press [ZOOM]. The calculator will zoom in the contents of the zoom box that you selected to fill the entire screen.

If you now press [ZOUT], the calculator will zoom out of the current box using the H- and V-Factors, which may not recover the graph view from which you started the zoom box operation.

ZDFLT, ZAUTO

Pressing ZDFLT re-draws the current plot using the default x- and y-ranges, i.e., -6.5 to 6.5 in x, and -3.1 to 3.1 in y. ZAUTO, on the other hand, uses a zoom window using the current independent variable (x) range, but adjusting the dependent variable (y) range to fit the curve (as when you use the function AUTO in the PLOT WINDOW ([←][WIN] - simultaneously)).

HZIN, HZOUT, VZIN and VZOUT

These functions zoom in and out the graphics screen in the horizontal or vertical direction according to the current H- and V-Factors.

CNTR

Zooms in with the center of the zoom window in the current cursor location. The zooming factors used are the current H- and V-Factors.

ZDECI

Zooms the graph so as to round off the limits of the x-interval to a decimal value.

ZINTG

Zooms the graph so that the pixel units become user-define units. For example, the minimum PICT window has 131 pixels. When you use ZINTG, with the cursor at the center of the screen, the window gets zoomed so that the x-axis extends from -64.5 to 65.5.

ZSQR

Zooms the graph so that the plotting scale is maintained at 1:1 by adjusting the x scale, keeping the y scale fixed, if the window is wider than taller. This forces a proportional zooming.

ZTRIG

Zooms the graph so that the x scale incorporates a range from about -3p to +3p, the preferred range for trigonometric functions.

> Note: None of these functions are programmable. They are only useful in an interactive way only. Do not confuse the command [ZFACT] in the ZOOM menu with the function ZFACTOR, which is used for gas dynamic and chemistry applications as described below. The only reason why we present this function at this point is because of its name.

Special function for gas dynamics: ZFACTOR(Tr, Pr)

> The function ZFACTOR(Tr, Pr) where Tr = reduced temperature ratio, and Pr = reduced pressure ratio. The function ZFACTOR returns the gas compressibility correction factor for non-ideal behavior of a hydrocarbon gas. In the formula, Tr = T/Tc, and Pr = P/Pc, Tc and Pc are known as the pseudocritical temperature and pressure, respectively. The function is valid for 1.05 < Tr < 3.0, and 0 < Pr < 30. For example, if Tr = 1.85 and Pr = 10, then the gas compressibility factor is calculated as follows:
>
> [1][.][8][5][ENTER] [1][0][ENTER] [CAT][ALPHA][Z] (find ZFACTOR using [▲] [▼]).
>
> When highlighted, press [OK]. The result is 1.13904784599.

Animating graphics

We have already presented a way to produce animation by using the Y-Slice plot. Let's look at a different example. Suppose that you want to animate the traveling wave, f(X,Y) = 2.5 sin(X-Y). We can treat the X as time in the animation producing plots of f(X,Y) vs. Y for different values of X. To produce this graph use the following:

- [↵][2D/3D] simultaneously. Select Y-Slice for TYPE. '2.5*SIN(X-Y)' for EQ. 't' for INDEP. Press [NXT][OK].

- [↵][WIN] simultaneously. Select:

```
    X-Left:  -5.        Y-Right:  5.
    Y-Near:  -5.        Y-Far:    5.
    Z-Low:   -2.5       Z-High:   2.5

    STEP: Indep: 20        Depnd: 20
```

- Press [ERASE][DRAW]. Allow some time for the calculator to generate all the needed graphics. When ready, it will show a traveling sinusoidal wave in your screen.

Animating a collection of graphics

The HP 49 G calculator provides the function ANIMATE to animate a number of graphics that have been placed in the stack. You can generate a graph in the graphics screen by using the commands in the PLOT and PICT menus. To place the generated graph in the stack, use PICT RCL. When you have n graphs in levels n through 1 of the stack, you can simply use the command n ANIMATE to produce an animation made of the graphs you placed in the stack.

Example 1 - Animating a ripple in a water surface

As an example, type in the following program that generates 11 graphics showing a circle centered in the middle of the graphics screen and whose radius increase by a constant value in each subsequent graph.

<<	Begin program
RAD	Set angle units to radians
131 R→B 64 R→B PDIM	Set PICT screen to 131×64 pixels
0 100 XRNG 0 100 YRNG	Set x- and y-ranges to 0-100
1 11 FOR j	Start FOR-NEXT loop with j = 1,2, ..., 11
ERASE	Erase current PICT
(50., 50.) '5*(j-1)' →NUM 0 '2*p' →NUM ARC	Draw a circle center (50,50), radius =5(j-1)
PICT RCL	Place current PICT on stack
NEXT	End FOR-NEXT loop
11 ANIMATE	When 11 pictures are in the stack, animate
>>	

Store this program in a variable called PANIM (Plot ANIMation). To run the program press [VAR] (if needed) [PANIM]. It takes the calculator more than one minute to generate the graphs and get the animation going. Therefore, be really patient here. You will see the hourglass symbol up in the screen for what seems a long time before the animation, resembling the ripples produced by a pebble dropped on the surface of a body of quiescent water, appears in the screen. To stop the animation, press [ON].

The 11 graphics generated by the program are still available in the stack. If you want to re-start the animation, simply use: [1][1][CAT][ALPHA][A], then use the up- and down-arrow keys ([▲][▼]) to highlight ANIMATE, then press [OK]. The animation will be re-started. Press [ON] to stop the animation once more. Notice that the number 11 will still be listed in stack level 1. Press [⇐] to drop it from the stack.

Suppose that you want to keep the figures that compose this animation in a variable. You can create a list of these figures, let's call it WLIST, by using:

    [1][1][↵][PRG][TYPE][→LIST] [→]['][ALPHA][ALPHA] [W][L][I][S][T] [ALPHA][STO▶]

Press [VAR] to recover your list of variables. The variable [WLIST] should now be listed in your soft-menu keys. To re-animate this list of variables you could use the following program:

<<	Start program
WLIST	Place list WLIST in stack
OBJ→	Decompose list in elements, stack level 1 = 11
ANIMATE	Start animation
>>	End program

Save this program in a variable called RANIM (Re-ANIMate). To run it, press [RANIM].

The following program will animate the graphics in WLIST forward and backwards:

<<	Start program
WLIST DUP	Place list WLIST in stack, make extra copy
REVLIST +	Reverse second copy's order, concatenate 2 lists
OBJ→	Decompose list in elements, stack level 1 = 22
ANIMATE	Start animation
>>	End program

Save this program in a variable called RANI2 (Re-ANImate version 2). To run it, press [RANI2]. The animation now simulates a ripple in the surface of otherwise quiescent water that gets reflected from the walls of a circular tank back towards the center. Press [OK] to stop the animation.

Example 2 - Animating the plotting of different power functions

Suppose that you want to animate the plotting of the functions $f(x) = x^n$, n = 0, 1, 2, 3, 4, in the same set of axes. You could use the following program:

<<	Begin program
RAD	Set angle units to radians
131 R→B 64 R→B PDIM	Set PICT screen to 131×64 pixels
0 2 XRNG 0 20 YRNG	Set x- and y-ranges
0 4 FOR j	Start FOR-NEXT loop with j = 0,1,...,4
'X^j' STEQ	Store 'X^j' in variable EQ
ERASE	Erase current PICT
DRAX LABEL DRAW	Draw axes, labels, and draw function
PICT RCL	Place current PICT on stack
NEXT	End FOR-NEXT loop
5 ANIMATE	When 5 pictures are in the stack, animate
>>	

Store this program in a variable called PWAN (PoWer function ANimation). To run the program press [VAR] (if needed) [PWAN]. You will see the calculator drawing each individual power function before starting the animation in which the five functions will be plotted quickly one after the other. To stop the animation, press [ON].

More information on the ANIMATE function

The ANIMATE function as used in the two previous examples utilized as input the graphics to be animated and their number. You can use additional information to produce the animation, such as the time interval between graphics and the number of repetitions of the graphics. The general format of the ANIMATE function in such cases is the following:

```
n-graphs   { n {#X #Y} delay rep }  ANIMATE
```

n represents the number of graphics, {#X #Y} stand for the pixel coordinates of the lower right corner of the area to be plotted (see figure below), delay is the number of seconds allowed between consecutive graphics in the animation, and rep is the number of repetitions of the animation.

Grabbing GROBs

The word GROB stands for GRaphics OBjects and is used in the HP 49 G environment to represent a pixel-by-pixel description of an image that has been produced in the calculator's screen. Therefore, when an image is converted into a GROB, it becomes a sequence of binary digits (binary digits = bits), i.e., 0's and 1's. To illustrate GROBs and conversion of images to GROBS consider the following exercise.

When we produce a graph in the calculator, the graph become the contents of a special variable called PICT. Thus, to see the last contents of PICT, you could type:

[ALPHA][ALPHA][P][I][C][T] [ENTER] [↶][RCL].

The display shows in stack level 1 the line Graphic 131×64 (if using the standard screen size) followed by a sketch of the top part of the graph. If you press [▼] then the graph contained in level 1 is shown in the calculator's graphics display. Press [CANCL] to return to normal calculator display.

The graph in level 1 is still not in GROB format, although it is, by definition, a graphics object. To convert a graph in the stack into a GROB, use:

[3][ENTER] [↶][PRG][NXT][GROB][→GROB]

Now we have the following information in level 1:

```
1: Graphic 13128 × 8
GROB 131 64 100000000
```

The first part of the description is similar to what we had originally, namely, Graphic 131×64, but now it is expressed as Graphic 13128 × 8. However, the graphic display is now replaced by a sequence of zeroes and ones representing the pixels of the original graph. Thus, the original graph as now been converted to its equivalent representation in bits.

You can also convert equations into GROBs. For example, using the equation writer type in the equation:
'X^2+3' into stack level 1, and then press

[1][ENTER] [↶][PRG][NXT][GROB][→GROB].

You will now have in level 1 the GROB described as: Graphic 28×6 'X^2-6'. As a graphic object this equation can now be placed in the graphics display. To recover the graphics display press [◄]. Then, move the cursor to an empty sector in the graph, and press [EDIT][NXT][NXT][REPL]. The equation 'X^2-5' is placed in the graph. Thus, GROBs can be used to document graphics by placing equations, or text, in the graphics display.

The GROB menu

The GROB menu, accessible through [←][PRG][NXT][GROB][→GROB], contains the following functions:

[→GROB][BLANK][GOR][GXOR][SUB][REPL]

Pressing [NXT] produces the second menu in GROB:

[→LCD][LCD→][SIZE][ANIMA][   ][PRG]

→GROB

Of these functions we have already used SUB, REPL, (from the graphics EDIT menu), ANIMATE [ANIMA], and →GROB. ([PRG] is simply a way to return to the programming menu.) While using →GROB in the two previous examples you may have noticed that I used a 3 while converting the graph into a GROB, while I used a 1 when I converted the equation into a GROB. This parameter of the function →GROB indicates the size of the object that is being converted into a GROB as 0 or 1 - for a small object, 2 - medium, and 3 - large. The other functions in the GROB menu are described following.

BLANK

The function BLANK, with arguments #n and #m, creates a blank graphics object of width and height specified by the values #n and #m, respectively. This is similar to the function PDIM in the GRAPH menu.

GOR

The function GOR (Graphics OR) takes as input $grob_2$ (a target GROB), a set of coordinates, and $grob_1$, and produces the superposition of $grob_1$ onto $grob_2$ (or PICT) starting at the specified coordinates. The coordinates can be specified as user-defined coordinates (x,y), or pixels {#n #m}. GOR uses the OR function to determine the status of each pixel (i.e., on or off) in the overlapping region between $grob_1$ and $grob_2$.

GXOR

The function GXOR (Graphics XOR) performs the same operation as GOR, but using XOR to determine the final status of pixels in the overlapping area between graphic objects $grob_1$ and $grob_2$.

Note: In both GOR and GXOR, when grob2 is replaced by PICT, these functions produce no output. To see the output you need to recall PICT to the stack by using either PICT RCL or PICTURE.

→LCD

Takes a specified GROB and displays it in the calculator's display starting at the upper left corner.

LCD→

Copies the contents of the stack and menu display into a 131 x 64 pixels GROB.

SIZE

The function SIZE, when applied to a GROB, shows the GROB's size in the form of two numbers. The first number, shown in stack level 2, represents the width of the graphics object, and the second one, in stack level 1, shows its height.

An example of a program using GROB

The following program produces the graph of the sine function including a frame - drawn with the function BOX - and a GROB to label the graph.   Here is the listing of the program:

<<	Begin program
RAD	Set angle units to radians
131 R→B 64 R→B PDIM	Set PICT screen to 131×64 pixels
-6.28 6.28 XRNG -2. 2. YRNG	Set x- and y-ranges
FUNCTION	Select the FUNCTION type for graphs
'SIN(X)' STEQ	Store the function sine into EQ
ERASE DRAX LABEL DRAW	Clear screen, draw axes, place labels, draw graph
(-6.28,-2.) (6.28,2.) BOX	Draw a frame around the graph
PICT RCL	Place contents of PICT on stack
"SINE FUNCTION"	Place graph label string in stack
1 →GROB	Convert string into a small GROB
(-6., 1.5) SWAP	Coordinates where to place the label GROB
GOR	Combine PICT with the label GROB
PICT STO	Save the combined GROB back into PICT
{ } PVIEW	Bring the PICT to the stack
>>	End program

Save the program under the name GRPR (GROB PRogram).   Press [GRPR] to run the program. The output will look like this

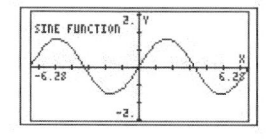

Strings attached

Strings are HP 49 G objects enclosed between double quotes. They are treated as text by the calculator. For example, the previous exercise used the string "SINE FUNCTION", transformed into a GROB, to label a graph. In the chapter on programming (Chapter 7) we learnt how to convert a number or algebraic expression into a string by using the function →STR, available in the TYPE menu: [↰][PRG][TYPE][→STR]. We also learnt how to use a string to tag a result through the function [→TAG], available in the TYPE menu. The function →OBJ, when applied to a string, separates the string into numbers and algebraic expressions. The function →OBJ is also available in the TYPE menu. Study the following examples:

[1][2] [↰][PRG][TYPE][→STR]   Produces the string "12"
"25.2" [↰][PRG][TYPE][→OBJ]   Produces the number 25.2
'x^2+y^2' [↰][PRG][TYPE][→STR]   Produces the string "'x^2+y^2'"
"23 ABD" [↰][PRG][TYPE][→OBJ]   Produces the number 23 and the algebraic 'ABD'
32.2 "g" [↰][PRG][TYPE][→TAG]   Produces the tagged quantity g: 32.2

Strings can be concatenated by using the function [+], for example:

"My " [ENTER] "dog ate " [ENTER] "the bone. " [+][+] results in the string "My dog ate the bone. "

In Chapter 7 we also learnt to use strings for input, using the function INPUT, and for output using MSGBOX. See Chapter 7 to review some of those examples.

The CHARS menu

The HP 49 G calculator provides the user with a number of additional functions for string manipulation under the CHARS menu. This menu is accessible through [↰][PRG][NXT][CHARS]. The functions available are:

[SUB][REPL][POS][SIZE][NUM][CHR]

Pressing [NXT] you get:

[OBJ→][→STR][HEAD][TAIL].

We are already familiar with the operation of OBJ→ and →STRING. We have also seen the functions SUB and REPL in relation to graphics earlier in this chapter. We have experience with the functions POS, SIZE, HEAD, and TAIL in relation with lists. These functions, SUB, REPL, POS, SIZE, HEAD, and TAIL have similar effects as in their graphics or lists counterparts. To see those effects on action try the following exercises:

Enter the string "MY NAME IS JOHN." In stack level 1. Then, press [ENTER] six times to make six copies of the string.

↓ [↰][PRG][NXT][CHARS][SIZE] returns the number 16, indicating that the string consists of 16 characters.

⬇ [4] [ENTER] [7] [ENTER] [⤶][PRG][NXT][CHARS][SUB], produces the subset "NAME", i.e., SUB extracts characters 4 through 7 from the original string. Press [⬅] to drop the result from the stack.

⬇ "IS" [⤶][PRG][NXT][CHARS][POS] returns the number 9, indicating that the sub-string "IS" within the original string starts at position 9. Press [⬅] to drop the result from the stack.

⬇ [⤶][PRG][NXT][CHARS][NXT][HEAD] returns the first character in the string "M". Press [⬅] to drop the result from the stack.

⬇ [⤶][PRG][NXT][CHARS][NXT][TAIL] returns the string with its first character removed. "Y NAME IS JOHN". Press [⬅] to drop the result from the stack.

⬇ [1][2] [ENTER] "Peter" [ENTER] [⤶][PRG][NXT][CHARS][REPL], results in "MY NAME IS PETER." In other words, the sub-string "PETER" replaces the characters starting at position 12.

The functions NUM and CHAR

The function NUM returns the numerical code corresponding to a single character or to the first character of a string. For example, "R" [NUM] produces the value 82, while "r" [NUM] results in the value 114. The function CHAR is the inverse function of NUM, i.e., it returns the character corresponding to the integer value in the stack. If you use a fractional value as input for CHAR, the number will be truncated to the lower integer, and then evaluated.

The characters list

The entire collection of characters available in the HP 49 G is accessible through the keystroke sequence [→][CHARS]. When you highlight any character, say they feed line character ↵ , you will see at the left side of the bottom of the screen the keystroke sequence to get such character (→. for this case) and the numerical code corresponding to the character (10 in this case).

Characters that are not define appear as a dark square in the characters list (·) and show (None) at the bottom of the display, even though a numerical code exists for all of them. Numerical characters show the corresponding number at the bottom of the display.

Letters show the code α (i.e., ALPHA) followed by the corresponding letter, for example, when you highlight M, you will see αM displayed at the lower left side of the screen, indicating the use of [ALPHA][M]. On the other hand, m shows the keystroke combination α⤶M, or [ALPHA][⤶][M].

Greek characters, such as σ, will show the code α→S, or [ALPHA][→][S]. Some characters, like ρ, do not have a keystroke sequence associated with them. Therefore, the only way to obtain such characters is through the character list by highlighting the desired character and pressing [ECHO1] or [ECHO].

Use [ECHO1] to copy one character to the stack and return immediately to normal calculator display. Use [ECHO] to copy a series of characters to the stack. To return to normal calculator display use

Parting thoughts about character strings

Unless you enjoy dealing with string characters to produce fancy input and output, all you really need to know about characters is how to enter them, concatenate them, and use them for tagging, and simple input and output. The emphasis of this book being on mathematical applications, we will use character strings sparingly through the remainder of this book.

A program with plotting and drawing functions - Generating Mohr's circle for two-dimensional stress

Background

In this section we develop a program to produce, draw and label Mohr's circle for a given condition of two-dimensional stress. The left-hand side figure below shows the given state of stress in two-dimensions, with σ_{xx} and σ_{yy} being normal stresses, and $\tau_{xy} = \tau_{yx}$ being shear stresses. The right-hand side figure shows the state of stresses when the element is rotated by an angle ϕ. In this case, the normal stresses are σ'_{xx} and σ'_{yy}, while the shear stresses are τ'_{xy} and τ'_{yx}.

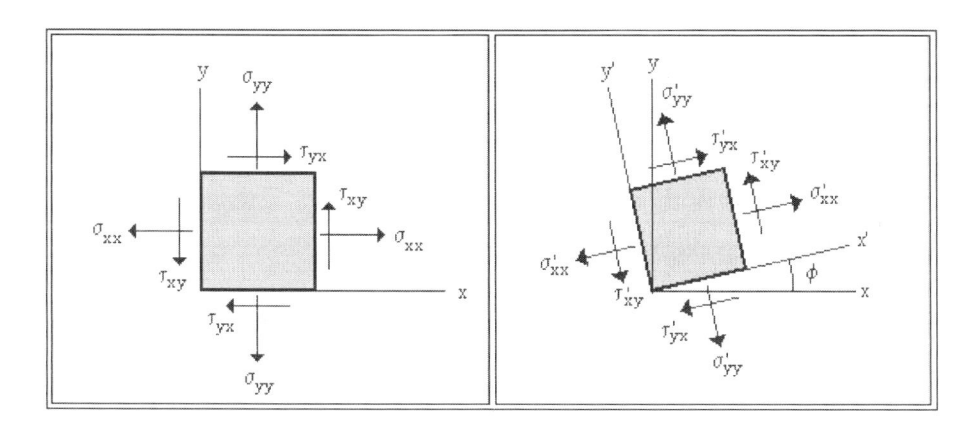

The relationship between the original state of stresses (σ_{xx}, σ_{yy}, τ_{xy}, τ_{yx}) and the state of stress when the axes are rotated counterclockwise by f (σ'_{xx}, σ'_{yy}, τ'_{xy}, τ'_{yx}), can be represented graphically by the construct shown in the figure below.

To construct Mohr's circle we use a Cartesian coordinate system with the x-axis corresponding to the normal stresses (σ), and the y-axis corresponding to the shear stresses (τ). Locate the points A(σ_{xx}, τ_{xy}) and B(σ_{yy}, τ_{xy}), and draw the segment AB. The point C where the segment AB crosses the σ_n axis will be the center of the circle. Notice that the coordinates of point C are (½·(σ_{yy} + σ_{xy}), 0). When constructing the circle by hand, you can use a compass to trace the circle since you know the location of the center C and of two points, A and B.

Let the segment AC represent the x-axis in the original state of stress. If you want to determine the state of stress for a set of axes x'-y', rotated counterclockwise by an angle ϕ with respect to the original set of axes x-y, draw segment A'B', centered at C and rotated clockwise by and angle 2ϕ with respect to segment AB. The coordinates of point A' will give the values $(\sigma'_{xx}, \tau'_{xy})$, while those of B' will give the values $(\sigma'_{yy}, \tau'_{xy})$.

The stress condition for which the shear stress, τ'_{xy}, is zero, indicated by segment D'E', produces the so-called principal stresses, σ^p_{xx} (at point D') and σ^p_{yy} (at point E'). To obtain the principal stresses you need to rotate the coordinate system x'-y' by an angle ϕ_n, counterclockwise, with respect to the system x-y. In Mohr's circle, the angle between segments AC and D'C measures $2\phi_n$.

The stress condition for which the shear stress, τ'_{xy}, is a maximum, is given by segment F'G'. Under such conditions both normal stresses, $\sigma'_{xx} = \sigma'_{yy}$, are equal. The angle corresponding to this rotation is ϕ_s. The angle between segment AC and segment F'C in the figure represents $2\phi_s$.

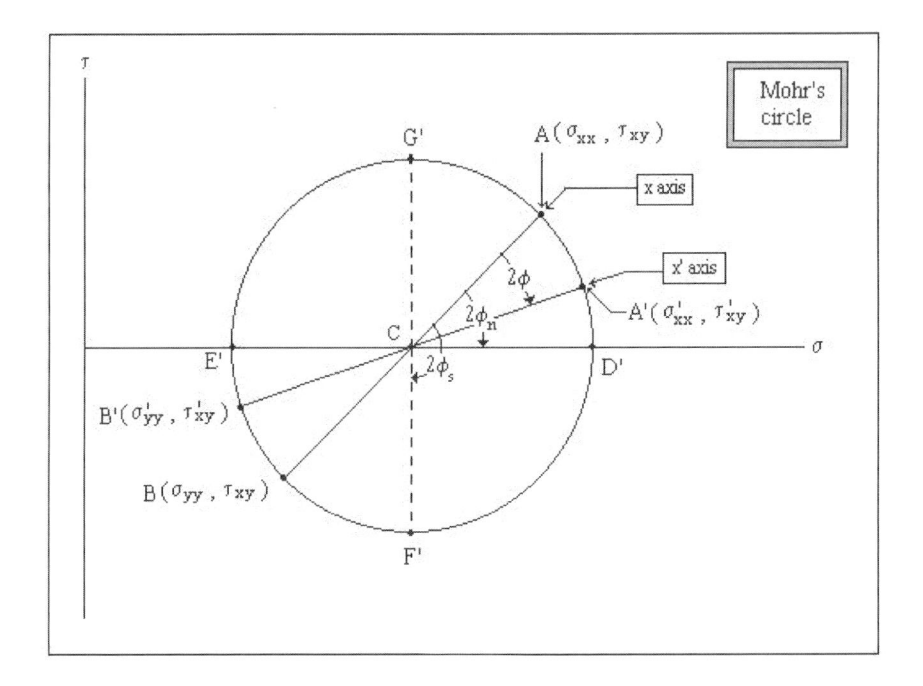

Modular programming

To develop the program that will plot Mohr's circle given a state of stress, we will use modular programming. Basically, this approach consists in decomposing the program into a number of sub-programs that are created as separate variables in the HP 49 G calculator. These sub-programs are then linked by a main program, that we will call MOHRCIRCL. We will first create a sub-directory called MOHRC within the HOME directory, and access it, as follows:

[→]['][ALPHA][ALPHA][M][O][H][R][C][ENTER]   Enter sub-directory name 'MOHRC'
[←][PRG][MEM][DIR][CRDIR]   Create sub-directory MOHRC

[VAR][MOHRC]   Access sub-directory

The next step is to create the main program and sub-programs within the sub-directory.

The main program MOHRCIRCL uses the following sub-programs:

- INDAT: Requests input of σx, σy, τxy from user, produces a list σL = {σx, σy, τxy} as output.
- CC&r: Uses σL as input, produces σc = ½(σx+σy), r = radius of Mohr's circle, φn = angle for principal stresses, as output.
- DAXES: Uses σc and r as input, determines axes ranges and draws axes for the Mohr's circle construct
- PCIRC: Uses σc, r, and φn as input, draw's Mohr's circle by producing a PARAMETRIC plot
- DDIAM: Uses σL as input, draws the segment AB (see Mohr's circle figure above), joining the input data points in the Mohr's circle
- σLBL: Uses σL as input, places labels to identify points A and B with labels "σx" and "σy".
- σAXS: Places the labels "σ" and "τ" in the x- and y-axes, respectively.
- PTTL: Places the title "Mohr's circle" in the figure.

Listings of the main program and sub-programs follow:

Main program MOHRCIRCL:

<<	Start main program
DEG	Set angular measures to degrees
INDAT	Call sub-program INDAT (input data)
→ σL	INDAT produces a list that is passed on to σL
<<	Start first sub-program within MOHRCIRCL
σL CC&r	With σL as input, call sub-program CC&r
→ σc r φn	Results of CC&r passed on as σc r φn
<<	Start second sub-program within MOHRCIRCL
σc r DAXES	With σc and r as input, call sub-program DAXES
σc r φn PCIRC	With σc, r, φn as input, call sub-program PCIRC
σL DDIAM	With σL as input, call sub-program DDIAM
σL DDIAM	With σL as input, call sub-program DDIAM
σL σLBL	With σL as input, call sub-program σLBL
σAXS	Call sub-program σAXS
PTTL	Call sub-program PTTL
{ } PVIEW	Recall contents of PICT to stack
>>	Close second sub-program in MOHRCIRCL
>>	Close first sub-program in MOHRCIRCL
>>	Close main program

Sub-program INDAT: gets input data, uses an input string with the INPUT function

<<	Start sub-program INDAT (Input DATa)
"Enter σx, σy, τxy:"	Prompt title for inputting data
{ "↵ :σx: ↵ :σy: ↵ :τxy:" {2 0} V }	Input string
INPUT	INPUT function using two previous lines
OBJ→	Decomposes input string into three tagged values
1 3 FOR j	Start FOR loop to de-tag values, j = 1,2,3
DTAG	De-tag last value in stack
3 ROLLD	Roll-down three elements in stack
NEXT	End of FOR loop
3 →LIST	Create list with the three de-tagged values

>>   End of sub-program INDAT

Sub-program CC&r:

<<   Start sub-program CC&r (Circle Center & radius)
EVAL   Decomposes list into values σx, σy, τxy
→ x y xy   Passes values σx, σy, τxy as x, y, xy
 <<   Start first sub-program within CC&r
 '(x+y)/2' →NUM   Calculates x-coordinate of Mohr's circle's center
 DUP x SWAP -   Calculates (σx-σc)
 xy R→C ATN2 2. /   Calculates ϕ_n = ½·tan^{-1}(xy/(σx-σc))
 '√(xy^2+((x-y)/2)^2)' →NUM   Calculates Mohr's circle's radius
  SWAP   Swap radius with angle
  >>   End first sub-program within CC&r
>>   End sub-program CC&r

Sub-program CC&r uses the sub-program ATN2 which determines the angle ϕ_n given as input the point (x,y) = x+y*i. The sub-program ATN2 was used instead of the function ATAN because the function ATAN(y/x) cannot handle very well situations such as when x =0, or when both y and x are negative. The listing of the sub-program ATN2 follows:

Sub-program ATN2:

<<   Start sub-program ATN2 (ATaN No. 2)
C→R   Converts input x+i*y (complex) into x and y (real)
1.E-10   Places the value 0.0000000001 in stack
→ X Y ε   Passes values in stack as X, Y, and ε (a tolerance)
 <<   Start first sub-program wihtin ATN2
  CASE   Start a CASE construct
   'X>0. AND ABS(Y)< ε   Corresponds to the case (X,Y) = (positive, 0),
     THEN 'ASIN(0.)' →NUM END   for which, ATN2 = ASIN(0.) = 0., i.e., ϕ_n = 0.
   'X>0. AND Y>0.   Corresponds to the case (X,Y) = (positive, positive),
     THEN 'ATAN(Y/X)' →NUM END   for which, ATN2 = ATAN(Y/X), i.e., 0 < ϕ_n <90°.
   'ABS(X)< ε AND Y>0   Corresponds to the case (X,Y) = (0, positive),
     THEN 'ACOS(0.)' →NUM END   for which, ATN2 = ACOS(0.) = 90°, i.e., ϕ_n = 90°.
   'X<0. AND Y>0.   Corresponds to the case (X,Y) = (negative, positive),
     THEN 'ACOS(0.) + ATAN(ABS(Y/X))'   for which, ATN2 = 90°+ATAN(|Y/X|),
       →NUM END   i.e., 90° < ϕ_n <180°.
   'X<0. AND ABS(Y)< ε   Corresponds to the case (X,Y) = (negative, 0),
     THEN '2*ACOS(0.)' →NUM END   for which, ATN2 = 2*ACOS(0.) , i.e., ϕ_n = 180°.
   'X<0. AND Y<0.   Corresponds to the case (X,Y) = (negative, negative),
     THEN '2*ACOS(0.) + ATAN(ABS(Y/X))'   for which, ATN2 = 180°+ATAN(|Y/X|),
       →NUM END   i.e., 180° < ϕ_n <270°.
   'ABS(X)< ε AND Y<0   Corresponds to the case (X,Y) = (0, negative),
     THEN '3*ACOS(0.)' →NUM END   for which, ATN2 = 3*ACOS(0.) , i.e., ϕ_n = 270°.
   'X>0. AND Y<0.   Corresponds to the case (X,Y) = (positive, negative),
     THEN '3*ACOS(0.) + ATAN(ABS(Y/X))'   for which, ATN2 = 270°+ATAN(|Y/X|),
       →NUM END   i.e., 270° < ϕ_n <360°.
   "ATN2(0,0) not defined." MSGBOX   Default result for the CASE structure, corresponds to
                                                   the case (X,Y) = (0,0), which is not defined.
  END   End CASE construct
 >>   End first sub-program within ATN2
>>   End sub-program ATN2

Note 1: Sub-program ATN2, a generalization of the HP 49 G function ATAN, can be used to determine the angle $\phi = \tan^{-1}(y/x)$, given the point (x,y) in the plane. The result is such that $0 \leq \phi < 360°$ when the angular measure is set to degrees (DEG). If the angular measure is set to radians (RAD), the result is such that $0 \leq \phi < 2\pi$.

Note 2: In general, to plot the Mohr's circle you will be working in Approximate mode. In such mode, the statement 'X == 0.' has no meaning. For that reason, such expression is replaced by 'ABS(X)<ε', where ε is a small number (ε = 1.0×10^{-10}, in this example). Thus, whenever $|X| < ε$, the number X is, for all practical purposes, equal to zero.

Sub-program DAXES:

<<	Start sub-program DAXES (Draw AXES)
→ c r	Take as input the values or c (σc) and r
<<	Start first sub-program within DAXES
'c-2.4*r' →NUM 'c+2.4*r' →NUM	Calculate min. & max. values of x for plot
'-1.2*r' →NUM DUP NEG	Calculate min. & max. values of y for plot
YRNG XRNG	Set x- and y-ranges for plot
ERASE DRAX	Erase current PICT and draw axes
>>	End first sub-program within DAXES
>>	End sub-program DAXES

Sub-program PCIRC:

<<	Start sub-program PCIRC (Plot CIRCle)
→ σc r φn	Take as input the value σc, r , and φn
<<	Start first sub-program within PCIRC
PARAMETRIC	Select a parametric plot to produce the circle
{ φ 0. 180. } INDEP	Select φ as indep. variable in range $0 < \phi < 180°$
1. RES	Select increment in φ to be equal to 1°
'σc+r*COS(2.*(φn- φ))+i*r*SIN(2.*(φn- φ))'	Equation that define the circle in parametric coords.
'σc' σc 'r' r 'φn' φn 6. →LIST	Create { 'σc' σc 'r' r 'φn' φ} to replace in eqn.
| STEQ	Replace list in eqn. with |, store eqn. in EQ.
DRAW	Draw EQ in plot, i.e., draw Mohr's circle
>>	End first sub-program within PCIRC
>>	End sub-program PCIRC

Sub-program DDIAM:

<<	Start sub-program DDIAM (Draw DIAMeter)
EVAL	Decompose input list
→ x y xy	Enter values from list as x, y, and xy
<<	Start first sub-program within DDIAM
x xy R→C DUP PPTS	Create A(σx, τxy), duplicate, plot point in PICT
y xy NEG R→C DUP PPTS	Create point B(σy, -τxy), duplicate, plot point
LINE	Draw a line between A and B
>>	End first sub-program within DDIAM
>>	End sub-program DDIAM

Sub-program DDIAM uses the sub-program PPTS which plots a point at a given location (x,y) in PICT, here is the listing of PPTS:

Sub-program PPTS:
```
<<                  Start sub-program PPTS (Plot PoinTS)
 C→PX DUP           Convert point (x,y) to pixels (hexadecimal)
 { 1 1 } ADD        Add 1 pixel to x and y
 SWAP { 1 1 } -     Subtract 1 pixel from x and y
 BOX                Create a box from (x+1,y+1) to (x-1,y-1) (pixels)
>>                  End subprogram PPTS
```

Sub-program σLBL:
```
<<                  Start sub-program σLBL (σ LaBeLs)
EVAL                Decompose input list
→ x y xy            Enter values from list as x, y, and xy
  <<                Start first sub-program within σLBL
  PICT DUP          Place PICT in stack and duplicate it
  "σx" 1 →GROB      Convert "σx" into a small graphics object
  x xy R→C PLPNT    Create point (x,xy) and locate place for label
  SWAP GOR          Swap place with GOR, combine PICT with GOR
  "σy" 1 →GROB      Convert "σy" into a small graphics object
  x xy R→C PLPNT    Create point (x,xy) and locate place for label
  SWAP GOR          Swap place with GOR, combine PICT with GOR
  >>                End first sub-program within σLBL
>>                  End sub-program σLBL
```

Sub-program σLBL uses sub-program PLPNT, which locates the place where to put the label given the coordinates of a point (x,y). The point where the label is placed is a few pixels (6 pixels) off the location of the point of interest. Here is a listing of sub-program PLPNT:

Sub-program PLPNT:
```
<<                  Start sub-program PLPNT (Place Label in a PoiNT)
DUP ATN2            Duplicate point and find angle (α) with ATN2
DUP COS             Get cosine of angle α
SWAP SIN            Swap levels 1 and 2, calculate sine of angle α
2 →LIST 6 *         Create list { cos α  sin α }, and multiply by 6
SWAP C→PX ADD       Convert list to pixels and add to original point
>>                  End sub-program PLPNT
```

Sub-program σAXS:
```
<<                  Start sub-program σAXS (s AXes labelS)
PICT                Recall contents of PICT to stack
{ # 0h # 0h }       At the origin of pixel coordinates,
"τ" 1 →GROB             take "τ" and convert it to a small graphics object
GOR                     and combine it with PICT creating new PICT
{ # 78h # 19h }     At the end of the σ-axis,
"σ" 1 →GROB             take "σ" and convert it to a small graphics object
GOR                     and combine it with PICT creating new PICT
>>                  End sub-program σAXS.
```

Sub-program PTTL:
```
<<
{ # 64h # 3h } { # A1h # 10h} BOX      Start sub-program PTTL (PloT TitLe)
                                        Create box near upper right corner of plot
PICT "Mohr's" 1 →GROB                   Create a small GROB for "Mohr's"
{ # 69h # 5h } SWAP REPL                Place GROB for "Mohr's at this location
PICT "circle" 1 →GROG                   Create a small GROB for "circle"
{ # 69h # Ah } SWAP REPL                Place the GROB for "circle" at this location
>>                                      End sub-program PTTL
```

Running the program

If you typed the programs in the order shown above, you will have in your sub-directory MOHRC the following variables:

[PTTL][σAXS][PLPNT][σLBL][PPTS][DDIAM]

Pressing [NXT] you find also:

[PCIRC][DAXES][ATN2][CC&r][INDAT][MOHRC]

Before re-ordering the variables, run the program once by pressing the soft-key labeled [MOHRC]. Use the following:

```
[MOHRC]                    Launches the main program MOHRCIRCL
[2][5][▼]                  Enter σx = 25
[7][5][▼]                  Enter σy = 75
[5][0][ENTER]              Enter τxy = 50, and finish data entry
```

At this point the program MOHRCIRCL starts calling the sub-programs to produce the figure. Be patient. The resulting Mohr's circle will look like this:

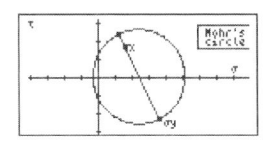

Because this view of PICT is invoked through the function PVIEW, we can not get any other information out of the plot besides the figure itself. To obtain additional information out of the Mohr's circle, end the program by pressing [ON]. Then, press [◄] to recover the contents of PICT in the graphics environment. The Mohr's circle now looks like this:

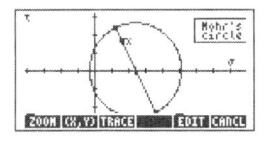

Press the soft-menu keys [TRACE] and [(x,y)]. At the bottom of the screen you will find the value of φ corresponding to the point A(σx, τxy), i.e.,

$\phi = 0$ (2.50E1, 5.00E1)

Press the right-arrow key ([▶]) to increment the value of ϕ and see the corresponding value of $(\sigma'_{xx}, \tau'_{xy})$. For example, for $\phi = 45°$, we have the values $(\sigma'_{xx}, \tau'_{xy}) = (1.00E2, 3.50E1) = (100, 35)$. The value of σ'_{yy} will be found at an angle 90° ahead, i.e., where $\phi = 45 + 90 = 135°$. Press the [▶] key until reaching that value of ϕ, we find $(\sigma'_{yy}, \tau'_{xy}) = (-1.00E-10, -2.5E1) = (0, 25)$.

To find the principal normal values press [◀] until the cursor returns to the intersection of the circle with the positive section of the σ-axis. The values found at that point are $\phi = 59°$, and $(\sigma'_{xx}, \tau'_{xy}) = (1.06E2, -1.40E0) = (106, -1.40)$. Now, we expected the value of $\tau'_{xy} = 0$ at the location of the principal axes. What happens is that, because we have limited the resolution on the independent variable to be $\Delta\phi = 1°$, we miss the actual point where the shear stresses become zero. If you press [◀] once more, you find values of are $\phi = 58°$, and $(\sigma'_{xx}, \tau'_{xy}) = (1.06E2, 5.51E0) = (106, 5.51)$. What this information tell us is that somewhere between $\phi = 58°$ and $\phi = 59°$, the shear stress, τ'_{xy}, becomes zero.

To find the actual value of ϕn, press [ON]. Then type the list corresponding to the values {σx σy τxy}, for this case, it will be

{ 25   75   50 } [ENTER]

Then, press [CC&r]. The last result in the output, 58.2825255885°, is the actual value of ϕn.

A program to calculate principal stresses

The procedure followed above to calculate ϕn, can be programmed as follows:

Program PRNST:

<<	Start program PRNST (PRiNcipal STresses)
INDAT	Enter data as in program MOHRCIRC
CC&r	Calculate σc, r, and fn, as in MOHRCIRC
"ϕn" →TAG	Tag angle corresponding to principal stresses
3 ROLLD	Move tagged angle to level 3
R→C DUP	Convert σc and r to a complex number (σc, r), duplicate
C→R + "σPx" →TAG	Calculate principal stress σPx, tag it
SWAP C→R - "σPy" →TAG	Swap levels 1 and 2, and calculate principal stress σPy, tag it.
>>	End program PRNST

To run the program use:

[VAR][PRNST]	Start program PRNST
[2][5][▼]	Enter σx = 25
[7][5][▼]	Enter σy = 75
[5][0][ENTER]	Enter τxy = 50, and finish data entry

The result is:

```
3:  φn:  58.2825255885
2:  σPx: 105.901699438
1:  σPy: -5.9016994375
```

Ordering the variables in the sub-directory

Running the program MOHRCIRCL for the first time produced a couple of new variables, PPAR and EQ. These are the Plot PARameter and EQuation variables necessary to plot the circle. I suggest that we re-order the variables in the sub-directory, so that the programs [MOHRC] and [PRNST] are the two first variables in the soft-menu key labels. This can be accomplished by creating the list { MOHRCIRCL PRNST } using:

[VAR][←][{}][MOHRC][PRNST][ENTER].

And then, ordering the list by using:

[←][PRG][MEM][DIR][NXT][ORDER].

After this call to the function ORDER is performed, press [VAR]. You will now see that we have the programs MOHRCIRCL and PRNST being the first two variables in the menu, as we expected.

A second example

Determine the principal stresses for the stress state defined by σ_{xx} = 12.5 kPa, σ_{yy} = -6.25 kPa, and τ_{xy} = - 5.0 kPa. Draw Mohr's circle, and determine from the figure the values of σ'_{xx}, σ'_{yy}, and τ'_{xy} if the angle ϕ = 35°.

To determine the principal stresses use the program [PRNST], as follows:

[VAR][PRNST]   Start program PRNST
[1][2][.][5][▼]   Enter σx = 12.5
[6][.][2][5][+/-][▼]   Enter σy = -6.25
[5][+/-][ENTER]   Enter τxy = -5, and finish data entry.

The result is:

```
3:   φn:  165.963756532
2:          σPx:  13.75
1:          σPy:  -7.5
```

To draw Mohr's circle, use the program [MOHRC], as follows:

[VAR][MOHRC]   Start program MOHRC
[1][2][.][5][▼]   Enter σx = 12.5
[6][.][2][5][+/-][▼]   Enter σy = -6.25
[5][+/-][ENTER]   Enter τxy = -5, and finish data entry.

The result is:

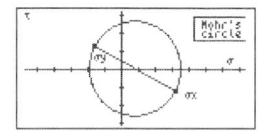

To find the values of the stresses corresponding to a rotation of 35° in the angle of the stressed particle, we use:

[ON][◄]	Clear screen, show PICT in graphics screen
[TRACE][(x,y)]	To move cursor over the circle showing values of ϕ and (x,y)

Next, press [►] until you read ϕ = 35. The corresponding coordinates are (1.63E0, -1.05E1), i.e., at

$$\phi = 35°, \sigma'_{xx} = 1.63 \text{ kPa, and } \sigma'_{yy} = -10.5 \text{kPa}.$$

12 Solution to equations

In this chapter we presents a diversity of methods for solving single equations of the form f(x) = 0, and a few for solving multiple equations of the form f(x,y,z,...) = 0, g(x,y,z,...) = 0, ..., h(x,y,z,....) = 0.

Symbolic solution of algebraic equations

The solution of algebraic equations in the HP 49 G calculator can be accomplished by a variety of methods. To begin with, we can simply isolate the variable we want to solve for from the algebraic equations. For example, to solve for m from the equation A·x+B·m = C, using the ISOL function, try:

[EQW][ALPHA][A][×][ALPHA][↵][X][+]	Type in A·x+
[ALPHA][B][×][ALPHA][↵][M]	Type in B·m
[→] [=] [ALPHA][C][ENTER]	Type in = C, place equation in stack
[ENTER]	Make additional copy of the equation
[→]['][ALPHA][↵][M][ENTER]	Type in 'm', the variable to isolate
[↵][S.SLV][ISOL]	Isolate variable

The result is: 'm=(C-x*A)/B'

The same solution can be obtained by using the function SOLVE, as follows. Press [←] to drop the later result, then type in 'm' once more, and press [SOLVE] (the second SOLVE key in the menu).

The functions ISOL and SOLVE

These two functions use as input an algebraic expression containing the variable to be solved for, and the name of the variable to be solved for, and provide the best expression possible that isolates or solves for the desired variable. To see another example of the application of these functions try solving the generic quadratic equation, $a \cdot x^2 + b \cdot x + c = 0$, for x, as follows:

[EQW]	Enter the equation writer
[ALPHA][↵][ALPHA]	Locks alphanumeric keyboard in lower case
[ALPHA][ALPHA]	Engage the alphanumeric keyboard
[A][×][X][ALPHA][y^x][2][▶][▶]	Type in ax^2
[ALPHA][ALPHA]	Engage alphanumeric keyboard once more
[+][B][×][X][▶][▶][+][C][▶][▶][→][=][0]	Type in +bx+c=0
[ENTER][ENTER]	Enter two copies equation in stack
[ALPHA][↵][X][ENTER]	Enter variable 'x' in stack
[↵][S.SLV][ISOL]	Use the function ISOL to solve for x

The result is the list: { 'x=-((b+√(b^2-4*a*c))/(2*a))' 'x=-((b-√(b^2-4*a*c))/(2*a))' }

[◄]   Drop the last result from the stack
[ALPHA][↰][X][ENTER]   Enter variable 'x' in stack once more
[↰][S.SLV][SOLVE] (second SOLVE key)   Use the function SOLVE to solve for x

The result is exactly the same as above.

The S.SLV menu

Since we accessed this menu (S.SLV, Symbolic SoLVer), we want to present some of its functions that can be used for solving algebraic equations. The S.SLV menu, accessible through [↰][S.SLV], presents the following functions:

[DSOL][ISOL][LDEC][LINSO][SOLVE][SOLVE]

Note: The first SOLVE function is SOLVEVX, the second one is SOLVE.

Press [NXT] to see the following menu:

[ZEROS][   ][   ][   ][   ][   ]

The operation of functions ISOL and SOLVE was already introduced. The function LINSO was introduced in relation with solving systems of linear equations in Chapter 10. The functions DSOL and LDEC are used for symbolic solution of differential equations, therefore, their introduction will be postponed for a later chapter on that subject. We are left to discuss the functions SOLVEVX and ZEROS.

The function SOLVEVX

The function SOLVEX solves for the default CAS variable contained in the reserved variable name VX. By default, this variable is set to be X. So, if you have that default value in VX, and your equation is written in terms of X, pressing the first [SOLVE] key in the S.SLV menu will produce the solution of the equation in terms of X. For example:

[EQW][ALPHA][A][×][ALPHA][X][+]   Type in A·x+
[ALPHA][B][×][ALPHA][↰][M]   Type in B·m
[→] [=] [ALPHA][C][ENTER]   Type in = C, place equation in stack
[↰][S.SLV][SOLVE](VX) (first [SOLVE] key)   Solve for default CAS variable VX = 'X'

The result is: 'X=-((m*B-C)/A)'

The function ZEROS

The function ZEROS finds the solutions of an equation of the form f(x) = 0, without showing their multiplicity. The function requires having as input the expression for the equation and the name of the variable to solve for. If the CAS is set to Exact, the ZEROS function will provide symbolic results, if possible. Otherwise, it will find numerical solutions. If the CAS is set to complex, ZEROS will find real and complex variables of the equation. Try the following example:

[EQW]   Enter the equation writer
[ALPHA][⤶][ALPHA]   Locks alphanumeric keyboard in lower case
[ALPHA][ALPHA]   Engage the alphanumeric keyboard
[A][×][X][ALPHA][y^x][2][▶][▶]   Type in ax^2
[ALPHA][ALPHA]   Engage alphanumeric keyboard once more
[+][B][×][X][▶][▶][+][C][▶][▶][⤶][=][0]   Type in +bx+c=0
[ENTER]   Enter equation in the stack
[ALPHA][⤶][X][ENTER]   Enter variable 'x' in stack
[⤶][S.SLV][NXT][ZEROS]   Use the function ISOL to solve for x

You will get the same result obtained earlier with ISOL and SOLVE, except that with ZEROS the results are not labeled as with SOLVE or ISOL.

The HP 49 G CAS (Calculator Algebraic System) is set up so that it provides symbolic solutions for equations that can be converted to a rational expression (i.e., the ratio of two polynomials). Expressions that cannot be converted to rational expressions need to be solved with the numerical solver, as shown below.

Numerical solver menu

The HP 49 G calculator provides a very powerful environment for the solution of single algebraic or transcendental equations. To access this environment we use the numerical solver (NUM.SLV) by using [→][NUM.SLV]. This produces a drop-down menu that includes the following options:

```
1. Solve equation..
2. Solve diff eq..
3. Solve poly..
4. Solve lin sys..
5. Solve finance..
```

Item 2. Solve diff eq.. is to be discussed in a later chapter on differential equations. Item 4. Solve lin sys.. was discussed in Chapter 10 (Matrices). Next, we present applications of items 3, 5, and 1, in that order.

Polynomial equations

We described briefly the use of item 3. Solve poly... in Chapter 8 (Algebra and related subjects). In this section we discuss the use of this solving environment in more detail.

Using the Solve poly... option in the HP48G/GX SOLVE environment you can:

(1) find the solutions to a polynomial equation;
(2) obtain the coefficients of the polynomial having a number of given roots; and,
(3) obtain an algebraic expression for the polynomial as a function of X.

Finding the solutions to a polynomial equation

A polynomial equation is an equation of the form:

$$a_n x^n + a_{n-1} x^{n-1} + \ldots + a_1 x + a_0 = 0.$$

The fundamental theorem of algebra indicates that there are n solutions to any polynomial equation of order n. Some of the solutions could be complex numbers, nevertheless. As an example, solve the equation:

$$3s^4 + 2s^3 - s + 1 = 0.$$

We want to place the coefficients of the equation in a vector [a_n a_{n-1}... a_1 a_0]. For this example, let's use the vector [3 2 0 -1 1]. To solve for this polynomial equation using the calculator, try the following:

[↱][NUM.SLV][▼][▼][OK]   Select Solve poly...
[↰][[ ]][3][SPC][2][SPC][0]][SPC][1][+/-][SPC][1][OK]   Enter the vector of coefficients
[SOLVE]   Solve for s
[ENTER]   Returns to stack.
[↰][PRG][TYPE][OBJ→][⇦][OBJ→] [⇦]   All solutions will be listed in the stack.

All the solutions are complex numbers: (0.432,-0.389), (0.432,0.389), (-0.766, 0.632), (-0.766, -0.632).

> Note: Recall that complex numbers in the HP 49 G calculator are represented as ordered pairs, with the first number in the pair being the real part, and the second number, the imaginary part. For example, the number (0.432,-0.389), a complex number, will be written normally as 0.432 - 0.389i, where i is the imaginary unit, i.e., $i^2 = -1$.
> Note: The <u>fundamental theorem of algebra</u> indicates that there are n solutions for any polynomial equation of order n. There is another theorem of algebra that indicates that if one of the solutions to a polynomial equation with real coefficients is a complex number, then the conjugate of that number is also a solution. In other words, complex solutions to a polynomial equation with real coefficients come in pairs. That means that polynomial equations with real coefficients of odd order will have at least one real solution.

Generating polynomial coefficients given the polynomial's roots

Suppose you want to generate the polynomial whose roots are the numbers [1 5 -2 4]. To use the HP 49 G calculator for this purpose,. follow the following steps:

[↱][NUM.SLV][▼][▼][OK]   Select Solve poly...
[▼][↰][[ ]][1][SPC][5][SPC][2][+/-][SPC][4][OK] Enter the vector of roots
[SOLVE]   Solve for the coefficients
[ENTER]   Returns to stack. Coefficients are listed in stack.

> Note: If you want to get a polynomial with real coefficients, but having complex roots, you must include the complex roots in pairs of conjugate numbers. To illustrate the point, generate a polynomial having the roots [1 (1,2) (1,-2)]. Verify that the resulting polynomial has only real coefficients. Also, try generating a polynomial with roots [1 (1,2) (-1,2)], and verify that the resulting polynomial will have complex coefficients.

Generating an algebraic expression for the polynomial

You can use the calculator to generate an algebraic expression for a polynomial given the coefficients or the roots of the polynomial. The resulting expression will be given in terms of the default CAS variable X. (The examples below shows how you can replace X with any other variable by using the function [|].)

To generate the algebraic expression using the coefficients, try the following example. Assume that the polynomial coefficients are [1 5 -2 4]. Use the following keystrokes:

[↱][NUM.SLV][▼][▼][OK]   Select Solve poly...
[↰][[ ]][1][SPC][5][SPC][2][+/-][SPC][4][OK]   Enter the vector of roots
[▲][SYMB]   Highlight coefficients; generate symbolic expression
[ENTER]   Returns to stack. Expression shown in stack.

The resulting expression is given as `'X^3+5*X^2-2*X+4'`.

To generate the algebraic expression using the roots, try the following example. Assume that the polynomial roots are [1 3 -2 1]. Use the following keystrokes:

[↱][NUM.SLV][▼][▼][OK]   Select Solve poly...
[▼][↰][[ ]][1][SPC][3][SPC][2][+/-][SPC][1][OK] Enter the vector of roots
[▼]   Highlight roots
[SYMB]   Generate algebraic expression.
[ENTER]   Returns to stack. Expression listed in stack.

The resulting expression is listed as `'(X-1)*(X-3)*(X+2)*(X-1)'`. To expand the products, you can use:
[↱][ALG][EXPA].

The resulting expression is: `'X^4+-3*X^3+-3*X^2+11*X-6'`.

A different approach to obtaining an expression for the polynomial is to generate the coefficients first, then generate the algebraic expression with the coefficients highlighted. For example, for this case try:

[↱][NUM.SLV][▼][▼][OK]   Select Solve poly...
[▼][↰][[ ]][1][SPC][3][SPC][2][+/-][SPC][1][OK] Enter the vector of roots
[SOLVE]   Solve for the coefficients
[SYMB]   Highlight roots, generate algebraic expression.
[ENTER]   Returns to stack. Expression listed in stack.

The resulting expression is: `'X^4+-3*X^3+-3*X^2+11*X+-6*X^0'`.
The coefficients are listed in stack level 2.

Evaluating a polynomial (or any) expression

Suppose that you want to evaluate the expression obtained above for X = 5. Keeping the expression in stack level 1, enter the following:

[↰][{}][ALPHA][X][SPC][5][ENTER]   Enter the list {X 5}
[↱][|]   Use the "evaluate at" expression

369

The result is 224. This procedure can be used to evaluate any HP48G/GX algebraic in level 2 using the values in the list in level 1.

Replacing the variable in the polynomial's algebraic expression

Press [→][UNDO] to recover the last operands. If instead of using {X 5} we use {X x}, the variable X will be replaced by x in the polynomial. Try the following:

[⇦][↵][{}][ALPHA][X][SPC][⇦][ALPHA][↵][X][ENTER]   Enter the list {X x}
[→][|]   Use the "evaluate at" expression

The result is the polynomial: : `x^4_3*x^3-3*x^2+11*x-6'`.

Financial calculations

The calculations in item 5. Solve finance.. in the Numerical Solver (NUM.SLV) are used for calculations of time value of money of interest in the discipline of engineering economics and other financial applications. Before discussing in detail the operation of this solving environment, we present some definitions needed to understand financial operations in the HP 49 G.

Definitions

In the development of engineering projects or projects of other type it is necessary to borrow money from a financial institution or from public funds. The amount of money borrowed is referred to as the Present Value (PV). This money is to be repaid through n periods (typically multiples or sub-multiples of a month) subject to an annual interest rate of I%YR. The number of periods per year (P/YR) is an integer number of periods in which the year will be divided for the purpose of repaying the loan money. Typical values of P/YR are 12 (one payment per month), 24 (payment twice a month), or 52 (weekly payments). The payment(PMT) is the amount that the borrower must pay to the lender at the beginning or end of each of the n periods of the loan. The future value of the money (FV) is the value that the borrowed amount of money will be worth at the end of n periods.

Payment at beginning or end of a period

Typically payment occurs at the end of each period, so that the borrower starts paying at the end of the first period, and pays the same fixed amount at the end of the second, third, etc., up to the end of the n-th period. Flag -14 in the HP 49 G calculator when not set uses the option *payment at the end* of each period. When Flat -14 is set, either by using

[1][4][+/-][↵][PRG][NXT][MODES][FLAG][SF],

or by using

[MODE][FLAGS] [▼][▼][▼][✓CHK][OK][OK],

then the option Payment at begin is set.

In the following exercises, let's keep FLAG -14 so that the option payment at the end is active.

> Note: Access to the financial calculation environment is also available directly from the keyboard by using [↰][FINANCE] (the [9] key).

Examples

Example 1

If $2 million are borrowed at an annual interest rate of 6.5% to be repaid in 60 monthly payments, what should be the monthly payment? For the debt to be totally repaid in 60 months, the future values of the loan should be zero. So, for the purpose of using the financial calculation feature of the HP 49 G we will use the following values: n = 60, I%YR = 6.5, PV = 2000000, FV = 0, P/YR = 12. To enter the data and solve for the payment, PMT, use:

[↰][FINANCE]	Start the financial calculation input form
[6][0][OK]	Enter n = 60
[6][.][5][OK]	Enter I%YR = 6.5 %
[2][0][0][0][0][0][0][OK]	Enter PV = 2,000,000 US$
[▼]	Skip PMT, since we will be solving for it
[0] [OK]	Enter FV = 0, the option End is highlighted
[▲][◀][SOLVE]	Highlight PMT and solve for it

The screen now shows the value of PMT as -39,132..., i.e., the borrower must pay the lender US $ 39,132.30 at the end of each month for the next 60 months to repay the entire amount. The reason why the value of PMT turned out to be negative is because the calculator is looking at the money amounts from the point of view of the borrower. The borrower has + US $ 2,000,000.00 at time period t = 0, then he starts paying, i.e., adding -US $ 39132.30 at times t = 1, 2, ..., 60. At t = 60, the net value in the hands of the borrower is zero. Now, if you take the value US $ 39,132.30 and multiply it by the 60 payments, the total paid back by the borrower is US $ 2,347,937.79. Thus, the lender makes a net profit of $ 347,937.79 in the 5 years that his money is used to finance the borrower's project.

Example 2

The same solution to the problem in Example 1 can be found by pressing [AMOR], which is stands for AMORTIZATION (and not for the Spanish word for "love"). This option is used to calculate how much of the loan has been amortized at the end of a certain number of payments. Suppose that we use 24 periods in the first line of the amortization screen, i.e., [2][4][OK]. Then, press [AMOR]. You will get the following result:

```
########## AMORTIZE ########
Payments:   24
Principal:  -723,211.43
Interest:   -215,963.68
Balance:    1,276,788.57
```

This screen is interpreted as indicating that after 24 months of paying back the debt, the borrower has paid up US $ 723,211.43 into the principal amount borrowed, and US $ 215,963.68 of interest. The borrower still has to pay a balance of $1,276,788.57 in the next 36 months.

Check what happens if you replace 60 in the Payments: entry in the amortization screen, then press [OK]
[AMOR]. The screen now looks like this:

```
########## AMORTIZE ########
Payments:   60
Principal:  -2,000,000,00
Interest:   -347,937.79
Balance:    -3.16E-6
```

This means that at the end of 60 months the US $ 2,000,000.00 principal amount has been paid, together with US $ 347,937.79 of interest, with the balance being that the lender owes the borrower US $ 0.000316. Of course, the balance should be zero. The value shown in the screen above is simply round-off error resulting from the numerical solution.

Press [ON] or [ENTER], twice, to return to normal calculator display.

Example 3

Let's solve the same problem as in Examples 1 and 2, but using the option that payment occurs at the beginning of the payment period. Use:

[←][FINANCE]	Start the financial calculation input form
[6][0][OK]	Enter n = 60
[6][.][5][OK]	Enter I%YR = 6.5 %
[2][0][0][0][0][0][0][OK]	Enter PV = 2,000,000 US$
[▼]	Skip PMT, since we will be solving for it
[0] [OK]	Enter FV = 0
[CHOOSE][▲][OK]	Change this option to Begin
[▲][◄][SOLVE]	Highlight PMT and solve for it

The screen now shows the value of PMT as -38,921..., i.e., the borrower must pay the lender US $ 38,921.48 at the beginning of each month for the next 60 months to repay the entire amount. Notice that the amount the borrower pays monthly, if paying at the beginning of each payment period, is slightly smaller than that paid at the end of each payment period. The reason for that difference is that the lender gets interest earnings from the payments from the beginning of the period, thus alleviating the burden on the lender.

Notes:

1. The financial calculator environment allows you to solve for any of the terms involved, i.e., n, I%YR, PV, FV, P/Y, given the remaining terms in the loan calculation. Just highlight the value you want to solve for, and press [SOLVE]. The result will be shown in the highlighted field.

> 2. The values calculated in the financial calculator environment are copied to the stack with their corresponding tag.
> 3. When you use the financial calculator environment for the first time within the HOME directory, or any sub-directory, it will generate the variables [n][I%YR][PV][PMT][PYR][FV] to store the corresponding terms in the calculations.

Solving equations with one unknown through NUM.SLV

The HP48G/GX calculator's NUM.SLV menu provides the item 1. Solve equation.. that can be used to solve different types of equations in a single variable, including non-linear algebraic and transcendental equations. For example, let's solve the equation:

$$e^x - \sin(\pi x/3) = 0.$$

Simply enter the expression as an HP48G/GX algebraic, i.e.,

['][↰][e^x] [ALPHA][↰][X] [▶] [-] [SIN] [↰][π] [×] [ALPHA][↰][X] [÷] [3] [▶] [↰][=][0][ENTER]]

and store it into variable EQ, i.e.,

['] [ALPHA][ALPHA][E][Q][ALPHA][STO]

Then, enter the SOLVE environment and select Solve equation..., by using:

[↱][NUM.SLV][OK]

The equation is already loaded, all you need to do is highlight the field in front of X: by using [▼], and press [SOLVE]. The solution shown is X: 4.5006... To obtain a negative solution, try entering a negative number in the X: field, for example, [3][+/-][OK] [▼] [SOLVE]. The solution is now X: -3.045.

How does the numerical solver for single-unknown equations work?

The numerical solver for single-unknown equations works as follows:

- It lets the user type in or CHOOSe an equation to solve.
- It creates an input form with input fields corresponding to all variables involved in equation EQ.
- The user needs to enter values for all variables involved, except one.
- The user then highlights the field corresponding to the unknown for which to solve the equation, and presses [SOLVE].
- The calculator uses a search algorithm to pinpoint an interval for which the function changes sign, which indicates the existence of a root or solution. It then utilizes a numerical method to converge into the solution.

The solution the calculator seeks is determined by the initial value present in the unknown input field. If no value is present, the calculator uses a default value of zero. Thus, you can search for more than one solution to an equation by changing the initial value in the unknown input field. Examples of the equations solutions are shown following.

Example 1 - Hooke's law for stress and strain.

The equation to use is Hooke's law for the normal strain in the x-direction for a solid particle subjected to a state of stress given by

$$\begin{bmatrix} \sigma_{xx} & \sigma_{xy} & \sigma_{xz} \\ \sigma_{yx} & \sigma_{yy} & \sigma_{yz} \\ \sigma_{zx} & \sigma_{zy} & \sigma_{zz} \end{bmatrix}.$$

The equation is

$$e_{xx} = \frac{1}{E}[\sigma_{xx} - n \cdot (\sigma_{yy} + \sigma_{zz})] + \alpha \cdot \Delta T,$$

where e_{xx} is the unit strain in the x-direction, σ_{xx}, σ_{yy}, and σ_{zz}, are the normal stresses on the particle in the directions of the x-, y-, and z-axes, E is Young's modulus or modulus of elasticity of the material, n is the Poisson ratio of the material, α is the thermal expansion coefficient of the material, and ΔT is a temperature increase.

Suppose that you are given the following data: σ_{xx}= 2500 psi, σ_{yy} =1200 psi , and σ_{zz} = 500 psi, E = 1200000 psi, n = 0.15, α = 0.00001/°F, ΔT = 60 °F. To calculate the strain e_{xx} use the following:

[↦][NUM.SLV][OK]   Access numerical solver to solve equations
[EQW]   Access the equation writer to enter equation

To enter the equation use:

[ALPHA][↵][E][ALPHA][↵][X] [↦][=][↵][( )] [ALPHA][↦][S][ALPHA][↵][X][-]
[ALPHA][↵][N][×][↵][( )][ALPHA][↦][S][ALPHA][↵][Y] [+]
[ALPHA][↦][S][ALPHA][↵][Z] [▶][▶][▶][▶][▶][▶][÷] [ALPHA][E] [▲][▲]
[+][ALPHA][↦][A] [×][ALPHA][↦][C][ALPHA][T]

[▶][2][5][0][0][OK]   Enter σ_{xx}= 2500
[0][.][1][5][OK]   Enter n = 0.15
[1][2][0][0][OK]   Enter σ_{yy} =1200
[5][0][0][OK]   Enter σ_{zz} =500
[1][2][0][0][0][0][0][OK]   Enter E = 1200000
[0][.][0][0][0][0][1][OK]   Enter α = 0.00001
[6][0][OK]   Enter ΔT = 60
[▼][SOLVE]   To solve for ex

The solution can be seen from within the SOLVE EQUATION input form by pressing [EDIT]. The resulting value is 2.470833333333E-3. Press [OK] to exit the EDIT feature.

Suppose that you want to determine the Young's modulus that will produce a strain of e_{xx} = 0.005 under the same state of stress neglecting thermal expansion. Use the following:

[◄][0][.][0][0][5][OK]   Enter e_{xx} = 0.005
[▼][▼][0][OK]   Enter ΔT = 0 to neglect thermal expansion
[▼][▼][►][►][SOLVE]   Select input field for E and solve for E
[EDIT]   To see resulting value

The solution is E = 449000 psi. Press [OK] [ON] to return to normal display.

Notice that the results of the calculations performed within the numerical solver environment have been copied to the stack. Also, you will see in your soft-menu key labels variables corresponding to those in EQ, i.e.,

[ex][ΔT][α][E][σz][σy]

Pressing [NXT] you will see:

[n][σx][other variables]

If you decide that you do not need to keep this equation and its variables, you can PURGE them by placing the variable names, and EQ, in a list, and using [TOOL][PURGE]. On the other hand, if you think this equation is to be used regularly, you may want to create a separate sub-directory to keep it on a permanent basis.

Example2 - Specific energy in an open channel flow

Specific energy in an open channel is defined as the energy per unit weight measured with respect to the channel bottom. Let E = specific energy, y = channel depth, V = flow velocity, g = acceleration of gravity, then we write

$$E = y + \frac{V^2}{2g}.$$

The flow velocity, in turn, is given by

$$V = Q/A,$$

where Q = water discharge, A = cross-sectional area. The area depends on the cross-section used, for example, for a trapezoidal cross-section, as shown in the figure below,

$$A = (b+m \cdot y) \cdot y,$$

where b = bottom width, and m = side slope of cross section.

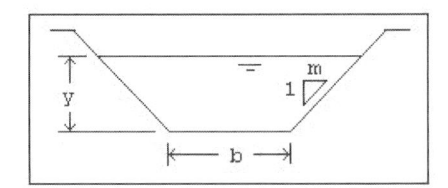

We can type in the equation for E as shown above and use auxiliary variables for A and V, so that the resulting input form will have fields for the fundamental variables y, Q, g, m, and b, as follows:

⬥ First, create a sub-directory called SPEN (SPecific ENergy) and work within that sub-directory.
⬥ Next, define the following variables:

[↵]['][ALPHA][Q][÷][ALPHA][A][ENTER] [↵]['][ALPHA][V] [ENTER] [STO▶]

[EQW][←][( )][ALPHA][←][B][+][ALPHA][←][M][×][ALPHA][←][Y][▶][▶][▶]
[×][ALPHA][←][Y][ENTER] [↵]['][ALPHA][A] [ENTER] [STO▶]

[↵]['][ALPHA][E][↵][=][ALPHA][←][Y][+]
[ALPHA][V][y^x][2][÷][←][( )][2][×][ALPHA][←][G][ENTER] [↵]
['][ALPHA][ALPHA][E][Q] [ENTER] [STO▶]

⬥ Launch the numerical solver for solving equations: [↵][NUM.SLV][OK]. Notice that the input form contains entries for the variables y, Q, b, m, and g.

⬥ Try the following input data: E = 10 ft, Q = 10 cfs (cubic feet per second), b = 2.5 ft, m = 1.0, g = 32.2 ft/s^2:

[▼][1][0][OK] [▶] [1][0][OK] [2][.][5][OK] [1][OK] [3][2][.][2][OK]

⬥ Solve for y: [▼][▶] [SOLVE]. The result is 0.149836.., i.e., y = 0.149836.

⬥ It is known, however, that there are actually two solutions available for y in the specific energy equation. The solution we just found corresponds to a numerical solution with an initial value of 0. To find the other solution, we need to enter a larger value of y, say [1][0][OK], return to the y input field with [◀], and press [SOLVE] again. The result is now 9.99990, i.e., y = 9.99990 ft.

This example illustrates the use of auxiliary variables to write complicated equations. When NUM.SLV is activated, the substitutions implied by the auxiliary variables are implemented, and the input screen for the equation provides input field for the primitive or fundamental variables resulting from the substitutions. The example also illustrates an equation that has more than one solution, and how choosing the initial guess for the solution may produce those different solutions.

In the next example we will use the DARCY function for finding friction factors in pipelines. Thus, we define the function in the following frame.

Special function for pipe flow: DARCY(ε/D, Re)

The Darcy-Weisbach equation is used to calculate the energy loss (per unit weight), h_f, in a pipe flow through a pipe of diameter D, absolute roughness ε, and length L, when the flow velocity in the pipe is V. The equation is written as

$$h_f = f \cdot \frac{L}{D} \cdot \frac{V^2}{2g}.$$

The quantity f is known as the friction factor of the flow and it has been found to be a function of the relative roughness of the pipe, ε/D, and a (dimensionless) Reynolds number, Re. The Reynolds number is defined as

$$Re = \rho VD/\mu = VD/\nu,$$

where ρ and μ are the density and dynamic viscosity of the fluid, respectively, and $\nu = \mu/\rho$ is the kinematic viscosity of the fluid.

The HP 49 G calculator provides a function called DARCY that uses as input the relative roughness ε/D and the Reynolds number, in that order, to calculate the friction factor f. For example, for ε/D = 0.0001, Re = 1000000, you can find the friction factor by using:

[.][0][0][0][1][ENTER] [1][0][0][0][0][0][0][ENTER] [CAT][ALPHA][D],

if needed use [▲][▼] to highlight DARCY, then press [OK]. The result is 1.3414320724E-2, i.e., f = 0.01341...

The function FANNING(ε/D, Re)

In aerodynamics applications a different friction factor, the Fanning friction factor, is used. The Fanning friction factor, f_F, is defined as 4 times the Darcy-Weisbach friction factor, f. The HP 49 G calculator also provides a function called FANNING that uses the same input as DARCY, i.e., ε/D and Re, and provides the FANNING friction factor.

Example 3 - Flow in a pipe

You may want to create a separate sub-directory to try this example.

The main equation governing flow in a pipe is, of course, the Darcy-Weisbach equation. Thus, type in the following equation into EQ:

[→][`][ALPHA][↵][H][ALPHA][↵][F][→][=][ALPHA][↵][F][×]
[↵][( )][ALPHA][L][÷][ALPHA][D] [▶] [×]
[↵][( )][ALPHA][V][y^x][2][÷][↵][( )][2][×][ALPHA][↵][G][ENTER]

['][ALPHA][ALPHA][E][Q] [ENTER] [STO▶]

Also, store the following variables:

[r→]['][CAT][ALPHA][D] (select DARCY) [ALPHA][r→][E]
[÷][ALPHA][D][r→][,] [ALPHA][R][ALPHA][←][E][ENTER]

['][ALPHA][←][F][ENTER] [STO▶]

[r→]['][←][π][×][ALPHA][D][y^x][2][÷][4][ENTER]

[r→]['][ALPHA][A][ENTER] [STO▶]

[r→]['][ALPHA][Q][÷][ALPHA][A][ENTER]

[r→]['][ALPHA][V] [ENTER] [STO▶]

[r→]['][ALPHA][V][×][ALPHA][D][÷][ALPHA][N][ALPHA][←][U][ENTER]

[r→]['][ALPHA][R][ALPHA][←][E][ENTER] [STO▶]

Now, launch the numerical solver, by using:

[r→][NUM.SLV][OK].

You will see the input screen for this equation containing input fields for the primitive variables hf, ε, Q, D, Nu, L, and g. Suppose that we use the values hf = 2 m, ε = 0.00001 m, Q = 0.05 m3/s, Nu = 0.000001 m^2/s, L = 20 m, and g = 9.806 m/s^2, find the diameter D. Enter the input values as follows:

[▼][2][OK]	Enter hf
[1][EEX][5][+/-][OK]	Enter ε
[▶]	Skip D
[.][0][5][OK]	Enter Q
[1][EEX][6][+/-][OK]	Enter Nu (i.e., ν = kinematic viscosity)
[2][0][OK]	Enter L
[9][.][8][0][6][OK]	Enter g

To solve for D, use: [▼][▶][▶] (highlight D)[SOLVE]. The solution is: 0.12, i.e., D = 0.12 m. Press [ENTER] or [ON] to return to normal calculator display. The solution for D will be listed in the stack.

To see the equation we are actually solving, try the following (set the CAS to Exact):

[←][{}]	Open list
[D][L][Nu][ε][hf][g][Q]	Place variable names D, L, Nu, ε, hf, g, and Q in list
[ENTER][TOOL][PURGE]	Purge variables D, L, Nu, ε, hf, g, and Q
[VAR][EQ]	Lists Darcy-Weisbach equation in stack
[r→]['] [f] [ENTER] [f]	Places the name 'f' and the contents of f in stack
[4][←][PRG][TYPE][→LIST]	Create a list with four elements
[r→][	][ENTER]	Replace variables in list (level 1) in equation (lev. 2)

If asked whether you want approximate mode, say NO, as many times as needed.

The resulting equation is

$$\text{'hf = DARCY}(\varepsilon,\text{Re})*(L/D)*((Q/A)^2/(2.*g))\text{'}$$

We now will replace the corresponding expressions for Re and Q, as follows:

[VAR][→]['] [Re] [ENTER] [Re]   Places the name 'Re' and the contents of Re in stack
[→]['] [A] [ENTER] [A]   Places the name 'A' and the contents of A in stack
[4][↰][PRG][TYPE][→LIST]   Create a list with four elements
[→][|][ENTER]   Replace variables in list (level 1) in equation (level 2)

If asked whether you want approximate mode, say NO, as many times as needed.

The resulting equation is:

$$\text{'hf= DARCY}(\varepsilon/D,\ V*D/\text{Nu})*(L/D)*((Q/(\pi*D^2/4))^2/(2*g))\text{'}$$

We need two more substitutions:

[VAR] [→]['] [V] [ENTER] [V]   Places the name 'V' and the contents of V in stack
[2][↰][PRG][TYPE][→LIST]   Create a list with four elements
[→][|][ENTER]   Replace variables in list (level 1) in equation (level 2)
[VAR] [→]['] [A] [ENTER] [A]   Places the name 'V' and the contents of V in stack
[2][↰][PRG][TYPE][→LIST]   Create a list with four elements
[→][|][ENTER]   Replace variables in list (level 1) in equation (level 2)

The resulting equation is:

$$\text{'hf= DARCY}(\varepsilon/D,\ Q/(\pi*D^2/4)*D/\text{Nu})*(L/D)*((Q/(\pi*D^2/4))^2/(2*g))\text{'}$$

which is the equation that is actually solved in NUM.SLV.

Functions UTPN, UTPT, UTPC, and UTPF

In Chapter 4 we introduced the functions UTPN (Upper-tail probability normal), UTPT(Upper-tail probability Student's t), UTPC(Upper-tail probability Chi square), and UTPF(Upper-tail probability F distribution). These functions provide values of the probability that a variable X, following a specified distribution, is larger than a particular value x. Utilizing the notation and definitions from Chapter 3, we have:

- $f(x)$ = probability density function or pdf
- $F(x) = P(X<x)$ = cumulative distribution function or CDF, where P( ) stands for "probability of."

An upper-tail probability is typically referred to as $\alpha = P(X>x)$. By the definition of probability,

$$\alpha = 1 - F(x).$$

Graphically, the CDF F(x) is the area under the pdf curve for X<x, and a is the area under the same curve for X>x, as illustrated in the figure below.

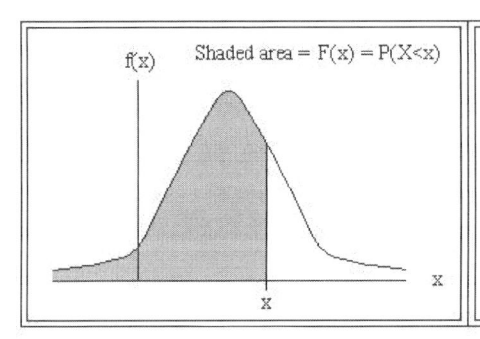

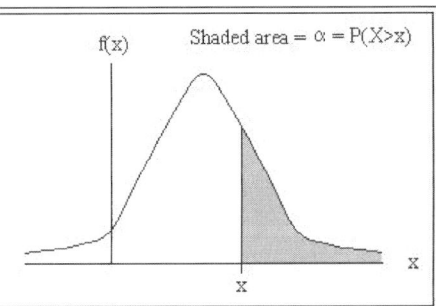

The upper-tail probability functions UTPN, UTPT, UTPC, and UTPF are available through the menu:

$$[\leftarrow][MTH][NXT][PROB][NXT].$$

The parameters associated with each of those probability distributions are:

- With UTPN (normal), μ = mean value, σ^2 = variance
- With UTPT (Student's t) and UTPC (Chi-square), ν = degrees of freedom
- With UTPF (F distribution), ν_N = degrees of freedom in the numerator, ν_D = degrees of freedom in denominator.

The corresponding parameters must be given as part of the input for the upper-tail probability functions as illustrated in the exercises in Chapter 4. In the following example we illustrate the use of the numerical solver to solve the equation defining inverse CDFs and upper tail probability distributions.

Example 3 - Upper tail probabilities for Normal, Student-t, χ^2, and F distributions

The expression defining the upper-tail probability function for the Normal variate is written as

$$\alpha = UTPN(\mu, \sigma^2, x),$$

or

$$UTPN(\mu, \sigma^2, x) - \alpha = 0.$$

The latter equation can be typed as a program:

$$\ll \mu\ \sigma 2\ x\ \text{UTPN}\ \alpha\ -\gg$$

and stored in the EQ variable: [`][ALPHA][ALPHA][E][Q] [ENTER] [STO▶].

> Note: The program showed above interpreted in RPN mode will produce the quantity UTPN(x,μ,σ2)-α. When this quantity is stored in EQ and used for an equation solution through NUM.SLV it is interpreted as UTPN(x,μ,σ2)-α= 0. As a general rule, if an equal sign is missing in the expression stored in EQ, then the expression is assumed to be equal to zero.

If we now launch the numerical solver, we obtain an input form involving the variables μ, σ2 (stands for σ^2), x , and α. For example, to find the value of x for which α = 0.05, when μ = 3.2, and σ^2 = 1.5, use:

[↱][NUM.SLV][OK]	Access numerical solver to solve equations
[▼][3][.][2][OK]	Enter μ
[1][.][5][OK]	Enter σ2
[▶]	Skip x
[.][0][5][OK]	Enter α
[▼] [▼][SOLVE]	Solve for x

The result is x = 5.21452.

If you want to find the value of α corresponding to x = 6, use:

[6][OK][SOLVE]

The result is (press [EDIT]) 1.112E-2, i.e., α = 0.01112.

You can create a sub-directory, call it UTPs, for Upper Tail probability functions, where you can store the program used above for the normal distribution (store it into the variable EQN), and programs corresponding to the other three distributions under consideration: t, χ^2, and F. The programs corresponding to the other distributions are shown below:

Program name	Program
EQT	$\ll \gamma\ x\ \text{UTPT}\ \alpha\ -\gg$
EQC	$\ll \gamma\ x\ \text{UTPC}\ \alpha\ -\gg$
EQF	$\ll \gamma N\ \gamma D\ x\ \text{UTPF}\ \alpha\ -\gg$

> Notice that I use the variable name γ (gamma) for ν (nu), because the HP 49 G calculator does not include the character ν.

Within the sub-directory you should create a variable named EQ, for example, using:

[0][`][ALPHA][ALPHA][E][Q][ENTER][STO]

Next, order the variables in your sub-directory using:

{ EQ EQN EQT EQC EQF } [↰][PRG][MEM][DIR][NEXT][ORDER].

At this point the sub-directory should have the following variables in its soft-menu key labels:

[EQ][EQN][EQT][EQC][EQF].

Whenever you want to use a particular equation you need to load it in EQ, and then launch the numerical solver. For example, if you want to use the Chi-square equation, use the following:

[→][EQC][↵][EQ][→][NUM.SLV][OK],

and you will be ready to solve for any term in the equation

$$UTPC(\gamma, x) - \alpha = 0.$$

A similar procedure can be used to solve for any term in the equations involving the UTPT and UTPF functions.

Example 4 - Universal gravitation

Newton's law of universal gravitation indicates that the magnitude of the attractive force between two bodies of masses m_1 and m_2 separated by a distance r is given by the equation

$$F = G \cdot \frac{m_1 \cdot m_2}{r^2}.$$

Here, G is the universal gravitational constant, whose value can be obtained through the use of the function CONST in the calculator by using:

[→][`][ALPHA][G][ENTER][CAT][ALPHA][C] (highlight CONST) [OK].

$G = 6.67259 \times 10^{-11}$ m^3/(s$^2 \cdot$kg).

We can solve for any term in the equation (except G) by entering the equation as:

[EQW][ALPHA][F][→][=] [ALPHA][G] [×] [ALPHA][↵][M][1][×]
[ALPHA][↵][M][2] [▶] [÷][ALPHA][↵][R][y^x][2][ENTER]

This equation is then stored in EQ: [`][ALPHA][ALPHA][E][Q] [ENTER] [STO▶].

Launching the numerical solver for this equation, i.e.,. [→][NUM.SLV][OK], results in an input form containing input fields for F, G, m1, m2, and r. Let's solve this problem using units with the following values for the known variables m1 = 1.0×10^6 kg, m2 = 1.0×10^{12} kg, r = 1.0×10^{11} m.

First, enter a value of 0_N in field F, by using:

[▼][0] [→][_][→][UNITS][NXT][FORCE][N][ENTER] to load the value of m1.

Next, use the following keystrokes to skip the field for F and enter the value of G from the stack:

[▶][NXT][CALC][→][`][ALPHA][ALPHA][C][O][N][S][T][↵] [( )] [G][ENTER][→][EVAL][OK]

Press [▶][1][EEX][6][→][_][→][UNITS][NXT][MASS][kg][ENTER] to load the value of m1.

Press [1][EEX][1][2][→][_][→][UNITS][NXT][MASS][kg][ENTER] to load the value of m1.

Press [1][EEX][1][1][→][_][→][UNITS][LENG][m][ENTER] to load the value of r.

Finally, to solver for F use: [▼][SOLVE]

Press [ENTER] to return to normal calculator display.

The solution is F : 6.67259E-15_N, or F = 6.67259×10^{-15} N.

> Note: When using units in the numerical solver make sure that all the variables have the proper units, that the units are compatible, and that the equation is dimensionally homogeneous.

Example 5 - Energy losses in three pipelines in series

Take the last version of the Darcy-Weisbach equation obtained at the end of example 1, above. Namely,

`'hf= DARCY(ε/D, Q/(π*D^2/4)*D/Nu)*(L/D)*((Q/(π*D^2/4))^2/(2*g))'`

Remove the terms `hf=` from this equation, so that all what is left is the right-hand side, i.e.,

`'DARCY(ε/D, Q/(π*D^2/4)*D/Nu)*(L/D)*((Q/(π*D^2/4))^2/(2*g))'`

To accomplish this use:

[▼][EDIT] [▶][▶][▶][▶][⇐][⇐][⇐][ENTER][ENTER]

Store this expression in a variable called RHDW (Right-Hand side of Darcy-Weisbach equation), use:

[→]['][ALPHA][ALPHA][R][H][D][W][ENTER][STO▶].

We are going to use this expression to create the equation relating the head loss (energy loss per unit weight) through three pipes in parallel. Basically, the total head loss hf is equal to the sum of head losses in pipes 1, 2, and 3, i.e., hf = hf$_1$ + hf$_2$ + hf$_3$, where the individual head losses can be calculated using the right-hand side of the Darcy-Weisbach equation, i.e., the expression you just stored in RHDW. Care should be taken to identify the variables L, D, and ε with an appropriate sub-index for each pipe. Let's proceed to obtain an expression for head loss in pipe 1, as follows:

[VAR][RHDW] will copy the expression in the stack. Enter the list {'D' 'D(1)' 'L' 'L(1)' 'ε' 'ε(1)' } [ENTER][ENTER]. Save a copy into a variable called VL. Then, press enter [→][|][ENTER], to replace D with D(1), L with L(1), and ε with ε (1). When asked if you want Approximate mode, say NO. Next, save this expression in a variable called 'EQT1'.

Now, press [VAR][RHDW][VL]. Edit the list so that all the 1's are replaced by 2's. To edit, press [▼]. Once you have the list {D 'D(2)' L 'L(2)' ε 'ε(2)' } in the stack, press [→][|][ENTER], to replace D with D(2), L with L(2), and ε with ε (2). When asked if you want Approximate mode, say NO. Next, save this expression in a variable called 'EQT2'.

Next, press [VAR][RHDW][VL]. Edit the list so that all the 1's are replaced by 3's. To edit, press [▼]. Once you have the list {D 'D(3)' L 'L(3)' ε 'ε(3)' } in the stack, press [→][|

][ENTER], to replace D with D(3), L with L(3), and ε with ε (3). When asked if you want Approximate mode, say NO. Next, save this expression in a variable called 'EQT3'.

Now, we will put together the equation as follows:

[EQT1][EQT2][+][EQT3][+] [↵]   ['] [ALPHA][←][H][ALPHA][←][F][ENTER]   [↵][=]

Now, solve this long expression into EQ: ['][ALPHA][ALPHA][E][Q] [ENTER] [STO▶].

We are not quite ready yet to solve a problem for we need to provide the data corresponding to diameters, lengths, and absolute roughnesses as vectors. For example, suppose that the pipes of interest have the following properties:

Pipe No.	L (ft)	D(ft)	ε(ft)
1	500	0.5	0.0001
2	700	0.25	0.00001
3	1000	0.75	0.001

We will need to load the following vectors in the variables L, D, and ε, i.e.,

[500 700 1000] [ENTER][↵]['][ALPHA][L][STO▶]
[0.5 0.25 0.75][ENTER][↵]['][ALPHA][D][STO▶]
[0.0001 0.00001 0.001][ENTER][↵]['][ALPHA][↵][E][STO▶]

Now, launch the numerical solver: [↵][NUM.SLV][OK]. The resulting input form has entries for e, D, Q, Nu, L, g, and hf. Due to the nature of the data, we would not be able to solve for D, L, or e, which are vectors. But we can solve for any of the other variables, i.e., Q, Nu, and hf. (g is a constant). Let's use Q = 0.1 cfs, and Nu = 1.0×10^{-6} ft^2/s, and solve for hf, as follows:

[▼][▶][▶]   Skip ε and D (already loaded)
[.][1][OK] [1][EEX][6][+/-][OK]   Enter Q and Nu
[▶][3][2][.][2][OK]   Skip L, enter g
[SOLVE]   Solve for hf

Press [ON] to return to normal calculator display. The solution is: hf: 2.557612..., i.e., h_f = 2.56 ft.

Changing the order of the variables

Try using the numerical solver again with the case of three pipes in series. Press [↵][NUM.SLV][OK]. You will see associated with soft-menu key F4 the menu [VARS]. Press that key and you get a list of variables { e D Q NU L g hf }, which reflects the way that the variables are listed in the input screen. You can manipulate this list to reorder the variables or remove variables (if you don't want to change its current value in future applications). After making any changes in the list, press [OK] to return to the input screen for the current equation.

Through these example we have shown you several applications of the numerical solver to obtain solutions of single-unknown equations. The following section is dedicated to the solution of simultaneous equations.

Graphical solution of single-unknown equations

Solution of single-unknown equations can be obtained graphically if other methods fail to provide a solution. The examples we show below can be solved by other methods, but are shown in here to illustrate the use of graphs for the solution of equations.

Example 1 - Solution of a cubic equation

Suppose that we want to find some of the solutions of the equation $x^3-7x^2+36 = 0$. We can plot the function $f(x) = x^3-7x^2+36$, and check where the graph intersects the x-axis. As presented in Chapter 10, to prepare this graph follow this procedure:

- Press [↰][2D/3D], simultaneously to access to the PLOT SETUP window.

- Change TYPE to FUNCTION, if needed, by using [CHOOS].

- Press [▼] and type in the equation 'X^3-7*X^2+36'.

- Make sure the independent variable is set to 'X'.

- Press [NXT][OK] to return to normal calculator display.

- Press [↰][WIN], simultaneously, to access the PLOT window (in this case it will be called PLOT -POLAR window).

- Change the H-VIEW range to -4 to 8, by using [4][+/-][OK][8][OK], then press [AUTO] to generate the vertical range.

- Press [ERASE][DRAW] to plot the function in polar coordinates. The result is the curve shown in the diagram below:

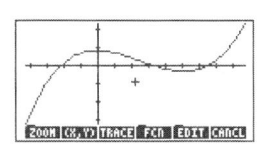

- Obviously, there are three points where the curve crosses the x-axis in the graph. To find those roots, we will use the functions in the [FCN] menu. Move the cursor near the leftmost root, and press [FCN][ROOT]. The first root is found to be Root: -2. Press [NXT]. Move the cursor near the second root, and press [ROOT]. The second root is found to be Root: 3. Press [NXT]. Finally, placing the cursor near the third root, and pressing [ROOT], produces the result Root: 6.

- Press [NXT][NXT][PICT][CANCL][ON] to return to normal calculator display.

Example 2- Clausius-Clapeyron equation

The Clausius-Clapeyron equation related the vapor pressures of a liquid, P_1 and P_2, at two temperatures T_1 and T_2, to the heat of vaporization ΔH_{vap} through

$$\ln\left(\frac{P_2}{P_1}\right) = \frac{\Delta H_{vap}}{R} \cdot \left(\frac{1}{T_2} - \frac{1}{T_1}\right)$$

where R is the universal gas constant. To find the value of this constant in the calculator use:

[↵]['][ALPHA][R][ENTER][CAT][ALPHA][C] (highlight CONST) [OK].

The calculator produces the result R = 8.31451_J/(gmol*K). This value can also be written as R = 8.31451_J/(mol*K).

The calculator recognizes the equivalence between gmol and mol, as shown here: enter gmol by using:

[ALPHA][↵][ALPHA] [ALPHA][ALPHA] [G][M][O][L] [ENTER]

The result is 1._mol.

Given P_1 = 33.86 kPa, and T_1 = 318 K, find the value of P_2 for T_2 = 298 K, if ΔH_{vap} = 36.9 kJ/mol. To make all units consistent, we will write P_1 = 33860 Pa, and ΔH_{vap} = 36900 J/mol. To obtain a graphical solution, we will have to plot the function f(X) = ln(X/33860) - (36900/8.31451)*(1/298-1/318), where X replaces P_2. The function to plot is, thus, f(X) = ln(X/33860) - 0.9366. To produce the graphical solution use the following:

- Press [↵][2D/3D], simultaneously to access to the PLOT SETUP window.

- Change TYPE to FUNCTION, if needed, by using [CHOOS].

- Press [▼] and type in the equation 'LN(X/33860)-0.9366'.

- Make sure the independent variable is set to 'X'.

- Press [NXT][OK] to return to normal calculator display.

- Press [↵][WIN], simultaneously, to access the PLOT window (in this case it will be called PLOT -POLAR window).

- Change the H-VIEW range to 80000 to 100000, by using [8][0][0][0][0][OK] [1][0][0][0][0][0][OK], then press [AUTO] to generate the vertical range.

- Press [ERASE][DRAW] to plot the function in polar coordinates. The resulting curve looks almost like a straight line crossing the X-axis at one point.

⬥ Move the cursor near the root, and press [FCN][ROOT]. The root is found to be Root: 86386.76, i.e., $P_2 = 86.386$ kPa.

⬥ Press [NXT][NXT][PICT][CANCL][ON] to return to normal calculator display.

Solving multiple equations

Many problems of science and engineering require the simultaneous solutions of more than one equation. The HP 49 G calculator provides several procedures for solving multiple equations as presented below.

Linear equation systems

Systems of linear equations were discussed in great detail in Chapter 10 (Matrices), therefore, we will not discuss them any further in this chapter.

Rational equation systems

Equations that can be re-written as polynomials or rational algebraic expressions can be solved directly by the calculator by using the function SOLVE. You need to provide the list of equations as elements of a vector. The list of variables to solve for also are provided as a vector. Make sure that the CAS is set to mode Exact before attempting a solution using this procedure. Also, the more complicated the expressions, the longer the CAS takes in solving a particular system of equations. Examples of this application follow:

Example 1 - Projectile motion

Enter the equations for two-dimensional projectile motion into a vector, i.e.,

['x = x0 + v0*COS(θ0)*t' 'y =y0+v0*SIN(θ0)*t - g*t^2/2'][ENTER]

Then enter the variables to solve for, say t and y, i.e.,

['t' 'y0'][ENTER]

To solve, first change CAS mode to Exact, then use: [↰][S.SLV][SOLVE] (second SOLVE key). After about 40 seconds, maybe more, you get as result a list:

{ 't = (x-x0)/(COS(θ0)*v0)'
'y0 = (2*COS(θ0)^2*v0^2*y+(g*x^2(2*x0*g+2*SIN(θ0))*COS(θ0)*v0^2)*x+
          (x0^2*g+2*SIN(θ0)*COS(θ0)*v0^2*x0)))/(2*COS(θ0)^2*v0^2)']}

Press [↱][EVAL] to remove the vector from the list, then [↰][PRG][TYPE][OBJ→][⇦] to get the equations listed separately in the stack.

Suppose that you were given the values: $x0 = 0$, $v0 = 10$ ft/s, $θ0 = 30°$, $g = 32.2$ ft/s^2, $x = 50$ ft, and $y = 20$ ft. (Press [ALPHA][ALPHA][D][E][G][ALPHA][ENTER] to change angular measure to degrees, if needed). You can obtain values for t and y0 from the equations in the stack, by replacing the values of the known variables using a list, as follows. Type the list, make a copy, and store the list into variable VL:

{'x0' 0 'v0' 10 'θ0' 30 'g' 32.2 'x' 50 'y' 20 }[ENTER][ENTER]
[→]['][ALPHA][ALPHA][V][L][ALPHA][STO▶]

Then, use [→][|][ENTER] to replace values in the y0 equation. (Approximate mode OK). Press [→][EVAL] to obtain a floating-point result. The result is 'y0 = 527.80'.

Press [⇦][VAR][VL] [→][|][ENTER] ([→][EVAL], if needed). The result is 't = 5.77.'

> Note: This method worked fine in this example because the unknowns t and y0 were algebraic terms in the equations. This method would not work for solving for θ0, since θ0 belongs to a transcendental term.

Example 2 - Stresses in a thick wall cylinder

Consider a thick-wall cylinder for inner and outer radius a and b, respectively, subject to an inner pressure P_i and outer pressure P_o. At any radial distance r from the cylinder's axis the normal stresses in the radial and transverse directions, σ_{rr} and $\sigma_{\theta\theta}$, respectively, are given by

$$\sigma_{\theta\theta} = \frac{a^2 \cdot P_i - b^2 \cdot P_o}{b^2 - a^2} + \frac{a^2 \cdot b^2 \cdot (P_i - P_o)}{r^2 \cdot (b^2 - a^2)},$$

$$\sigma_{rr} = \frac{a^2 \cdot P_i - b^2 \cdot P_o}{b^2 - a^2} - \frac{a^2 \cdot b^2 \cdot (P_i - P_o)}{r^2 \cdot (b^2 - a^2)}.$$

Notice that the right-hand sides of the two equations differ only in the sign between the two terms. Therefore, to write these equations in the HP 49 G calculator, I suggest you type the first term and store in a variable T1, then the second term, and store it in T2. Writing the equations afterwards will be matter of recalling the contents of T1 and T2 to the stack and adding and subtracting them. Here is how to do it with the equation writer:

Enter and store term T1:

[EQW] [↵][( )] [ALPHA][↵][A] [y^x][2] [▶] [×] [ALPHA][P][ALPHA][↵][I] [-]
[ALPHA][↵][B] [y^x][2] [▶] [×] [ALPHA][P][ALPHA][↵][O] [▶][▶][▶][▶]
[÷] [↵][( )] [ALPHA][↵][A] [y^x][2] [▶] [-][ALPHA][↵][B] [y^x][2] [▶] [ENTER]

[→]['][ALPHA][T][1][STO▶]

Enter and store term T2:

[EQW] [ALPHA][↵][A] [y^x][2] [▶] [×] [ALPHA][↵][B] [y^x][2] [▶] [×]
[↵][( )] [ALPHA][P][ALPHA][↵][I] [−] [ALPHA][P][ALPHA][↵][O] [▶] [▶][▶][▶][▶]
[÷][ALPHA][↵][R] [y^x][2] [▶] [×] [↵][( )] [ALPHA][↵][A] [y^x][2] [▶] [-][ALPHA][↵][B] [y^x][2]
[▶]
[ENTER]

[→]['][ALPHA][T][2][STO▶]

Create the equation for $\sigma_{\theta\theta}$:

[VAR][T1][T2][+] [→]['][ALPHA][→][S][ALPHA][→][T] [ENTER] [▶] [→][=]

Create the equation for σ_{rr}:

[VAR][T1][T2][-] [→]['][ALPHA][→][S][ALPHA][←][R] [ENTER] [▶] [→][=]

Put together a vector with the two equations:

[2] [←][PRG][TYPE][→ARRY]

Now, suppose that we want to solve for P_i and P_o, given a, b, r, σ_{rr}, and $\sigma_{\theta\theta}$. Thus, we enter a vector with the unknowns:

[←] [[ ]] [→]['] [ALPHA][P][ALPHA][←][I] [▶][SPC] [→]['] [ALPHA][P][ALPHA][←][O] [ENTER]

To solve for P_i and P_o, use:

[←][S.SLV][SOLVE] (second SOLVE key)

It may take the calculator a minute to produce the result:

```
{ ['Pi=-(((σθ-σr)*r^2-(σθ+σr)*a^2)/(2*a^2))'
  'Po=-(((σθ-σr)*r^2-(σθ+σr)*b^2)/(2*b^2))' ] }.
```

Using the Multiple Equation Solver (MES)

The multiple equation solver is an environment where you can solve a system of multiple equations by solving for one unknown from one equation at a time. It is not really a solver to simultaneous solutions, rather, it is a one-by-one solver of a number of related equations. To illustrate the use of the MES for solving multiple equations we present an application related to trigonometry in the next section.

Solution of triangles using the MES

In this section we use one important application of trigonometric functions: calculating the dimensions of a triangle. As a matter of fact, the word 'trigonometric' implies solution of triangles for it is composed of the Greek roots: 'Tri' (three), 'gonos' (sides), and 'metron' (measurement). Unlike the presentation in this book, trigonometric functions were first defined in terms of the dimensions of a right triangle as presented below. Solution of triangles using trigonometric functions can be extended to any type of triangle through the so-called sine and cosine laws. The solution is implemented in the calculator using the Multiple Equation Solver, or MES.

Trigonometric functions in a right triangle

Consider the right triangle ABC shown in the figure below.

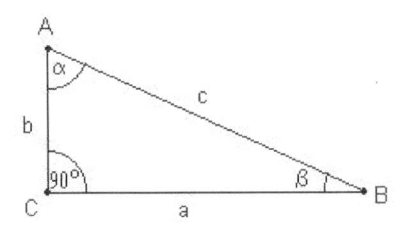

The triangle has sides of lengths a, b, c, opposite, respectively, to the angles α, β, and $\gamma = 90°$. The side opposite to the 90o angle is known as the hypotenuse of the triangle. For all triangles is it known that the sum of the interior angles is always equal to 180°. For the triangle above, then, $\alpha + \beta + 90° = 180°$, or

$$\alpha + \beta = 90°.$$

Also, from the Pythagorean theorem, we know that

$$a^2 + b^2 = c^2.$$

In terms of the lengths of the sides of this right triangle, the trigonometric functions of the angle α are defined as follows:

$$\sin\alpha = \frac{a}{c} = \frac{opposite\ side}{hypotenuse}\ ;\quad \cos\alpha = \frac{b}{c} = \frac{adjacent\ side}{hypotenuse}\ ;\quad \tan\alpha = \frac{a}{b} = \frac{opposite\ side}{adjacent\ side}\ ;$$

$$\csc\alpha = \frac{c}{a} = \frac{hypotenuse}{opposite\ side}\ ;\quad \sec\alpha = \frac{c}{b} = \frac{hypotenuse}{adjacent\ side}\ ;\quad \cot\alpha = \frac{b}{a} = \frac{adjacent\ side}{opposite\ side}\ .$$

Using the general definitions in terms of opposite side, adjacent side, and hypotenuse, we can also write, for the angle β:

$$\sin\beta = \frac{b}{c}\ ;\quad \cos\beta = \frac{a}{c}\ ;\quad \tan\beta = \frac{b}{a}\ ;\quad \cot\beta = \frac{a}{b}\ ;\quad \sec\beta = \frac{c}{a}\ ;\quad \csc\beta = \frac{c}{b}.$$

It follows from these definitions that

$$\sin\alpha = \cos\beta,\ \cos\alpha = \sin\beta,\ \tan\alpha = \cot\beta,\ \cot\alpha = \tan\beta,\ \sec\alpha = \csc\beta,\ \csc\alpha = \sec\beta.$$

Or, since, $\beta = 90° - \alpha = (\pi/2 - \alpha)^{rad}$:

$\sin \alpha = \cos (\pi/2 - \alpha)$, $\cos \alpha = \sin (\pi/2 - \alpha)$, $\tan \alpha = \cot (\pi/2 - \alpha)$,

$\cot \alpha = \tan (\pi/2 - \alpha)$, $\sec \alpha = \csc (\pi/2 - \alpha)$, $\csc \alpha = \sec (\pi/2 - \alpha)$.

Solution of triangles

Consider the triangle ABC shown in the figure below.

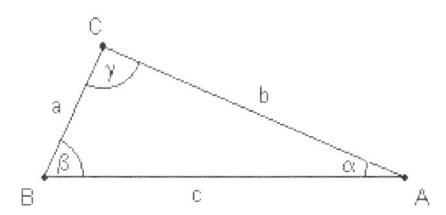

The sum of the interior angles of any triangle is always 180°, i.e.,

$$\alpha + \beta + \gamma = 180°.$$

The sine law indicates that:

$$\frac{\sin \alpha}{a} = \frac{\sin \beta}{b} = \frac{\sin \gamma}{c}.$$

The cosine law indicates that:

$$a^2 = b^2 + c^2 - 2 \cdot b \cdot c \cdot \cos \alpha,$$
$$b^2 = a^2 + c^2 - 2 \cdot a \cdot c \cdot \cos \beta,$$
$$c^2 = a^2 + b^2 - 2 \cdot a \cdot b \cdot \cos \gamma.$$

In order to solve any triangle, you need to know at least three of the following six variables: a, b, c, α, β, γ. Then, you can use the equations of the sine law, cosine law, and sum of interior angles of a triangle, to solve for the other three variables. In the solution of triangles, therefore, we distinguish the following cases:

Case I. Given three sides, find the three angles. In this case, you can find the three angles using the cosine law:

$\alpha = \cos^{-1}[(b^2+c^2-a^2)/(2 \cdot b \cdot c)]$, $\beta = \cos^{-1}[(a^2+c^2-b^2)/(2 \cdot a \cdot c)]$, and $\gamma = \cos^{-1}[(a^2+b^2-a^2)/(2 \cdot a \cdot b)]$.

Case II: Given two sides and the angle in between, find the other side and the two remaining angles. Say that we know a, b, and γ. Using the cosine law, we can find c from

$$c^2 = a^2 + b^2 - 2 \cdot a \cdot b \cdot \cos \gamma.$$

Then, using the sine law, we can get

$\alpha = \sin^{-1}(a \cdot \sin\gamma / c)$, and $\beta = \sin^{-1}(b \cdot \sin\beta / a)$, or $\beta = 180° - (\alpha + \gamma)$.

Case III: Given two sides and an angle opposite to one of the sides. Say that we know a, b, and α. From the sine law we can get,

$$\beta = \sin^{-1}(b \cdot \sin \alpha / a); \text{ and } \gamma = 180° - (\alpha + \beta).$$

Finally, we can find c from the cosine law

$$c^2 = a^2 + b^2 - 2 \cdot a \cdot b \cdot \cos \gamma.$$

Calculating the triangle's area with Heron's formula

Once the three sides are known, then the area of the triangle can be calculated with Heron's formula

$$A = \sqrt{s \cdot (s-a) \cdot (s-b) \cdot (s-c)},$$

where s is known as the semi-perimeter of the triangle, i.e.,

$$s = \frac{a+b+c}{2}.$$

Triangle solution using the HP 49 G's Multiple Equation Solver

The Multiple Equation Solver (MES) is a feature that can be used to solve two or more coupled equations. It must be pointed out, however, that the MES does not solve the equations simultaneously. Rather, it takes the known variables, and then searches in a list of equations until it finds one that can be solved for one of the unknown variables. Then, it searches for another equation that can be solved for the next unknowns, and so on, until all unknowns have been solved for.

Creating a working directory

We will use the MES to solve for triangles by creating a list of equations corresponding to the sine and cosine laws, the law of the sum of interior angles, and Heron's formula for the area. First, create a sub-directory within HOME that we will call TRIANG, and move into that directory:

[→]['][ALPHA][ALPHA][T][R][I][A][N][G][ENTER]   Enter name of directory in the stack]
[←][PRG][MEM][DIR][CRDIR]   Create directory TRIANG
[VAR][TRIAN]   Move into directory TRIANG

Entering the list of equations

Within TRIANG, enter the following list of equations either by typing them directly on the stack or by using the equation writer. (Recall that [ALPHA][→][A] produces the character α, and [ALPHA][→][B] produces the character β. The character γ needs to be [ECHO]ed from [→][CHARS].):

'SIN(α)/a = SIN(β)/b'
'SIN(α)/a = SIN(γ)/c'
'SIN(β)/b = SIN(γ)/c'
'c^2 = a^2+b^2-2*a*b*COS(γ)'
'b^2 = a^2+c^2-2*a*b*COS(β)'
'a^2 = b^2+c^2-2*b*c*COS(α)'
'α+β+γ = 180'
's = (a+b+c)/2'
'A =√ (s*(s-a)*(s-b)*(s-c))'

Then, enter the number [9], and create a list of equations by using: [←][PRG][LIST][→LIST]. Store this list in the variable EQ:

[VAR][→]['][ALPHA][ALPHA][E][Q][ENTER] [STO▶].

The variable EQ contains the list of equations that will be scanned by the MES when trying to solve for the unknowns.

Entering a window title

Next, we will create a string variable to be called TITLE to contain the string "Triangle Solution", as follows:

[→][" "]   Open double quotes in stack
[ALPHA][←][ALPHA] [ALPHA][ALPHA]   Locks keyboard into lower-case alpha.
[←][T][R][I][A][N][G][L][E][SPC]   Enter text: Triangle_
[←][S][O][L][U][T][I][O][N]   Enter text: Solution
[ENTER]   Enter string "Triangle Solution" in stack

[→][']	Open single quotes in stack
[ALPHA][ALPHA][T][I][T][L][E][ENTER]	Enter variable name 'TITLE'
[STO▶]	Store string into 'TITLE'

Creating a list of variables

Next, create a list of variable names in the stack that will look like this:

$$\{ \ a \ b \ c \ \alpha \ \beta \ \gamma \ A \ s \ \}$$

and store it in variable LVARI (List of VARIables). The list of variables represents the order in which the variables will be listed when the MES gets started. It must include all the variables in the equations, or it will not work with [MITM] (see below). Here is the sequence of keystrokes to use to prepare and store this list:

[←][{ }]	Open list brackets
[ALPHA][←][ALPHA] [ALPHA][ALPHA]	Locks keyboard into lower-case alpha
[A][SPC][B][SPC][C][SPC]	Enter a b c
[→][A][SPC][→][B][SPC][ALPHA]	Enter α β and
[→][CHARS]	Open the character list
(find γ) [ECHO][ENTER] [ENTER]	Enter γ, and enter list in stack
[→][']	Open single quotes in stack
[ALPHA][ALPHA][L][V][A][R][I][ENTER]	Enter variable name 'TITLE'
[STO▶]	Store string into 'TITLE'

Press [VAR], if needed, to get your variables menu. Your menu should show the variables

[LVARI][TITLE][EQ].

Preparing to run the MES

The next step is to activate the MES and try one sample solution. Before we do that, however, we want to set the angular units to DEGrees, if they are not already set to that, by typing:

[ALPHA][ALPHA][D][E][G][ENTER].

Next, we want to keep in the stack the contents of TITLE and LVARI, by using:

[TITLE][LVARI].

To access the MES functions, you need to access menu 113.01. This menu was accessible in the HP 48 G/G+/GX calculators through a direct keystroke sequence, but in the HP 49 G, it needs to be accessed through:

[1][1][3][.][0][1] [SPC] [ALPHA][ALPHA] [M][E][N][U] [ENTER]

You will get the soft menu keys:

[EQLIB][COLIB][MES][UTILS][   ][   ]

Press the [MES] key. The MES menu includes the following functions:

[MSOLV][MINIT][MITM][MUSER][MCALC][MROOT].

We will use only the functions MINIT, MITM, and MSOLV. These stand for:

- MINIT: MES INITialization: initializes the variables in the equations stored in EQ.

- MITM: MES' Menu Item: Takes a title from stack level 2 and the list of variables from stack level 1 and places the title atop of the MES window, and the list of variables as soft menu keys in the order indicated by the list. In the present exercise, we already have a title ("Triangle Solution") and a list of variables ({ a b c α β γ A s }) in stack levels 2 and 1, respectively, ready to activate [MITM].

- MSOLV: MES SOLVEr; activates the Multiple Equation Solver (MES) and awaits for input by the user.

Running the MES interactively

To get the MES started, press:

[MINIT]	Initialize MES variables
[MITM]	Load title and list of equations.
[MSOLV]	Starts MES solution

The MES is launched with the following list of variables available:

[a][b][c][α][β][γ]

Press [NXT] to see the next list of variables. You should have:

[A][s][   ][   ][   ][ALL]

Press [NXT] to see the third list of variables. You should see:

[MUSER][MCALC][   ][   ][   ][   ].

Press [NXT] once more to recover the first variable menu.

Let's try a simple solution of Case I, using a = 5, b = 3, c = 5. Use the following entries:

[5][a]   a:5 is listed in the top left corner of the display. Soft key label [a] changes.
[3][b]   b:3 is listed in the top left corner of the display. Soft key label [b] changes.
[5][c]   c:5 is listed in the top left corner of the display. Soft key label [c] changes.

To solve for the angles use:

[↰][α]   Calculator reports Solving for α, and shows the result α: 432.542396876.

This value is, however, unreasonable, since the sum of the interior angles of a triangle must be 180°. Because the solution is obtained numerically, it will depend on the initial value of the variable that we are solving for. This value is somewhat arbitrarily set by the calculator, and, in this case, produced too large a solution. Let's try changing the initial value to obtain a more reasonable solution as follows:

[1][0][α]   Re-initialize a to a smaller value.
[↰][α]   Calculator reports Solving for α, and shows the result α: 72.5423968763.
Next, we calculate the other two values:

[↰][β]   The result is β: 34.9152062474.
[↰][γ]   The result is γ: 72.5423968763.

You should have the values of the three angles listed in stack levels 3 through 1. Press [+] twice to check that they add indeed to 180°.

Press [NXT] to move to the next variables menu. To calculate the area use: [↰][A]. The calculator first solves for all the other variables, and then finds the area as A: 7.15454401063.

> Note: When a solution is found, the calculator reports the conditions for the solution as either Zero, or Sign Reversal. Other messages may occur if the calculator has difficulties finding a solution.

Press [VAR] to exit the MES environment.

Organizing the variables in the sub directory

Your variable menu will now contain the variables:

[A][s][g][b][a][c]

Press [NXT] to see the second variables menu:

[b][a][Mpar][LVARI][TITLE][EQ].

Variables corresponding to all the variables in the equations in EQ have been created. There is also a new variable called Mpar (MES parameters), which contains information regarding the setting up of the MES for this particular set of equations. If you use [↦][Mpar] to see the contents of the variable Mpar, you will get the cryptic message: Library Data. The meaning of this is that the MES parameters are coded in a binary file, which cannot be accessed by the editor.

Next, we want to place them in the menu labels in a different order than the one listed above, by following these steps:

1. Create a list containing { EQ Mpar LVARI TITLE }, by using:

   [↰][{ }][EQ][Mpar][LVARI][TITLE][ENTER].

2. Place contents of LVARI in the stack, by using: [LVARI].

3. Join the two lists by pressing [+].

4. Use the following keystroke sequence to order the variables as shown in the list in stack level 1:

   [↰][PRG][MEM][DIR][NXT][ORDER]

5. Press [VAR] to recover your variables list. It should now look like this:

[EQ][Mpar][LVARI][TITLE][a][b]

6. Press [NXT], to see:

[c][α][β][γ][A][s]

7. Press [NXT] to recover the first variable menu.

Programming the MES triangle solution using User RPL

To facilitate activating the MES for future solutions, we will create a program that will load the MES with a single keystroke. The program should look like this:

<< DEG MINIT TITLE LVARI MITM MSOLVR >>

and can be typed in by using:

[↱][<< >>]	Opens the program symbol
[ALPHA][ALPHA]	Locks alphanumeric keyboard
[D][E][G][SPC]	Type in DEG (to ensure that angular units are set to DEGrees)
[M][I][N][I][T][SPC]	Type in MINIT_
[ALPHA]	Unlocks alphanumeric keyboard
[TITLE][SPC]	List the name of variable TITLE in the program
[LVARI]	List the name of variable LVARI in the program
[ALPHA][ALPHA]	Locks alphanumeric keyboard
[M][I][T][M][SPC]	Type in MITM_
[M][S][O][L][V][R]	Type in MSOLVR
[ENTER]	Enter program in stack

Store the program in a variable called TRISOL, for TRIangle SOLution, by using:

[↱]['] [ALPHA][ALPHA] [T][R][I][S][O][L][ENTER] [STO▶].

Press [VAR], if needed, to recover your list of variables. A soft key label [TRISO] should be available in your menu.

Running the program - solution examples

To run the program, press the [TRISO] soft menu key. You will now have the MES menu corresponding to the triangle solution. Let's try examples of the three cases listed earlier for triangle solution.

Example 1 - Case I - Right triangle: Use a = 3, b = 4, c = 5. Here is the solution sequence:

[4][a][4][b][5][c]   To enter data
[↰][α]   The result is α: 36.8698976458
[↰][β]   The result is β: 53.1301023541.

[←][γ]	The result is γ: 90.
[NXT]	To move to the next variables menu.
[←][A]	The result is A: 6.
[NXT][NXT]	To move to the next variables menu.

Example 2 - Case I - Any type of triangle: Use a = 3, b = 4, c = 6. The solution procedure used here consists of solving for all variables at once, and then recalling the solutions to the stack:

[VAR][TRISO]	To clear up data and re-start MES
[3][a][4][b][5][c]	To enter data
[NXT]	To move to the next variables menu.
[←][ALL]	The calculator takes its time to solve for all the unknowns.
[→][ALL]	The calculator shows the solution in the following screen:

```
#### Triangle Solution ####
γ: 117.279612736
β: 36.3360575147
α: 26.3843297495
s: 6.5
A: 5.33268225193

[VALU■][EQNS][PRINT]…[EXIT]
```

At the bottom of the screen, you will have the soft menu keys:

[VALU■][EQNS][PRINT][   ][   ][EXIT]

The square dot in [VALU■] indicates that the values of the variables, rather than the equations from which they were solved, are shown in the display. To see the equations used in the solution of each variable, press the [EQNS] soft menu key. The display will now look like this:

```
#### Triangle Solution ####
γ: 'c^2=a^2+b^2-2*a*…
β: 'b^2=a^2+c^2-2*a*…
α: 'a^2=b^2+c^2-2*b*…
s: 's=(a+b+c)/2'
A: 'A=√(s*(s-a)*(s-b…'

[VALU][EQNS■][PRINT]…[EXIT]
```

The soft menu key [PRINT] is used to print the screen in a printer, if available. And [EXIT] returns you to the MES environment for a new solution, if needed. To return to normal calculator display, press [VAR].

The following table of triangle solutions shows the data input in bold face and the solution in italics. Try running the program with these inputs to verify the solutions. Please remember to press [VAR][TRISO] at the end of each solution to clear up variables and start the MES solution again. Otherwise, you may carry over information from the previous solution that may wreck havoc with your current calculations.

a	b	c	$\alpha(°)$	$\beta(°)$	$\gamma(°)$	A
2.5	6.9837	7.2	20.299	75	84.7707	8.6933
7.2	8.5	14.26	22.6162	27	130.3837	23.3086
21.92	17.5	13.2	90	52.97	37.03	115.5
41.92	23	29.6	75	32	73	328.81
10.27	3.26	10.5	77	18	85	16.66
17	25	32	31.79	50.78	97.44	210.71

Adding an INFO button to your directory

An information button can be useful for your directory to help you remember the operation of the functions in the directory. In this directory, all we need to remember is to press [TRISO] to get a triangle solution started. You may want to type in the following program: <<"Press [TRISO] to start." MSGBOX >>, and store it in a variable called INFO. As a result, the first variable in your directory will be the INFO button.

Velocity and acceleration in polar coordinates - solution using the calculator's MES

Two-dimensional particle motion in polar coordinates often involves determining the radial and transverse components of the velocity and acceleration of the particle given r, r' = dr/dt, r" = d²r/dt², θ, θ' = dθ /dt, and, θ" = d²θ/dt². The following equations are used:

$$v_r = r' \qquad v_\theta = r\theta' \qquad a_r = r'' - r\theta'^2 \qquad a_\theta = r\theta'' + 2r'\theta'$$

Create a subdirectory called POLC (POLar Coordinates), which we will use to calculate velocities and accelerations in polar coordinates. Within that subdirectory, enter the following variables:

Program or value variable:	store into
<< PEQ STEQ MINIT NAME LIST MITM MSOLVR >>	SOLVEP
"vel. & acc. polar coord."	NAME
{ r rD rDD θD θDD vr vθ v ar aθ a }	LIST
{ 'vr = rD' 'vθ = r*θD' 'v = √(vr^2 + vθ^2)' 'ar = rDD - r*θD^2' 'aθ = r*θDD + 2*rD*θD' 'a = √(ar^2 + aθ^2)' }	PEQ

An explanation of the variables follows:

SOLVEP = a program that triggers the HP48G or GX multiple equation solver for the particular set of equations stored in variable PEQ;

NAME = a variable storing the name of the multiple equation solver, namely, "vel. & acc. polar coord.";

LIST = a list of the variable used in the calculations, placed in the order we want them to show up in the multiple equation solver environment;

PEQ = list of equations to be solved, corresponding to the radial and transverse components of velocity (v_r, $v\theta$) and acceleration (a_r, $a\theta$) in polar coordinates, as well as equations to calculate the magnitude of the velocity (v) and the acceleration (a) when the polar components are known.

r, rD, rDD = r (radial coordinate), r-dot (first derivative of r), r-double dot (second derivative of r).

θD, θDD = θ-dot (first derivative of θ), θ-double dot (second derivative of θ).

Suppose you are given the following information:

$$r = 2.5, rD = 0.5, rDD = -1.5, \theta D = 2.3, \theta DD = -6.5.$$

and you are asked to find vr, vθ, ar, aθ, v, and a.

⬇ Start the multiple equation solver by pressing [VAR][SOLVE]. The calculator produces a screen labeled , "vel. & acc. polar coord.", that looks as follows:

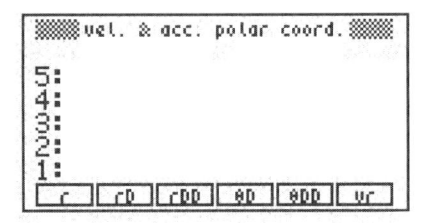

⬇ To enter the values of the known variables, just type the value and press the button corresponding to the variable to be entered. Use the following keystrokes:

[2][.][5] [r] [.][5][rD] [1][.][5][+/-][rDD] [2][.][3][θD][6][.][5][+/-][θDD].

Notice that after you enter a particular value, the calculator displays the variable and its value in the upper left corner of the display.

✦ We have now entered the known variables. To calculate the unknowns we can proceed in two ways:

a). Solve for individual variables, for example, [←][vr] gives vr: 0.500. Press [NXT] [←][vθ] to get vθ : 5.750 , and so on. The remaining results are v: 5.77169819031; ar: -14.725; aθ: -13.95; and a: 20.2836911089.;

or,

b). Solve for all variables at once, by pressing [←][ALL]. The calculator will flash the solutions as it finds them. When the calculator stops, you can press [→][ALL] to list all results. For this case we have:

Pressing the soft-menu key [EQNS] will let you know the equations used to solve for each of the values in the screen:

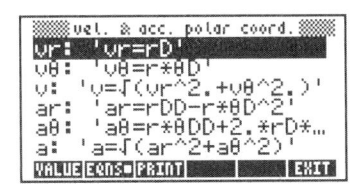

To use a new set of values press, either [EXIT][ALL][NEXT][NEXT], or [VAR][SOLVE].

Let's try another example using r = 2.5, vr = rD = -0.5, rDD = 1.5, v = 3.0, a = 25.0. Find, θD , θDD, vθ, ar, and aθ. Enter the following values: [2][.][5][r][.][5][+/-][ENTER][ENTER][rD][vr]
[1][.][5][rDD][3][NXT] [v] [2][5][a]   To solve and show results, press [←][ALL] (wait) [→][ALL]. The results are given in the following screen shot:

Using the SOLVESYS library for simultaneous equations

Systems of non-linear equations can be solved using the library SOLVESYS, developed by Sune Bredahl, a computer scientist from Denmark, and available in the Internet at: http://www.hp.org

Installing the SOLVESYS library

1. Make sure you are in the HOME directory. Download the variable that contains the library SOLVESYS from the INTERNET, and place it in your calculator in a variable that you could call SOLVES. To install SOLVESYS use these keystrokes:

Keystrokes	Display shows
[→][SOLVES] [0][STO]	Library 1550: SOL.... (*) (blank display)

    (*) 1550 is an ID number for your library selected by the library's author.

2. To check if installation was successful, enter: [←][LIB][:0:]. There should be a variable labeled [1550] in port 0.

3. The next step is to load the library and attach it to the HOME directory, by pressing, simultaneously, [ON][F4 D]. A menu will be provided. Press [Q] to reboot the calculator.

Example 1 - two conic equations

Let's use SOLVESYS to solve the following system of non-linear equations given as a list:

$$\{ \, '(X-1)^2+(Y-2)^2=3' \quad 'X^2/4+Y^2/3=1' \, \}.$$

Now, store this list into variable [EQ] by using:

$$[←][\, EQ \,]$$

Next, start the SOLVESYS library by using:

$$[→][LIBRARY][SOLVE][SOLVE]$$

The system is already loaded, as shown in the display.

⚓ Press [OK] to initialize values. The next screen (CURRENT VALUES) is used to enter initial values of X and Y, with default values of 1.0. Let's use those values to solve the system.
⚓ Press [SOLVE]. The bottom of the screen shows a value eq: that keeps changing (hopefully becoming smaller). This value is the error in the solution. After a while, you get a <!>Zero message.
⚓ Press [OK] to see the values of X and Y shown that form the solution. X = 1.906, Y = 0.524. You need to press [EDIT] to see any of the highlighted values in the solution.

⬥ Press [INFO] to obtain information about the solution process. You may get a message box similar to this:

```
m/n:  2/2
Δx:   5.62E-4
eq:   1.28E-6
lsq:
```

m/n stands for the number of equations (m) and the number of unknowns (n). For exact solutions m = n. Δx represents the increment in the x, y values obtained in the last step of the solution. Eq represents the error in the solution. And lsq is used when finding a least-square solution (also included with SOLVESYS).

⬥ Press [OK][ON] to quit the SOLVESYS library. The most current solution will be shown in the stack.

To get a different solution you may want to start by entering different values in the START VALUES screen. For example, try using X = -1, Y = -1.

⬥ Start the SOLVESYS library by using: [↪][LIBRARY][SOLVE][SOLVE][OK]
⬥ Enter [1][+/-][OK] [1][+/-][OK].
⬥ Press [SOLVE]. Press [OK] when advised that a solution has been found. This solution converged to X = -0.6910, Y = 1.6253, with an error ε = 5.62E-8.
⬥ Press [OK] to leave the SOLVE screen.
⬥ Press [ON] to quit the SOLVESYS library.

Press [VAR] to recover your variable menu.

Example 2 - Manning's equation for circular cross-section

Manning's equation is used to solve for uniform flow (constant-depth flow) in an open channel. The equation is written as,

$$Q = (C/n)(A^{5/3}/P^{2/3})S^{1/2},$$

where Q = flow rate or volumetric discharge [m^3/s, ft^3/s = cfs], A = cross-sectional area [m^2, ft^2], P = wetted perimeter [m, ft], S = channel bed slope [m/m, ft/ft, i.e., dimensionless], C = units coefficient [dimensionless, C = 1.0 for the S.I. system, C = 1.489 for the English system], n = Manning's resistance coefficient [dimensionless, a function of the channel surface roughness].

For a circular cross-section, if Y = flow depth [m, ft], D = diameter of the circular cross-section [m, ft], and β = central half-angle [radians], then

$$Y = (D/2)(1 - \cos β),$$

$$A = (D^2/4)(β - \cos β \sin β),$$

and,

$$P = \beta D.$$

Replacing the expressions for A and P into Manning's equation and raising both sides to the third power results in:

$$((D^2/4)(\beta - \cos\beta \sin\beta))^5/(\beta D)^2 = (Q \cdot n/C \cdot \sqrt{S})^3.$$

To solve for values of Y and β, given Q = 1 cfs, C = 1.489, n = 0.012, D = 3 ft, and S = 0.00001, enter the following equations:

'Y=(D/2)*(1-cos(ß))'

and

'(D^2/4*(ß-cos(ß)*sin(ß))^5/(ß*D)^2=(Q*n/C*√S)^3'

in levels 1 and 2 of the stack. Then enter:

[2][PRG][TYPE][→LIST]

to create the list:

{ 'Y=(D/2)*(1-cos(ß))' '(D^2/4*(ß-cos(ß)*sin(ß))^5/(ß*D)^2=(Q*n/C*√S)^3'}.

Store the list in variable EQ:

['][α][E][α][Q] [STO].

To use SOLVESYS, use the following keystrokes: [→][LIBRARY][SOLVE][SOLVE][OK].

Enter the values of the constants in the problem within square brackets:

[←][[ ]][1][.][4][8][9][OK]   [←][[ ]][3][OK]   [←][[ ]][1][OK]   [←][[ ]][0][.][0][0][0][0][1][OK]

Leave the value of 1 for Y and skip to the N: field to enter

[←][[ ]][0][.][0][1][2][OK].

Leaving the value of 1 for ß, also, press [SOLVE]. The first screen shows that there are two equations, with two unknowns and 5 constants. Next, you get a screen showing the values of the variables as they change, as well as the current value of the error, ε. The iterations stop when ε = 1.9200×10^{-9}. Press [OK] to see the solution as:

Y = 1.3869, ß = 1.4953.

Press [QUIT] to end the library operation. The solution will be listed in the stack. Also, the library would have created variables corresponding to the constant values used in the solution, namely, N, S, D, Q, and C. These variables are always created in the HOME directory, regardless of which directory was used to launch the library. Press [←] [HOME] [VAR] to see these variables. To delete the variables, use:

[←][{}][N][S][D][Q][C][ENTER].

The following list will be shown in stack level 1: { n S D Q C }. To purge the variables enter: [TOOL][PURGE].

Graphical solutions of two simultaneous equations

You can use the graphics environment to find the solution of two simultaneous equations. The solution will be simple the coordinates of the point of intersection of two curves. For example, to solve

$$\tan(x) = x,$$

you could plot the functions $f_1(X) = \text{TAN}(X)$, and $f_2(X) = X$, and find their intersections. Here is how:

- Press [←][2D/3D], simultaneously to access to the PLOT SETUP window.

- Change TYPE to FUNCTION, if needed, by using [CHOOS].

- Press [▼] and type in the equation list { 'TAN(X)' 'X' }.

- Make sure the independent variable is set to 'X'.

- Press [NXT][OK] to return to normal calculator display.

- Press [←][WIN], simultaneously, to access the PLOT window (in this case it will be called PLOT -POLAR window).

- Change both the H-VIEW and V-VIEW ranges to -5 to 5.

- Press [ERASE][DRAW] to plot the function in polar coordinates. The resulting plot looks as follows:

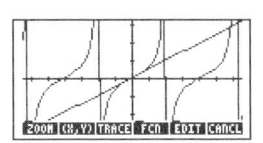

- Notice that there are vertical lines that represent asymptotes. These are not part of the graph, but show up because TAN(X) goes to at certain values of X. There are at least three roots in this interval, however. The one in the center is X = 0. To find the one to the right, move the cursor near the intersection of the two lines, and press [FCN][ISECT]. The result is I-sect: (4.4934, 4.4934). Thus, the first positive solution of $\tan(x) = x$ is $x = 4.4934$.

- Press [NXT][NXT][PICT][CANCL][ON] to return to normal calculator display.

REFERENCES – Vol. I only

Devlin, Keith, 1998, "The Language of Mathematics," W.H. Freeman and Company, New York.

Heath, M. T., 1997, "Scientific Computing: An Introductory Survey," WCB McGraw-Hill, Boston, Mass.

Newland, D.E., 1993, "An Introduction to Random Vibrations, Spectral & Wavelet Analysis - Third Edition," Longman Scientific and Technical, New York.

Tinker, M. and R. Lambourne, 2000, "Further Mathematics for the Physical Sciences," John Wiley & Sons, LTD., Chichester, U.K.

REFERENCES – Vol. II only

Farlow, Stanley J., 1982, "Partial Differential Equations for Scientists and Engineers," Dover Publications Inc., New York.

Friedman, B., 1956, "Principles and Techniques of Applied Mathematics," (reissued 1990), Dover Publications Inc., New York.

Kottegoda, N. T., and R. Rosso, 1997, "Probability, Statistics, and Reliability for Civil and Environmental Engineers," The Mc-Graw Hill Companies, Inc., New York.

Kreysig, E., 1983, "Advanced Engineering Mathematics - Fifth Edition," John Wiley & Sons, New York.

REFERENCES – For both Vols. I and II

Gullberg, J., 1997, "Mathematics - From the Birth of Numbers," W. W. Norton & Company, New York.

Harris, J.W., and H. Stocker, 1998, "Handbook of Mathematics and Computational Science," Springer, New York.

Hewlett Packard Co., 1999, HP 49 G GRAPHING CALCULATOR USER'S GUIDE.

Hewlett Packard Co., 2000, HP 49 G GRAPHING CALCULATOR ADVANCED USER'S GUIDE

INDEX

!

!, 51

#

∂, 24, 28
→ARRY, 172
→COL, 219
→DIAG, 214
→GROB, 350, 351
→LCD, 352
→NUM, 29
→OBJ, 353
→ROW, 221
→STR, 353
→TAG, 109
→V2, 175
→V3, 175
←WID, 170

%

%, 50
%CH, 50
%T, 50

::

::, 28

|

|, 21, 146

+

+COL, 171
+ROW, 171

1

√x, 25
10^x, 25

3

3D menu, 321

A

ABCUV, 159
ABS, 25, 50, 60, 175, 229
ACOS, 24
ACOS2S, 150
ACOSH, 49
ADD, 68
ADDTM, 156
adjacent side, 390
ALG, 27, 144
Algebra, 134
algebraic, 142
Algebraic, 10, 16
algebraic equations, 365
algebraic expressions for polynomials, 369
algebraic mode, 7, 9, 10, 12, 15
algebraic object, 15, 142
algorithm, 275
ALPHA, 26
Alphabetic characters, 18
alphabetic key, 18
amortization, 372
AND, 118
angle mode, 35
ANIMATE, 348, 349
annual interest rate, 370
ANS, 29
anti-symmetric matrix, 223
ARC, 336
ARG, 25, 60
ARITH, 27
arrays, 15
ASIN, 24
ASIN2C, 150
ASIN2T, 150
ASINH, 49
Assembler language, 9
ATAN, 24
ATAN2S, 150
ATANH, 49
ATICK, 320, 321
augmented matrix, 242
AUTO, 316

auxiliary variables, 376
AXES, 321, 326
AXL, 198, 237
AXM, 237
AXQ, 258

B

backward substitution, 241
bar plots, 302
BASE, 28
BASE menu, 51, 343
Base-10 logarithm, 48
BEGIN, 19
BESTFI, 326
beta distribution, 82
BIN, 343
binary digits, 350
Binary integers, 15
binary system, 342
Binomial distribution, 79
BIT and BYTE menus, 344
bits, 350
BLANK, 351
BOX, 332, 336
BOXZ, 346
Branching, 119
Bredahl, Sune, 402

C

C→PX, 336
CALC, 27
Calculator Algebraic System, 45
Cartesian representation of complex numbers, 59
CAS, 20
CAS settings, 46
CASE, 124
CASINFO, 163
CAT, 22
cdf, 54, 78
CDF, 379
CEIL, 50
center of mass, 183
CENTR, 320
Central Processing Unit, 7
CHAR, 354
characteristic polynomial of a matrix, 253
characters list, 354
CHARS, 22, 354
CHARS menu, 353
Chinese Remainder Theorem, 160
CHINREM, 160
Chi-squared (χ^2) distribution, 57

CIRCL, 333
Clausius-Clapeyron equation, 386
CLEAR, 21, 24
CMD, 22
CMPLX, 27
CMPLX menus, 60
CNCT, 327
CNRM, 230
CNTR, 346
COL, 218
-COL, 171
COL-, 220
COL→, 219
COL+, 219
column norm of a matrix, 230
column vectors, 196
COMB, 51
combinations, 52
Combinatorics, 52
command catalog, 41
complex number, 14, 15

complex numbers, 59

Complex numbers, 15
CON, 212
concatenate two lists, 68
COND, 231
condition number of a matrix, 231
congruence, 154
conic curve plots, 295
CONJ, 60
CONST, 41
CONT, 29
continuous random variables, 54
CONVERT, 27
coordinate system, 35
COPY, 21
correlation coefficient, 277
COS, 24
COSH, 49
cosine law, 391
Cramer's rule, 235
CROSS, 175
cross product, 178
CSWP, 220
cubic equation, 385
cumulative distribution function, 78, 379
Cumulative Distribution Function, 54
CUSTOM, 20
CUT, 22
CYLIN, 176

D

D→R, 50

DARCY(ε/D, Re), 377
Darcy-Weisbach equation, 377
DATA menu, 325
DBUG, 103
Debugging a program, 103
DEC, 343
decimal system, 342
DEF, 27
DEL, 24, 171, 333
Deleting variables, 30
DET, 233
De-tagging, 110
determinant of a matrix, 180, 233
DIAG→, 214
diagonal matrix, 200
differential equation plots, 299
Dimensional analysis, 263
dimensional homogeneity, 263
Dirac's delta function, 269
directed segment, 176
direction cosines, 266
directories, 9, 16
Directories, 15
Discrete probability distributions, 78
discussion group, 8
DISP, 20
DIV2, 160
DIV2M, 156
DIVMO, 156
DO, 130
DOLIST, 93
DOSUBS, 94
DOT, 175
DOT-, 332
dot vector product, 178
DOT+, 332
DRAW, 317
Drawing commands, 334
DRAX, 317
DROP, 23
DUP, 21

E

ECHO, 354
ECHO1, 354
EDIT, 20
EEX, 25
EGCD, 160
EGV, 254
EGVL, 253
eigenvalue equation, 253
eigenvalues, 231, 253
eigenvectors, 231, 253
Einstein's summation convention, 202

electrical circuit, 260
elements of a matrix, 200
END, 20
engineering economics, 370
Engineering number format, 33
ENTER, 29
EPS, 167
EPSX0, 167
EQ, 76, 317, 328
equation writer, 11, 142
Equation Writer, 12
EQW, 23, 142
ERASE, 317, 333
Euclidean norm of a matrix, 230
Euler formula, 59
EVAL, 23
Evaluating a polynomial or algebraic, 369
e^x, 24
EXP&LN, 26
EXPAN, 144
EXPFI, 326
EXPLN, 147
EXPM, 49, 147
Exponential, 47
exponential distribution, 82
EYEPT, 323

F

FACTOR, 144
Factorials, 52
FACTORMOD, 158
FANNING(ε/D, Re), 377
Fast 3D plots, 306
FCOEF, 153
FILES, 19
FINANCE, 26, 372
Financial calculations, 370
finite arithmetic, 154
flag, 13, 14
FLAG menu, 326
FLAGS, 20
FLOOR, 50
flow in a pipe, 377
FOR...NEXT, 129
FOR...STEP, 130
forward elimination, 241
FP, 50
Frobenius norm, 229
FROOTS, 154
full pivoting, 244
function for pipe flow, 377
function of one variable, 69
FUNCTION plots, 284
functions, 67

fundamental theorem of algebra, 368
FV, 373

G

gamma distribution, 81
Gamma function, 53
GAUSS, 259
Gaussian elimination, 241
Gauss-Jordan elimination, 243
GCD, 161
GCDMO, 156
General solutions, 13
Geometric Distribution, 80
GET, 85
GETI, 85
Global, 15
global variable, 72
GO→, 170
GOR, 351
GOTO, 171
Graphical solution of equations, 385
graphics animation, 347
Graphics objects, 15
Graphics OR, 351
Graphics XOR, 351
Greek letters, 32
Gridmap plots, 311
GROB, 350
GROB menu, 351
GXOR, 351

H

HADAMARD, 237
HALFTAN, 150
HEAD, 85
HERMITE, 161
Heron's formula, 392
HEX, 343
hexadecimal, 343
HILBERT, 216
HIST], 22
histogram plots, 302
HOME directory, 7, 16, 17
Hooke's law for stress and strain, 374
HORNER, 162
HP Basic, 9
HYP menu, 49
hyperbolic functions, 49
Hypergeometric Distribution, 80
hypotenuse, 390
HZIN, 346
HZOUT, 346

I

i, 21
I%YR, 370
ideal gas law, 43, 106
identity matrix, 200
IDN, 212
IF...THEN...ELSE...END, 121
IF...THEN...END, 119
IFTE, 74
IM, 60
imaginary part, 59
imaginary unit, 59
INDEP, 319
INFO, 317, 322, 325
input string, 102
integrals, 82
Interactive drawing, 331
Interactive input, 100
interactive plots, 328
internal vector product, 178
inverse function, 289
inverse matrix, 208
inverse trigonometric functions, 135
INVMO, 156
IP, 50
ISOL, 365

J

JORDAN, 254

K

Kronecker's delta function, 201

L

LABEL, 316, 333
LAGRANGE, 163
LCD→, 352
LCM, 164
LCXM, 240
least-square method, 225
LEGENDRE, 164
LIB, 27
LibMaker, 9, 10
library, 9, 10
LIN, 147
LINE, 332, 335
linear equation systems, 224
LINFI, 326
LINSOLVE, 251
list, 13, 14, 15, 16, 17
Lists, 67

LN, 24
LNCOL, 147
LNP1, 49, 147
Local, 15
local variable, 72
LOG, 25
logarithm, 136
LOGFL, 326
LOGIC menu, 344
Logical operators, 118
LQ, 257
LQ factorization of a matrix, 257
LSQ, 238, 251
LU, 256
LU matrix decomposition, 229

M

MAD, 239, 255
main diagonal of a matrix, 200
Manning's equation, 97
MANT, 50
manuals, 8
MARK, 332
mass distribution function, 78
mathematical constants, 38
MATRICES, 27
matrix, 14, 15
matrix CREATE menu, 216
matrix *decomposition*, 229
matrix MAKE menu, 209
matrix writer, 169
MAX, 50
mdf, 78
mean of a sample, 86
MENU, 333
MES, 389
message box, 112
MIN, 50
MOD, 50, 156
MODE, 19
MODL, 326
MODST, 156
modular inverse, 158
modular programming, 356
MODULO, 156, 163
modulus, 156
Mohr's circle, 355
moment of a force, 188
motion in polar coordinates, 399
moving average, 94
MTH, 23
MTH menu, 49
MTRW, 169
MTWR, 23
multiple linear fitting, 272

MULTM, 156

N

Natural logarithm, 47
natural river cross section, 337
NEG, 60
Nested IF, 122
non-singular matrix, 231
NORM, 228
Normal distribution, 54
normal stress, 267
normal unit vector, 266
NOT, 118
Notation, 7, *10*
NUM, 354
NUM.SLV, 26, 224
NUM.SLV menu, 367
Number Format, 33
number of periods per year, 370
numerical solver, 224
numerical solver menu, 367
NUMX, 324
NUMY, 324
NXT, 22

O

OBJ→], 69
object-oriented programming, 132
Objects, 15
OCT, 343
OFF, 29
OOP, 132
OPER, 237
operands, 11, 12
Operating Mode, 10, 13
operating system, 7, 8
operator, 11, 12
opposite side, 390
OR, 118
Orthogonal matrices, 230

P

Parametric plots in the plane, 296
PARTFRAC, 153
partial pivoting, 244
PASTE, 22
Payment at begin, 371
Payment at beginning, 371
payment at the end, 370
PCAR, 253
PCOEF, 165
pdf, 54, 379

PDIM, 335
periods for PVM, 370
PERM, 51
permutation matrix, 244
permutations, 52
PEVAL, 167
Physical constants, 39
PICT, 334, 335, 350
PICT→, 334
PICTURE, 336
pivot, 244
PIX?, 336
Pixels coordinates, 345
PIXOFF, 336
PIXON, 336
planar motion, 191
PLOT, 282
PLOT menu, 315
PLOT SETUP, 282
PLOT WINDOW, 283
Poisson Distribution, 80
POLAR plots, 294
polar representation of complex numbers, 59
polynomial coefficients, 368
Polynomial fitting, 273
Ports, 10
POS, 85
position vector, 182
Power, 47
Powers of 10, 48
powers of ten, 35
POWMO, 156
PPAR, 288
Present Value, 370
PRG, 22
principal stresses, 356
Principal stresses, 270
Principal value, 13, 14
PROB menu, 51
probability density function, 54, 379
probability functions, 52
program, 9, 10, 12, 13, 15, 17
Program loops, 125
program-generated plots, 329
programs, 67
Projectile motion, 387
projection of a vector, 177
PROOT, 166
PROPFAC, 153
Pr-Surface plots, 312
PRV, 22
Ps-Contour plots, 309
pseudo-random number generator, 53
PTAYL, 166
PTYPE, 318
PTYPE m, 322

PTYPE menu, 325
PURGE, 21
PUT, 85
PUTI, 85
PV, 370
PVIEW, 336
PWRFI, 326
PX→C, 336
Pythagorean theorem, 390

Q

QR, 257
QR factorization of a matrix, 257
QUADF, 258
quadratic equation, 13, 14
quadratic form, 258
QUOTIENT, 167
QXA, 258

R

R→D, 50
RAND, 51
random numbers, 52, 53
RANK, 232
rank of a matrix, 232
RANM, 213
Rational equation systems, 387
RCI, 222
RCIJ, 222
RCL, 20, 22
RDM, 212
RE, 60
REAL menu, 50
real numbers, 15, 45
real part, 59
REALASSUME, 163
RECT, 176
Re-dimensioning, 213
register, 11, 12, 13
Relational operators, 117
relative acceleration, 191
relative position vector, 186
relative velocity, 191
REMAINDER, 167
REPL, 214, 334
RES, 320
RESET, 321, 324, 326
resultant, 184
REVLI, 68
right-hand rule for vector product, 178
rigid body, 191
RND, 50
RNRM, 230

ROM, 7, 8
ROW, 220
-ROW, 171
ROW-, 221
row norm of a matrix, 230
row vector, 199
row vectors, 196
ROW→, 221
ROW+, 221
Row-Reduced Echelon Form of a matrix, 252
RPN, 7, 10, 11, 12, 13, 15
RREF, 252
RSD, 239
RSWP, 222
RUN, 103

S

S.SLV, 26
S.SLV menu, 366
SAME, 119
Saving a graph, 286
SCALE, 320
SCALEH, 320
SCALEW, 320
scatter plots, 302
SCHUR, 257
Schur *decomposition* of a matrix, 257
Scientific number format, 33
SEQ, 95
sequential programming, 96
shear stress, 267
SIGN, 50, 60
significative figures, 34
SIMP2, 152
SIMU, 327
SIN, 24
SINCOS, 151
sine law, 391
singular matrix, 231
Singular Value Decomposition, 230, 256
singular values, 230
singular vectors, 230
SINH, 49
SIZE, 85, 352
slope field plots, 305
SNRM, 230
soft menu keys, 7, 16, 17
solutions to a polynomial equation, 368
SOLVE, 365
Solve finance..., 370
Solve poly..., 367
Solvesys, 9
SOLVESYS library, 402
SOLVEX, 366
solving multiple equations, 387

SORT, 68
Specific energy in open channel flow, 375
spectral norm of a matrix, 230
spectral radius of a matrix, 231
SPHER, 176
square matrix, 200
Square power, 47
Square root, 47
SRAD, 231
SST↓, 104
STACK, 21
stack levels, 11, 12
standard deviation of a sample, 86
standard format, 33
START...NEXT, 126
START...STEP, 128
STAT, 27
STAT menu, 324
statistics from grouped data, 87
STOP], 22
STREAM, 94
Stress at a point in a solid, 266
stress tensor, 268
Stresses in a thick wall cylinder, 388
string concatenation, 113, 353
Strings, 353
Structural mechanics, 261
Student-t distribution, 56
SUB, 213, 333
sub-directories, 16
SUBS, 145
Substitution of algebraic expressions, 145
SUBTM, 156
sum of square errors, 277
sum of squared totals, 277
summations, 79
SVD, 230, 256
SVL, 257
SWAP, 21
SYLVESTER, 259
SYMB, 23
Symbolic Solver menu, 366
symmetric matrix, 223
system flag, 13, 14
System RPL, 9
Systems of linear equations, 387

T

table of values, 299
TAIL, 85
TAN, 24
TAN2SC, 151
TAN2SC2, 151
TANH, 49
TCHEBYCHEFF, 167

TCOLLECT, 151
tensor, 201
TEXPA, 147
textbook-like appearance, 48
The arrow keys, 19
The F distribution, 57
The multiple equation solver, 389
three-dimensional graphics, 327
TIME, 26
time value of money, 370
titled FILE MANAGER, 16
TLIN, 151
TLINE, 332, 336
TOOL, 20
TPAR, 76
TRACE, 236
trace of matrix, 236
TRAN, 237
transcendental functions, 134
transpose of a matrix, 201
Triangle solution, 392
triangles, 391
tridiagonal matrix, 201
TRIG, 26, 151
TRIGCOS, 152
trigonometric functions, 134
Trigonometric functions, 48
Trigonometric identities, 135
TRIGSIN, 152
TRIGTAN, 152
TRN, 211
TRNC, 50
Truth plots, 301
truth table, 118
TSIMP, 147
two-dimensional graphics, 327
TYPE, 109
TYPE menu, 353

U

UNDO, 22
unit imaginary number, 14, 15
Unit objects, 15
unit vector, 177
units, 42
UNITS, 27
universal gravitation, 382
UPDIR, 21
upper-tail probability, 380
Upper-tail probability Chi square, 379
Upper-tail probability F distribution, 379
Upper-tail probability normal, 379
Upper-tail probability Student's t, 379
User RPL, 9, 10, 15
Utility keys, 19

UTPC, 51, 379
UTPF, 51, 379
UTPN, 51, 379
UTPT, 51, 379

V

V→, 175
VANDERMONDE, 215
VAR, 21
Variable Scope, 73
variance of a sample, 86
Vector, 15
vector product, 178
vector triple product, 178
vectors, 169
VECTR, 174
VERSION, 8
VIEW, 20
visualize a polynomial fitting, 341
VPAR, 313, 324
VPAR menu, 322
VX, 18, 163
VZIN, 346
VZOUT, 346

W

Weibull distribution, 82
WHILE…REPEAT…END, 131
WID→, 170
wide rectangular channel, 97
Wireframe plots, 307

X

X, 25
X,Y→, 334
x^2, 24
XCOL, 326
XOR, 118
XPON, 50
XRNG, 319
XROOT, 149
XVOL, 323
XXRNG, 323

Y

YCOL, 326
YRNG, 319
Y-Slice plots, 310
YVOL, 323
y, 24

YYRNG, 323
$^y\sqrt{x}$, 24

Z

ZAUTO, 346
ZDECI, 346
ZDFLT, 346
ZEROS, 366
ZFACT, 345, 346
ZFACTOR(Tr, Pr), 347
ZIN, 345
ZINTG, 346
ZLAST, 346
ZOOM menu, 345
ZTRIG, 347
ZVOL, 323

→

→STK, 171

Δ

ΔLIST, 68

Π

ΠLIST, 68

Σ

Σ, 24
ΣLIST, 68
ΣPAR, 326
ΣPAR menu, 325

π

π, 28

∞

∞, 28

√

$\sqrt{x}$, 24

ABOUT THE AUTHOR

Gilberto E. Urroz is an Associate Professor of Civil and Environmental Engineering and a researcher at the Utah Water Research Laboratory, both at Utah State University, in Logan, Utah. He has been a teacher of engineering disciplines for more than 15 years both in his native Nicaragua and in the United States.

His teaching experience includes courses on introductory physics, engineering mechanics, probability and statistics for engineers, computer programming, fluid mechanics, hydraulics, and numerical methods. His research interests include mathematical and numerical modeling of fluid systems, hydraulic structures, and erosion control applications.

Dr. Urroz is an expert on the HP 48 G and HP 49 G series calculator and has written several books on applications of these computing devices to disciplines such as engineering mechanics, hydraulics, and science and engineering mathematics. His personal interests include reading, music, opera, theater, and taijiquan.

ABOUT GREATUNPUBLISHED.COM

greatunpublished.com is a website that exists to serve writers and readers, and remove some of the commercial barriers between them. When you purchase a greatunpublished.com title, whether you receive it in electronic form or in a paperback volume or as a signed copy of the author's manuscript, you can be assured that the author is receiving a majority of the post-production revenue. Writers who join greatunpublished.com support the site and its marketing efforts with a per-title fee, and a portion of the site's share of profits are
channeled into literacy programs.

So by purchasing this title from greatunpublished.com, you are helping to revolutionize the publishing industry for the benefit of writers and readers.
And for this we thank you.

1825846

Made in the USA